VOLUME SIXTY NINE

ADVANCES IN
MARINE BIOLOGY
Marine Managed Areas and Fisheries

ADVANCES IN MARINE BIOLOGY

VOLUME SIXTY NINE

ADVANCES IN
MARINE BIOLOGY
Marine Managed Areas and Fisheries

Edited by

MAGNUS L. JOHNSON
*Centre for Environmental and Marine Sciences,
University of Hull, Scarborough,
North Yorkshire, United Kingdom*

JANE SANDELL
*Scottish Fishermen's Organisation, Peterhead,
Aberdeenshire, United Kingdom*

AMSTERDAM • BOSTON • HEIDELBERG • LONDON
NEW YORK • OXFORD • PARIS • SAN DIEGO
SAN FRANCISCO • SINGAPORE • SYDNEY • TOKYO
Academic Press is an imprint of Elsevier

Academic Press is an imprint of Elsevier
32 Jamestown Road, London NW1 7BY, UK
525 B Street, Suite 1800, San Diego, CA 92101-4495, USA
225 Wyman Street, Waltham, MA 02451, USA
The Boulevard, Langford Lane, Kidlington, Oxford OX5 1GB, UK

First edition 2014

Notices
Knowledge and best practice in this field are constantly changing. As new research and
experience broaden our understanding, changes in research methods, professional practices,
or medical treatment may become necessary.

Practitioners and researchers must always rely on their own experience and knowledge in
evaluating and using any information, methods, compounds, or experiments described
herein. In using such information or methods they should be mindful of their own safety and
the safety of others, including parties for whom they have a professional responsibility.

To the fullest extent of the law, neither the Publisher nor the authors, contributors, or editors,
assume any liability for any injury and/or damage to persons or property as a matter of
products liability, negligence or otherwise, or from any use or operation of any methods,
products, instructions, or ideas contained in the material herein.

ISBN: 978-0-12-800214-8
ISSN: 0065-2881

For information on all Academic Press publications
visit our website at store.elsevier.com

Working together
to grow libraries in
developing countries

www.elsevier.com • www.bookaid.org

CONTRIBUTORS TO VOLUME 69

Richard Appeldoorn
Department of Marine Sciences, University of Puerto Rico, Mayagüez, Puerto Rico

Fabio Badalamenti
CNR-IAMC Sede di Castellammare del Golfo, Castellammare del Golfo, Italy

David K.A. Barnes
British Antarctic Survey, Natural Environment Research Council, Cambridge, United Kingdom

Ivonne Bejarano
Department of Marine Sciences, University of Puerto Rico, Mayagüez, Puerto Rico

Mark Belchier
British Antarctic Survey, Natural Environment Research Council, Cambridge, United Kingdom

Helen Bloomfield
School of Environmental Sciences, University of Liverpool, Liverpool, United Kingdom

Louis W. Botsford
Department of Wildlife, Fish, and Conservation Biology, University of California, Davis, California, USA

Judith Brown
Government of South Georgia and the South Sandwich Islands, Government House, Stanley, Falkland Islands

Ben Carr
Boston University, Department of Biology, Boston, Massachusetts, USA

Mark H. Carr
Department of Ecology and Evolution, University of California, Santa Cruz, California, USA

Jennifer E. Caselle
Marine Science Institute, University of California, Santa Barbara, California, USA

Alex J. Caveen
Seafish Industry Authority, Grimsby, and School of Marine Science and Technology, Newcastle University, Newcastle upon Tyne, United Kingdom

Martin A. Collins
Government of South Georgia and the South Sandwich Islands, Government House, Stanley, Falkland Islands

Giovanni D'Anna
CNR-IAMC Sede di Castellammare del Golfo, Castellammare del Golfo, Italy

Jeffrey C. Drazen
Department of Oceanography, University of Hawaii, Honolulu, HI, USA

Euan Dunn
Royal Society for the Protection of Birds, Bedfordshire, United Kingdom

Richard P. Dunne
West Briscoe, Barnard Castle, Durham, United Kingdom

Graham Farebrother
Sydney Fish Market, Pyrmont, NSW, Australia and Fisheries Aquaculture and Coasts Centre, Institute for Marine and Antarctic Studies, University of Tasmania, Hobart, Tasmania, Australia

Clare Fitzsimmons
School of Marine Science and Technology, Newcastle University, Newcastle upon Tyne, United Kingdom

Alan M. Friedlander
Fisheries Ecology Research Laboratory, Department of Biology, University of Hawaii, Honolulu, HI, and National Geographic Society, Washington, DC, USA

Susie M. Grant
British Antarctic Survey, Natural Environment Research Council, Cambridge, United Kingdom

Tim S. Gray
School of Geography Politics and Sociology, Newcastle University, Newcastle upon Tyne, United Kingdom

Edwin A. Hernández-Delgado
Center for Applied Tropical Ecology and Conservation, Coral Reef Research Group, and Department of Biology, University of Puerto Rico, San Juan, Puerto Rico

Ray Hilborn
School of Aquatic and Fishery Sciences, University of Washington, Box 355020, Seattle, Washington, 98195, USA

Regen Jamieson
New England Aquarium, 1 Central Wharf, Boston, MA, 02110, USA

Magnus L. Johnson
Centre for Environmental and Marine Sciences, University of Hull, Scarborough, United Kingdom

Estelle V. Jones
School of Marine Science and Technology, Newcastle University, Newcastle upon Tyne, United Kingdom

Les Kaufman
New England Aquarium, 1 Central Wharf, Boston, MA, 02110; Boston University, Department of Biology, Boston, Massachusetts, and Conservation International, Arlington, Virginia, USA

Bob Kearney
Emeritus Professor in Fisheries Management, Institute for Applied Ecology, University of Canberra, Bruce ACT, Australia

John N. Kittinger
Conservation International, Betty and Gordon Moore Center for Science and Oceans, Honolulu, HI, USA

Paula Lightfoot
School of Marine Science and Technology, Newcastle University, Newcastle upon Tyne, United Kingdom

Sangeeta Mangubhai
New England Aquarium, 1 Central Wharf, Boston, MA, 02110, USA

Daniel Mateos-Molina
Department of Marine Sciences, University of Puerto Rico, Mayagüez, Puerto Rico, and Departamento de Ecología e Hidrología, Universidad de Murcia, Campus de Espinardo, Murcia, Spain

Michael I. Nemeth
Department of Marine Sciences, University of Puerto Rico, Mayagüez, Puerto Rico

Richard S. Nemeth
University of the Virgin Islands, Center for Marine and Environmental Studies, St. Thomas, U. S. Virgin Islands

David Obura
New England Aquarium, 1 Central Wharf, Boston, MA, 02110, USA, and CORDIO East Africa, P.O. Box 1013, Mombasa, Kenya

Margherita Pieraccini
Law School, University of Bristol, Bristol, United Kingdom

Ray Pierce
EcoOceania, Speewah, Queensland, Australia

Carlo Pipitone
CNR-IAMC Sede di Castellammare del Golfo, Castellammare del Golfo, Italy

Nicholas V.C. Polunin
School of Marine Science and Technology, Newcastle University, Newcastle upon Tyne, United Kingdom

Betarim Rimon
Phoenix Island Protected Area Office, Ministry of Environment, Lands and Agriculture Development, P.O. Box 234, Tarawa, Kiribati

Bud Ris
New England Aquarium, 1 Central Wharf, Boston, MA, 02110, USA, and Phoenix Islands Protected Area Conservation Trust, P.O. Box 366, Tarawa, Kiribati

Randi Rotjan
New England Aquarium, 1 Central Wharf, Boston, MA, 02110, USA

Peter H. Sand
Ludwig-Maximilians-Universität München, München, Germany

Stuart Sandin
Scripps Institution of Oceanography, UC San Diego, La Jolla, California, USA

Michelle T. Schärer-Umpierre
Interdisciplinary Center for Coastal Studies, and Department of Marine Sciences, University of Puerto Rico, Mayagüez, Puerto Rico

Peter Shelley
Conservation Law Foundation, Boston, Massachusetts, USA

Tyler B. Smith
University of the Virgin Islands, Center for Marine and Environmental Studies, St. Thomas, U. S. Virgin Islands

Kostantinos A. Stamoulis
Fisheries Ecology Research Laboratory, Department of Biology, University of Hawaii, Honolulu, HI, USA

Iain J. Staniland
British Antarctic Survey, Natural Environment Research Council, Cambridge, United Kingdom

Selina M. Stead
School of Marine Science and Technology, Newcastle University, Newcastle upon Tyne, United Kingdom

Greg Stone
Conservation International, Arlington, Virginia, USA, and Phoenix Islands Protected Area Conservation Trust, P.O. Box 366, Tarawa, Kiribati

U. Rashid Sumaila
The University of British Columbia Fisheries Centre, Vancouver, British Columbia, Canada

Christopher J. Sweeting
School of Marine Science and Technology, Newcastle University, Newcastle upon Tyne, United Kingdom

Sue Taei
Conservation International Pacific Islands and Oceans Programme, P.O. Box 2035, Apia, Samoa

Heather Tausig
New England Aquarium, 1 Central Wharf, Boston, MA, 02110, USA

Tukabu Teroroko
Phoenix Island Protected Area Office, Ministry of Environment, Lands and Agriculture Development, P.O. Box 234, and Phoenix Islands Protected Area Conservation Trust, P.O. Box 366, Tarawa, Kiribati

Simon Thorrold
Woods Hole Oceanographic Institution, Woods Hole, Massachusetts, USA

Brian N. Tissot
Marine Laboratory, Humboldt State University, Trinidad, CA, USA

Teuea Toatu
Phoenix Islands Protected Area Conservation Trust, P.O. Box 366, Tarawa, Kiribati

Philip N. Trathan
British Antarctic Survey, Natural Environment Research Council, Cambridge, United Kingdom

Manuel Valdés-Pizzini
Department of Social Sciences, University of Puerto Rico, Mayagüez, Puerto Rico

Tomás Vega Fernández
CNR-IAMC Sede di Castellammare del Golfo, Castellammare del Golfo, Italy

J. Wilson White
Department of Biology and Marine Biology, University of North Carolina Wilmington, Wilmington, North Carolina, USA

Brooke Wikgren
New England Aquarium, 1 Central Wharf, Boston, MA, 02110, USA

CONTENTS

SERIES CONTENTS FOR LAST FIFTEEN YEARS*

*The full list of contents for volumes 1–37 can be found in volume 38

POEM

THE SEA ADDRESSES AN MPA

You have drawn a shape in my mind
and want me to think myself into it
like an amnesiac gazing
at a photograph of childhood.

Must I sing in this garden
while earth's guts billow and muffle
square miles the sun's crystalline
music once struck into song?

Thin as the paper I'm written on,
my bones cannot bear the weight of their losses:
pincers shatter, exoskeletons buckle,
swaying worlds vanish.

Your harbours are empty; you grow
old beside me, hungering
for a picture of our past you can look into
through a glass-bottomed boat,

as if you were not there, your plastic pollen
not in my every breath. You have
drawn a box in my mind
to hold the story of us – it may yet unfold.

John Wedgwood Clarke

PREFACE

Magnus L. Johnson*, Jane Sandell†

*Centre for Environmental and Marine Sciences, University of Hull, Scarborough, North Yorkshire, United Kingdom
†Scottish Fishermen's Organisation, Peterhead, Aberdeenshire, United Kingdom

> *Conservation will either contribute to solving the problems of the rural people who live day to day with wild animals, or those animals will disappear, Jonathon Adams & Thomas McShane, WWF.*
>
> **Brockington (2009)**
>
> *What must be understood about the future land rights in the world is that most of the Earth's remaining natural resources and most of its high biodiversity ecosystems are currently occupied by people, most of who are indigenous. So whether the ultimate quarry is gold, oil, timber, tin or tigers, human inhabitants are going to be placed in conflict with other interests. And to them it doesn't really matter much whether the conflict is with an extractive transnational, the World Bank, a BINGO [Big International NGO] or the Brazilian military; the end result is pretty much the same – loss of livelihood, food security, freedom and culture.*
>
> **Dowie (2009)**

This volume comprises a series of case studies of large marine protected areas (MPAs) around the world. We were stimulated to encourage its development, as is pointed out in the introduction, by the polarised views and often advocacy, as opposed to science, led proposals for large areas that are designated as "protected". We sought authors who would base their studies on evidence from areas where spatial management has delivered very different outcomes; areas where it has been relatively successful, such as South Georgia, Hawaii and California, where it has the potential to be successful, such as the Phoenix Islands Protected Area and those where it has been less effective or is not meeting the stated aims such as the Great Barrier Reef, the North Sea and the Mediterranean. We have also included a chapter on one of the most recently declared, controversial, possibly illegal and, it seems, least likely to offer any significant benefit in its current form; the Chagos Archipelago.

Themes that emerge repeatedly in the studies are the lack of science underpinning the development of MPAs, a lack of clear objectives or indicators for monitoring performance, an inability to marry the need for protection with the level of protection, with the exception of South Georgia, the lack of financial support for enforcement and the lack of ongoing study or biological monitoring in the areas after they have been established on paper. Even one of the staunchest advocates for MPAs recently admitted that they are failing for the waters around the United Kingdom:

> *None of the 27 conservation zones declared in 2013 have yet received any new protection. My students and I have probed Department for Environment, Food and Rural Affairs, the Marine Management Organisation and various inshore fisheries and conservation authorities and it seems that virtually no new protection is on offer.*
>
> **Roberts (2014)**

In many cases, designation is not supported by those who depend upon resources from the MPAs—indigenous peoples who have fished for centuries before the arrival of the western conservation ideal. These are the peoples who really understand the environment through their daily use and contact with it. They are not supporting them, not only because exclusion from their traditional grounds, in many cases, threatens rather than enhances their way of life but also because of the perverse effects often resulting from protection. Nor is the MPA approach favoured by fishermen from more developed countries, who see fishery conservation measures, rather than total bans, as a more relevant, comprehensible and effective alternative. And as Ray Hilborn points out in the introduction to this volume, dividing the world into protected and unprotected areas also increases the vulnerability of those areas outside MPAs.

In addition, it is clear that static MPAs are not able to deal with changes in distributions of species, particularly for highly migratory stocks, and the naturally huge fluctuations in numbers over time. The "wicked problem" of natural resource management on the marine commons is a function of known unknowns and diverse desires of stakeholders (Balint et al., 2011). The legislation required to deal with such situations would have to be so complex that it would be unworkable. In fisheries, we need to be seeking solutions that harness the incredible capacity of human societies to deal with complex and ever-changing situations. Harnessing this ability requires the development of management frameworks that recognise the propensities of human beings and facilitate community management of common pool resources (Ostrom et al., 1999) rather than imposes exclusion.

REFERENCES

Balint, P.J., et al., 2011. Wicked Environmental Problems: Managing Uncertainty and Conflict. Island Press, Washington.
Brockington, D., 2009. Celebrity and the Environment. Zed Books, London.

Dowie, M., 2009. Conservation Refugees. MIT Press, London.

Ostrom, E., et al., 1999. Revisiting the commons: local lessons, global challenges. Science 284 (9), 278–282.

Roberts, C., 2014. England's marine conservation network is worse than useless. The Guardian. Available at: http://www.theguardian.com/environment/2014/jun/17/england-marine-conservation-zones.

CHAPTER ONE

Introduction to Marine Managed Areas

Ray Hilborn[1]
School of Aquatic and Fishery Sciences, University of Washington, Box 355020, Seattle,
Washington, 98195, USA
[1]Corresponding author: e-mail address: hilbornr@gmail.com

Contents

Abstract

No issue in marine conservation and management seems to have generated as much interest, and controversy as marine protected areas (MPAs). In the past 30 years, a substantial scientific literature on the subject has developed, international agreements have set targets for proportion of the sea to be protected, and hundreds of millions of dollars have been spent on research and advocacy for MPA establishment. While the objectives of MPAs are diverse, few studies evaluate the success of MPAs against stated objectives. It is clear that well-enforced MPAs will protect enough fish from exploitation that within reserves abundance increases, fish live to be larger, and measures of diversity are higher. What is much more poorly understood is the impacts of reserve establishment on areas outside reserves. Theory suggests that when stocks are seriously overfished outside reserves, the yield and abundance outside the reserves may be increased by spillover from the reserve. When stocks are not overexploited, reserve establishment will likely decrease the total yield. The chapters in this volume explore a broad set of case studies of MPAs, their objectives and their outcomes.

Keywords: Marine protected areas, MPAs, Closed areas, Spatially explicit management, Fisheries management, Marine reserves

Advances in Marine Biology, Volume 69
ISSN 0065-2881
http://dx.doi.org/10.1016/B978-0-12-800214-8.00001-3

1. INTRODUCTION

This chapter is intended to review a bit of history of MPAs and set the stage for the case studies in this volume. It is very much a personal perspective and is certainly not intended to be a thorough literature review.

1.1. Background history of MPAs

Closed areas have always been a part of fisheries management. Traditional fishing practices in the Western Pacific as documented by Johannes (1978) and others almost always included areas that were closed to protect spawning or juveniles. Johannes highlights marine tenure as the underlying concept of traditional fisheries management in the Pacific, and later, in his list of methods used, closures are mentioned as the first and most ubiquitous method. McClenachan and Kittinger (2013) describe traditional fishing practices in Hawaii and also found that closed areas were one of the most common elements of fisheries management by the Polynesians.

In western fisheries management, too, closed areas have been an integral part of the toolkit. Areas closed to protect such critical habitats as spawning or juvenile-rearing areas are documented as far back as the nineteenth century (Fishery Board of Scotland, 1895). Management of salmon in Alaska has relied primarily on permanently closing most areas to fishing and regulating harvest through short-term openings of some specific areas (Clark et al., 2006). However, most other western fisheries management has relied on closing areas for specific species with specific gears only, while few areas were designated as permanently closed to all fishing—that is, there were few no-take areas.

However, beginning in the 1990s, a movement began to set aside significant areas of the ocean as permanent no-take areas. A review of the scientific literature shows that papers published with 'marine reserve' or 'marine protected area' in the title or abstract (Fig. 1.1) were rare before the 1990s, became more frequent around 2000, and rose to 270 in 2013.

The development of the MPA literature in the 1990s and 2000s was often supported by major funding from US Foundations and NGOs advocating the establishment of MPAs and reflected an increasing concern about the state of marine ecosystems:

> In light of the declining catches and failure of many marine fisheries, biologists dissatisfied with the effectiveness of current management practices have recently advocated the use of harvest refuges as a potentially effective strategy for protecting and/or enhancing harvestable stocks.
>
> **Carr and Reed (1993)**

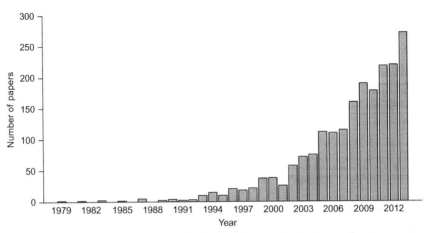

Figure 1.1 The number of papers published each year with 'marine reserve' or 'marine protected area' in the title or abstract.

However, it was not only the perceived failure of the management of marine fisheries that provided impetus for the MPA movement but also a broader concern with the state of the oceans in general. At the AAAS meeting in 2001, a group of 161 marine scientists released a consensus statement on the need for marine reserves that began

> *The declining state of the oceans and the collapse of many fisheries have created a critical need for new and more effective management of marine biodiversity, populations of exploited species and overall health of the oceans. Marine reserves are a highly effective but under-appreciated and under-utilized tool that can help alleviate many of these problems.*

The idea that MPAs are an effective way to manage fisheries was widely accepted. For instance, 'What reserves offer that other management tools cannot is an ability to control fishing rates in a manner that is relatively easy to enforce and requires relatively little scientific information (Nowlis and Friedlander, 2005)'. Clearly, any tool that is effective and easy, and, best of all, requiring little scientific information would be irresistible.

Between 2007 and 2009, US Foundations and NGOs spent $250,000,000 per year on marine conservation with much of this supporting research on MPAs and advocating for their establishment (California Environmental Associates, 2012; Fig. 4.1). This is the only period where such numbers are available but such funding has been going on since the 1990s. These efforts have indeed been very successful in many countries. International agreements, through the Convention on Biodiversity, accepted a target of 10% of the ocean to be designated as MPAs by 2020.

The United States and Australia in particular declared large areas of their economic zones to be so protected. In the United States, primarily in Alaska and in the Western Pacific, over two-thirds of the economic zones were qualified as MPAs because of the prohibition of bottom trawling.

A key characteristic of the MPA movement is that it has largely been driven by concerns of biodiversity, and that its advocates have come almost exclusively from the marine biology community. Little attention has been paid, however, to the broader implications of potentially restricting food production and food security or how it might affect people (Hilborn, 2013). Perhaps most worrying, though, is that establishing MPAs seems to have become an end in itself, the benefits being so self-evident that any careful statement of objectives or evaluation was clearly unnecessary. This self-evident nature of benefits is clearly reflected in the second sentence of the 2001 consensus statement: 'Marine reserves are a highly effective but under-appreciated and under-utilized tool that can help alleviate many of these problems'. The assertion that marine reserves are 'highly effective … (and) can help alleviate many of these problems' needs to be critically assessed, and to do that we need to look at their objectives.

1.2. Objectives

Protected areas can provide many benefits to marine ecosystems that can include (1) protection of biodiversity, (2) more tourism, (3) more fish production and better yields, (4) providing reference sites for evaluating human impacts and (5) providing a safety net of resources for times of need. The traditional practices discussed earlier were generally meant to enhance fish production and provide a safety net. Protecting spawning and/or juvenile habitat can have clear benefits through preserving the fish stocks and therefore were commonly included as part of fisheries management systems; they can be 'traditional' or associated with a centralised governmental management of fish resources.

The oceans are indeed under a broad range of threats including climate change, ocean acidification, pollution, loss of coastal habitat for breeding grounds, land-based runoff of sediments, overfishing and destructive fishing practices (Sutherland et al., 2012). MPAs, because they can only address overfishing and destructive fishing practices, have their limits as a management tool. The great threats of climate change, ocean acidification, pollution and land-based impacts are immune to MPAs. Furthermore, where the MPAs are being established is usually related to the current distribution of habitats, and those appear to be changing along with the global climate.

A key driver of the MPA movement was the litany of clear fisheries management failures leading to the collapse of a number of fisheries in the 1990s, most dramatically that of the northern cod (*Gadus morhua*) fishery of Newfoundland. The fisheries management systems of the 1990s were obviously failing to protect many fish stocks, and the advocates of MPAs suggested that fish yields would be improved by their establishment. The document 'The Science of Marine Reserves' (Partnership for Interdisciplinary Studies of Coastal Oceans, 2002) stated that reserves boost the productivity of fisheries outside their boundaries and argued that benefits to fisheries would be maximised if 40% of the total area was closed to fishing.

The push for MPAs was centred in the west coast of the United States, where the David and Lucile Packard Foundation provided tens of millions of dollars in support for research and advocacy. Most of the academics who led the MPA movement were located on the west coast, the National Science Foundation's National Center for Ecological Analysis and Synthesis (NCEAS) in Santa Barbara provided a nexus for the scientific and advocacy work, and two programmes to establish MPAs were implemented in California. In the first of these programmes, a set of marine reserves in the Channel Islands of Southern California was established in 2003, and it was argued that MPAs would provide strong benefits to fisheries. Later, a much more extensive set of reserves was established along the entire California coast, and I participated in the science advice teams for part of that process.

By the late 2000s, the arguments had fundamentally changed. The science teams were explicitly told that the establishment of reserves was not to be considered part of the fisheries management system, but that the MPAs were established primarily for the purpose of biodiversity. This reflects, I believe, a major change in the focus of much of the MPA advocacy movement, at least in those countries that have effective fisheries management systems, and stems from a general recognition that in many countries fisheries management has changed significantly since the 1990s and that the current management systems can, and do, effectively protect fish stocks.

As context for the chapters in this volume, I would like to review what we now know about the impact of MPAs on a range of objectives including the abundance of fish, the functioning of ecosystems and the provision of food security. First, I will discuss what we know from models then what we have learned from empirical analysis.

2. MODELS AND THEORY

Models have been used to evaluate the impacts of closed areas on abundance and potential fish yield at least since Beverton and Holt's classic 1957 book (Beverton and Holt, 1957). In section 10.2, they explore the impact on yield and abundance of protecting portions of the population from fishing. The subject appeared to lie dormant until the 1990s with the exception of Sluczanowski (1984), who explored rotational harvest strategies using models similar to Beverton and Holt. In the 1990s and early 2000s, a substantial number of modelling papers were published including Polacheck (1990), DeMartini (1993), Sumaila (1998), Guenette and Pitcher (1999), Maury and Gascuel (1999), Dahlgren and Sobel (2000), Jennings (2000), Mangel (2000), Pezzey et al. (2000) and Stockhausen et al. (2000).

Gerber et al. (2003) provided a review of 34 models that had been published. They found that all but one were single-species models, most assumed uniform distribution of larvae between closed and open areas, most assumed that density dependence happened post-dispersal, most assumed no age structure and sedentary adults and most were deterministic models with permanent closed areas.

Emerging from the models available up to 2003 were four conclusions (Botsford et al., 2003). (1) The effects on yield per recruit of adding reserves are essentially the same as increasing the size limit. (2) The effect on yield of adding reserves is practically the same as decreasing fishing mortality. (3) Reserves for preserving biodiversity are most effective for species with low rates of juvenile and adult movement, while reserves for fishery management are most effective for species with intermediate rates of adult movement. (4) Larger fractions of coastline in reserves are required for species with longer dispersal.

The key result in terms of benefiting fisheries yield was that when fishing mortality rates were in the range that produced maximum sustainable yields, there were no fishery benefits. When fishing mortality rates were significantly above those that produce maximum sustainable yields, then fisheries yields can be improved by protected areas. The key missing elements from models of up to 2003 were multispecies evaluations, explicit models of fleet movement and profitability and spatially explicit models that could represent habitat variation.

Hilborn et al. (2006) provided a spatially explicit model of a coastline with vessel movement and consideration of fisheries regulations and demonstrated that appropriate catch regulation always provided better yield outcomes than any MPA pattern. Walters et al. (2007) provided the spatially explicit models that could be used to evaluate proposed MPA designs that included site-specific habitat information and a dynamic fleet movement model. This approach was extended by Costello et al. (2010) who included multispecies models and explicit larval dispersal models based on models of oceanic currents and larval behaviour. Both the Walters and Costello models have the potential for source–sink dynamics—that is where particularly good habitats generate much of the production of larvae that can then migrate to habitats suitable for growth but not for reproduction. The source–sink dynamics provide a circumstance where even when fishing mortality rates are well regulated outside reserves, fish yields can be improved by specific patterns of MPA location—specifically by closing the most productive 'source' habitats. The historical use of closed areas for protection of spawning and juvenile rearing was, in essence, the recognition of source–sink dynamics.

These more recent models demonstrate several results that are perhaps surprising. While it is almost always the case that abundance of fish will increase inside the reserves if fishing is indeed stopped, depending on how fisheries are regulated outside the reserves, total abundance may actually decline when reserves are implemented when the displaced effort ends up overexploiting stocks outside the reserves. Much of the early literature emphasised that MPAs, if properly sized, could provide the same fishery benefits as catch regulation. However, the size has to be exactly tuned to the dispersal of the fish, and with multiple species, it is usually impossible for the reserve(s) to be the right size for each species. For species that move little, fisheries benefits will be maximised with small reserves, but for highly dispersive species, reserves must be much larger. Large reserves essentially lock up biomass of low-dispersal species and reduce potential fishery yield. Fisheries management tools that are species specific can be much more effective at managing mixed fisheries than no-take areas. For instance, Branch and Hilborn (2008) showed that the individual vessel quotas in the British Columbia demersal fishery provided incentives for the fishing vessels to be very site specific in their fishing locations, thereby allowing them to avoid species for which there was little quota.

A key feature of no-take areas is that they will increase fishing pressure outside reserves, often called displacement. The naïve vision of lots of large fish inside reserves did not consider the consequences of more fishing pressure elsewhere and the negative consequences of this displacement. A key question in any reserve design or evaluation is whether the benefits inside reserves are more than offset the negative impacts of extra fishing pressure outside reserves (Fogarty and Botsford, 2007). When stocks are seriously overfished, the extra pressure outside the reserves appears to be far more than compensated by the benefits inside reserves, but when stocks are not over-fished, the opposite appears to be the case. I believe that the MPA advocacy movement strongly believed that almost all fisheries were overexploited so that it would be highly likely that the fisheries would benefit from the reserves.

3. EMPIRICAL DATA

As mentioned earlier, NCEAS provided the nexus of activity among marine ecologists in MPA science and advocacy, and especially with a series of papers regarding what is known about MPA impact on abundance inside reserves. Halpern and Warner (2002) provided the first meta-analysis of changes within reserves, showing that density of individuals, total fish bio-mass, average size and diversity indices were higher inside reserves than out-side. An obvious concern about such comparisons is that the abundance outside the reserves will be affected by effort displacement. Later papers (Halpern, 2003; Lester et al., 2009) expanded the analysis to include before and after comparisons. A recent paper by authors largely outside the NCEAS network (Edgar et al., 2014) identified five conditions under which reserves were most likely to increase abundance of fish inside reserves, and these include (1) the reserves are no-take, (2) there is effective enforcement, (3) the MPAs have been in place for a significant period of time, (4) the MPAs are large and (5) the MPAs are isolated either by distance or by habitat bar-riers. Of course increased abundance inside reserves is a necessary condition for almost all objectives of reserves, but it does not indicate that there are any benefits outside.

I believe we can say with confidence that the empirical data show that MPAs will increase the abundance of targeted fish species inside the reserves if most of the above conditions are met. But empirical data are lacking on the impact of reserve implementation on abundance of fish outside reserves and the impact of reserves on fisheries yields. While there are some individual studies, there has not yet been a meta-analysis of those impacts, and the dif-ficulties of experimental design may be very hard to overcome. If abundance rises both inside and outside reserves, was this due to the MPA spillover or was there a change in environmental conditions? For instance, Roberts et al.

(2001) showed data on abundance inside and outside, before and after, from St. Lucia. Abundance increased after reserve implementation both inside and outside, but more inside roughly tripling on the inside and doubling on the outside. However, abundance on the outside and on the inside increased in the first year, something that could not have been due to increased abundance inside and spillover when dealing with fish species that do not mature until they are several years old (Hilborn, 2002).

One of the most convincing studies for fishery benefits is found in Kerwath et al. (2013) who have data not only from inside and outside reserves and before and after implementation, but also from other, similar sites that are at quite a distance and where no reserves have been established—presumably the best control possible. Catch and catch per unit effort (CPUE) near the MPA increased after MPA establishment, but did not increase at more distant sites, and unlike the Roberts study, the increase in outside the reserves was not instantaneous but took several years to develop.

Hamilton et al. (2010) presented preliminary evaluations of the Channel Islands marine reserves established in California, and these results were very consistent with models. The fisheries outside the reserves are well regulated, and the abundance of target species inside increased after the reserves were established, whereas they decreased outside. Their analysis did not compute total abundance, but it appears that total abundance of target species likely decreased or stayed the same after reserve establishment. This paper only had data for the first few years after reserve establishment so caution is need in interpreting the results.

4. CONCLUSIONS

Closed areas, both permanent and temporary, are a significant tool in marine resource management but, before reserves are implemented, objectives need to be clearly defined and an evaluation framework established. For those who see MPAs as an end in itself, it is important that they be permanent. But if we consider MPAs as a tool to achieve social objectives, we must evaluate what has happened against the objectives of the programme and be prepared to change the MPA design which might involve moving the MPA, or expanding or reducing their size.

We need to also consider the cost of implementing, enforcing and monitoring reserves as opposed to other possible expenditures that can yield the same benefits. A key lesson from the Marine Life Protection Act in

California and much experience in the developing world is that you must consult with user groups (Weible, 2008). This consultation is very expensive, the MLPA process in California, for example, costs tens of millions of dollars. Indeed, the State of California alone could not have afforded the process, and in the end, the many years of meetings, consultation and analysis were largely funded by the same foundations that had funded the initial advocacy that led to the passage of the law. The programme officer of the US Foundation told me recently that after reviewing the expenditures on MPAs, it was clear that the money would have been more effectively spent on improving fisheries management.

We need to abandon the acceptance of MPAs as an end in itself with the sole objective to increase abundance inside reserves. More case studies of reserves are needed to evaluate the range of objectives and the performance of the MPAs against those objectives. Peter Kareiva and others (Kareiva and Marvier, 2012; Kareiva et al., 2011) have argued that conservation needs to move beyond protected areas as the central tool and protects biodiversity in areas that people use by working with resource-dependent communities. The vehement rejection of these ideas by many in the conservation community (Soule, 2013) reflects a divide found in both marine and terrestrial conservation.

What follows in this collection is a major step in this direction. These papers consider the range of possible objectives and most attempt to evaluate the performance of the MPAs accordingly.

I would like to close this introduction with two quotes that, while more than a decade old, seem just as relevant today.

The rush to implement MPAs has set the stage for paradoxical differences of opinions in the marine conservation community. The enthusiastic prescription of simplistic solutions to marine conservation problems risks polarization of interests and ultimately threatens bona fide progress in marine conservation. The blanket assignment and advocacy of empirically unsubstantiated rules of thumb in marine protection creates potentially dangerous targets for conservation science. Clarity of definition, systematic testing of assumptions, and adaptive application of diverse MPA management approaches are needed so that the appropriate mix of various management tools can be utilized, depending upon specific goals and conditions. Scientists have a professional and ethical duty to map out those paths that are most likely to lead to improved resource management and understanding of the natural world, including the human element, whether or not they are convenient, politically correct or publicly magnetic. The use of MPAs as a vehicle for promoting long-term conservation and sustainable use of marine biodiversity is in need of focus, and both philosophical and applied tune ups.

Agardy et al. (2003)

Marine reserves, together with other fishery management tools, can help achieve broad fishery and biodiversity objectives, but their use will require careful planning and evaluation. Mistakes will be made, and without planning, monitoring and evaluation, we will not learn what worked, what did not, and why. If marine reserves are implemented without case by case evaluation and appropriate monitoring programs, there is a risk of unfulfilled expectations, the creation of disincentives, and a loss of credibility of what potentially is a valuable management tool.

Hilborn et al. (2004)

REFERENCES

Agardy, T., Bridgewater, P., Crosby, M.P., Day, J., Dayton, P.K., Kenchington, R., Laffoley, D., Mcconney, P., Murray, P.A., Parks, J.E., Peau, L., 2003. Dangerous targets? unresolved issues and ideological clashes around marine protected areas. Aquat. Conserv. 13, 353–367.

Beverton, R.J.H., Holt, S.J., 1957. On the Dynamics of Exploited Fish Populations. Her Majesties Stationary Office, London.

Botsford, L.W., Micheli, F., Hastings, A., 2003. Principles for the design of marine reserves. Ecol. Appl. 13(Suppl.), S25–S31.

Branch, T.A., Hilborn, R., 2008. Matching catches to quotas in a multispecies trawl fishery: targeting and avoidance behavior under individual transferable quotas. Can. J. Fish. Aquat. Sci. 68, 1435–1446.

California Environmental Associates, 2012. Charting a Course to Sustainable Fisheries. San Francisco California. Available at, http://www.packard.org/what-were-learning/resource/charting-a-course-to-sustainable-fisheries/.

Carr, M.H., Reed, D.C., 1993. Conceptual issues relevant to marine harvest refuges— examples from temperate reef fishes. Can. J. Fish. Aquat. Sci. 50, 2019–2028.

Clark, J.H., Mcgregor, A., Mecum, R.D., Krasnowski, P., Carroll, A.M., 2006. The commercial salmon fishery in Alaska. Alaska Fish. Res. Bull. 12, 1–146.

Costello, C., Rassweiler, A., Siegel, D., De Leo, G., Micheli, F., Rosenberg, A., 2010. The value of spatial information in MPA network design. Proc. Natl. Acad. Sci. U. S. A. 107, 18294–18299.

Dahlgren, C.P., Sobel, J., 2000. Designing a Dry Tortugas ecological reserve: how big is big enough? . . . to do what? Bull. Mar. Sci. 66, 707–719.

Demartini, E.E., 1993. Modeling the potential for fishery reserves for managing pacific coral reef fishes. Fish. Bull. 91, 414–427.

Edgar, G.J., Stuart-Smith, R.D., Willis, T.J., Kininmonth, S., Baker, S.C., Banks, S., Barrett, N.S., Becerro, M.A., Bernard, A.T.F., Berkhout, J., Buxton, C.D., Campbell, S.J., Cooper, A.T., Davey, M., Edgar, S.C., Forsterra, G., Galvan, D.E., Irigoyen, A.J., Kushner, D.J., Moura, R., Parnell, P.E., Shears, N.T., Soler, G., Strain, E.M.A., Thomson, R.J., 2014. Global conservation outcomes depend on marine protected areas with five key features. Nature 506, 216–220.

Fishery Board of Scotland, 1895. Fourteenth Annual Report of the Fishery Board of Scotland. House of Commons Parliamentary Papers Online. ProQuest Information and Learning Company. http://www.proquest.com/products-services/House-of-Commons-Parliamentary-Papers.html.

Fogarty, M.J., Botsford, L., 2007. Population connectivity and spatial management of marine fisheries. Oceanography 20, 112.

Gerber, L.R., Botsford, L.W., Hastings, A., Possingham, H.P., Gaines, S.D., Palumbi, S.R., Andelman, S.J., 2003. Population models for marine reserve design: a retrospective and prospective synthesis. Ecol. Appl. 13, S47–S64.

Guenette, S., Pitcher, T.J., 1999. An age-structured model showing the benefits of marine reserves in controlling overexploitation. Fish. Res. 39, 295–303.

Halpern, B.S., 2003. The impact of marine reserves: do reserves work and does reserve size matter? Ecol. Appl. 13, S117–S137.

Halpern, B.S., Warner, R.R., 2002. Marine reserves have rapid and lasting effects. Ecol. Lett. 5, 361–366.

Hamilton, S.L., Caselle, J.E., Malone, D.P., Carr, M.H., 2010. Incorporating biogeography into evaluations of the channel islands marine reserve network. Proc. Natl. Acad. Sci. U. S. A. 107, 18272–18277.

Hilborn, R., 2002. Marine reserves and fisheries management. Science 295, 1233–1234.

Hilborn, R., 2013. Environmental cost of conservation victories. Proc. Natl. Acad. Sci. U. S. A. 110, 9187.

Hilborn, R., Stokes, K., Maguire, J.J., Smith, T., Botsford, L.W., Mangel, M., Orensanz, J., Parma, A., Rice, J., Bell, J., Cochrane, K.L., Garcia, S., Hall, S.J., Kirkwood, G.P., Sainsbury, K., Stefansson, G., Walters, C., 2004. When can marine reserves improve fisheries management? Ocean Coast. Manage. 47, 197–205.

Hilborn, R., Micheli, F., De Leo, G.A., 2006. Integrating marine protected areas with catch regulation. Can. J. Fish. Aquat. Sci. 63, 642–649.

Jennings, S., 2000. Patterns and prediction of population recovery in marine reserves. Rev. Fish Biol. Fish. 10, 209–231.

Johannes, R.E., 1978. Traditional marine conservation methods in Oceania and their demise. Annu. Rev. Ecol. Syst. 9, 349–364.

Kareiva, P., Marvier, M., 2012. What is conservation science. Bioscience 62, 962–969.

Kareiva, P., Lalasz, R., Marvier, M., 2011. Conservation in the anthropocene: beyond solitutude and fragility. In: Shellenberger, M., Nordhaus, T. (Eds.), Love Your Monster: Postenvironmentalism and the Anthropocene. Breakthrough Institute, Oakland, CA.

Kerwath, S.E., Winker, H., Gotz, A., Attwood, C.G., 2013. Marine protected area improves yield without disadvantaging fishers. Nat. Commun. 4. http://dx.doi.org/10.1038/ncomms3347.

Lester, S.E., Halpern, B.S., Grorud-Colvert, K., Lubchenco, J., Ruttenberg, B.I., Gaines, S.D., Airame, S., Warner, R.R., 2009. Biological effects within no-take marine reserves: a global synthesis. Mar. Ecol. Prog. Ser. 384, 33–46.

Mangel, M., 2000. Irreducible uncertainties, sustainable fisheries and marine reserves. Evol. Ecol. Res. 2, 547–557.

Maury, O., Gascuel, D., 1999. SHADYS ('simulateur halieutique de dynamiques spatiales'), a GIS based numerical model of fisheries. Example application: the study of a marine protected area. Aquat. Living Resour. 12, 77–88.

Mcclenachan, L., Kittinger, J.N., 2013. Multicentury trends and the sustainability of coral reef fisheries in Hawai'i and Florida. Fish Fish. 14, 239–255.

Nowlis, J.S., Friedlander, A., 2005a. Marine reserve function and design for fisheries management. In: Norse, E.A., Crowder, Larry B. (Eds.), In: Marine Conservation Biology: The Science of Maintaining the Sea's Biodiversity, vol. 22. Island Press, Washington, DC, pp. 280–301.

Partnership for Interdisciplinary Studies of Coastal Oceans, 2002. Science of Marine Reserves. http://www.piscoweb.org.

Pezzey, J.C.V., Roberts, C.M., Urdal, B.T., 2000. A simple bioeconomic model of a marine reserve. Ecol. Econ. 33, 77–91.

Polacheck, T., 1990. Year round closed areas as a management tool. Nat. Resour. Model. 4, 327–354.

Roberts, C.M., Bohnsack, J.A., Gell, F., Hawkins, J.P., Goodridge, R., 2001. Effects of marine reserves on adjacent fisheries. Science 294, 1920–1923.

Sluczanowski, P.R., 1984. A management-oriented model of an abalone fishery whose sub-stocks are subject to pulse fishing. Can. J. Fish. Aquat. Sci. 41, 1008–1014.

Soule, M.E., 2013. The "New Conservation". Conserv. Biol. 27, 895–897.

Stockhausen, W.T., Lipcius, R.N., Hickey, B.M., 2000. Joint effects of larval dispersal, population regulation, marine reserve design, and exploitation on production and recruitment in the Caribbean spiny lobster. Bull. Mar. Sci. 66, 957–990.

Sumaila, U.R., 1998. Protected marine reserves as fisheries management tools: a bio-economic analysis. Fish. Res. 37, 287–296.

Sutherland, W.J., Aveling, R., Bennun, L., Chapman, E., Clout, M., Côté, I.M., Depledge, M.H., Dicks, L.V., Dobson, A.P., Fellman, L., 2012. A horizon scan of global conservation issues for 2012. Trends Ecol. Evol. 27, 12–18.

Walters, C.J., Hilborn, R., Parrish, R.H., 2007. An equilibrium model for predicting the efficacy of marine protected areas in coastal environments. Can. J. Fish. Aquat. Sci. 64, 1009–1018.

Weible, C.M., 2008. Caught in a maelstrom: implementing California marine protected areas. Coast. Manag. 36, 350–373.

CHAPTER TWO

The South Georgia and the South Sandwich Islands MPA: Protecting A Biodiverse Oceanic Island Chain Situated in the Flow of the Antarctic Circumpolar Current

Philip N. Trathan*,[1], Martin A. Collins[†], Susie M. Grant*,
Mark Belchier*, David K.A. Barnes*, Judith Brown[†], Iain J. Staniland*

*British Antarctic Survey, Natural Environment Research Council, Cambridge, United Kingdom
[†]Government of South Georgia and the South Sandwich Islands, Government House, Stanley, Falkland Islands
[1]Corresponding author: e-mail address: pnt@bas.ac.uk

Contents

Advances in Marine Biology, Volume 69
ISSN 0065-2881
http://dx.doi.org/10.1016/B978-0-12-800214-8.00002-5

Abstract

South Georgia and the South Sandwich Islands (SGSSI) are surrounded by oceans that are species-rich, have high levels of biodiversity, important endemism and which also support large aggregations of charismatic upper trophic level species. Spatial management around these islands is complex, particularly in the context of commercial fisheries that exploit some of these living resources. Furthermore, management is especially complicated as local productivity relies fundamentally upon biological production transported from outside the area. The MPA uses practical management boundaries, allowing access for the current legal fisheries for Patagonian toothfish, mackerel icefish and Antarctic krill. Management measures developed as part of the planning process designated the whole SGSSI Maritime Zone as an IUCN Category VI reserve, within which a number of IUCN Category I reserves were identified. Multiple-use zones and temporal closures were also designated. A key multiple-use principle was to identify whether the ecological impacts of a particular fishery threatened either the pelagic or benthic domain.

Keywords: Pelagic protection, Benthic protection, High biodiversity, Fisheries, Antarctic krill, South Georgia, MPA

1. INTRODUCTION TO THE AREA

1.1. Background and ecological characteristics

The Scotia Sea is located in the southwest Atlantic (Figure 2.1) where the marine ecosystem is dominated by the eastward-flowing Antarctic Circumpolar Current (ACC) and waters of the Weddell–Scotia Confluence. The flow of the ACC is constrained by bathymetry, particularly where it passes over the Scotia Arc (Orsi et al., 1995). Topographic steering pushes cold polar waters to lower latitudes (Murphy et al., 2004; Thorpe et al., 2004, 2007); this also results in associated intense eddy activity and mixing. South Georgia and the South Sandwich Islands (SGSSI) lie within the Scotia Sea, so their marine ecosystems are dominated by the strong advective flow of the ACC (see Atkinson et al., 2001; Murphy et al., 2007a for a description of the ecology of the Scotia Sea ecosystem). Much of the Scotia Sea also lies within that region that experiences strong seasonality, with seasonally changing irradiance and seasonal sea-ice cover that annually extends northwards in winter inundating many of the South Sandwich Islands, but rarely reaching as far north as South Georgia itself. Intense summer phytoplankton blooms

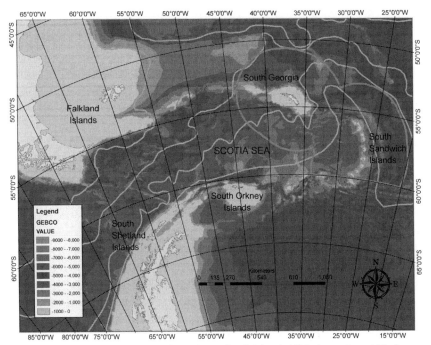

Figure 2.1 The South Georgia and South Sandwich Islands MPA shown in red (dark grey in the print version), with the eastward-flowing Antarctic Circumpolar Current (ACC) fronts in green (white in the print version): south to north, Southern ACC Boundary, Southern ACC Front, Polar Front, Sub-Antarctic Front. *Bathymetry data from GEBCO.*

probably result from the mixing of micronutrients into surface waters through the flow of the ACC over the Scotia Arc (Korb et al., 2004). This intense, seasonal production is critical to the Scotia Sea marine system with much of it eventually consumed by a range of species including Antarctic krill (*Euphausia superba*); krill can comprise up to 50% of zooplankton biomass at South Georgia (Atkinson et al., 2001) and are themselves a major prey item for the abundant seabird and marine mammal populations that use the Scotia Sea. Climate variability has also long been recognised as a feature of the Scotia Sea (e.g. Trathan et al., 2003), which has been linked to the El Niño-Southern Oscillation (e.g. Trathan et al., 2006a) and the Southern Annular Mode (e.g. Murphy et al., 2007b). This variability affects the physical properties of the marine system as well as primary and secondary production. It also affects krill population dynamics and dispersal (Murphy et al., 2007b), which in turn impacts higher trophic level predator foraging, breeding performance and population dynamics (Trathan et al., 2006a, 2007).

The Scotia Sea ecosystem has been highly perturbed as a result of harvesting over the past two centuries (e.g. Everson, 1977; Kemp, 1929; Kock, 1992), while significant ecological changes have also occurred in response to rapid, regional warming during the latter part of the 20th century (e.g. Whitehouse et al., 2008). The combination of historical perturbation together with recent, rapid regional change suggests that the Scotia Sea ecosystem is likely to show further significant change over the coming decades, which may also result in major ecological shifts.

Today, and despite such concerns for the future, the waters within the South Georgia and South Sandwich Islands Maritime Zone (SGSSIMZ) remain among the most productive in the Southern Ocean, supporting a great diversity and abundance of wildlife (Trathan et al., 1996), including benthos (Hogg et al., 2011; Rogers et al., 2012), pelagic organisms (Atkinson et al., 2001; Murphy et al., 2007a), and marine predators, including land-based seabirds and marine mammals (Clarke et al., 2012). The biological diversity and abundance within the waters of the SGSSIMZ also mean that the area continues to be attractive for the commercial harvesting of marine living resources (Agnew, 2004). Consequently, in 2012, the Government of South Georgia and the South Sandwich Islands (GSGSSI) established a Marine Protected Area (MPA) including a number of fishery no-take zones within the SGSSIMZ (Figure 2.1). The management plan for the MPA[1] lists in detail an inventory of species present in the SGSSIMZ.

1.2. Fisheries pre-MPA

Fishing for finfish and Antarctic krill has been taking place at South Georgia for over 40 years (Figure 2.2A). Within that period, major changes in management practices have occurred, including the international agreement to form the Commission for the Conservation of Antarctic Marine Living Resources (CCAMLR) in 1982 and the declaration of the SGSSIMZ in 1993. Management of fisheries prior to these dates was much less conservative and a number of stocks and species were harvested to very low levels following extensive industrial-scale fishing (Kock, 1992), especially by east European fleets.

In the SGSSIMZ, there are now legal licenced fisheries (Figure 2.2B) for Patagonian toothfish (*Dissostichus eleginoides*), mackerel icefish (*Champsocephalus gunnari*) and Antarctic krill, with a small fishery for Antarctic toothfish (*Dissostichus mawsoni*) at the South Sandwich Islands. Fisheries within the SGSSIMZ are currently managed within the international framework of

[1] See http://www.sgisland.gs/download/MPA/MPAManagementPlanv2.0.pdf (accessed 25 April 2014).

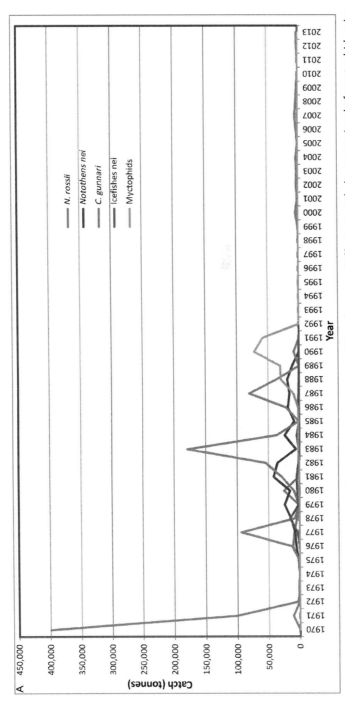

Figure 2.2 (A) Historical fisheries at South Georgia; the large catches of *Notothenia rossii*, *Champsocephalus gunnari* and of myctophids prior to 1993 were taken by vessels from the former Soviet Union.

(Continued)

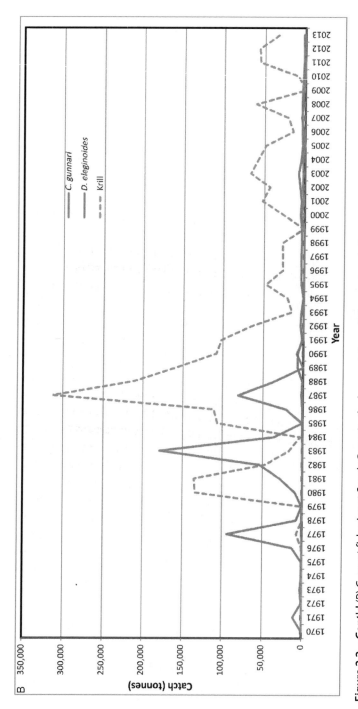

Figure 2.2—Cont'd (B) Current fisheries at South Georgia; the large catches of *Champsocephalus gunnari* and *Euphausia superba* prior to 1993 were taken by vessels from the former Soviet Union.

CCAMLR. Quotas and regulations are set by CCAMLR, but the GSGSSI often imposes lower quotas and stricter regulations than CCAMLR. In 2013/2014, the CCAMLR catch limit for Patagonian toothfish at South Georgia was 2400 tonnes, while the corresponding catch limits for Patagonian toothfish and Antarctic toothfish at the South Sandwich Islands were 45 and 24 tonnes, respectively. However, GSGSSI set a lower catch limit for toothfish than did CCAMLR, setting a limit of 2200 tonnes at South Georgia. In 2013/2014, the catch limit for icefish was 4635 tonnes, while the limit for krill was 372,000 tonnes (279,000 tonnes around South Georgia and 93,000 tonnes around the South Sandwich Islands).

The fishery for Patagonian toothfish began in the late 1980s and expanded rapidly with considerable amounts taken that were illegal, unregulated and unreported (IUU). The fishery uses demersal longlines (Agnew, 2004; Collins et al., 2010), deploying large numbers of hooks baited with sardine or squid. In the early years, it had major problems with seabird by-catch, with large numbers of albatrosses and petrels caught and drowned after being attracted to the baited hooks. In response to this mortality pressure, CCAMLR introduced strict mitigation measures designed to prevent seabird by-catch. These regulations, which include seasonal closures, line-weighting regimes (to ensure baited hooks sink rapidly) and night setting requirements, have virtually eliminated seabird by-catch in the SGSSIMZ. The fishery is now restricted to night setting in the winter months to minimise interactions with foraging seabirds during their breeding season. Since 2010, CCAMLR and GSGSSI have permitted a gradual extension to the fishing season, with the season starting 5 days earlier each year; in the 2013/2014 season, fishing started on April 6. The season extension has not led to increased levels of seabird by-catch, which still remains negligible (between 2008 and 2013 the number of seabirds caught has been 0, 2, 2, 1, 2, 1; this compares with 5755 in 1997[2]).

In 2004, a minimum fishing depth of 500 m was introduced by GSGSSI to protect smaller toothfish. The minimum depth was increased to 550 m in 2010 and 700 m in 2011. Finfish by-catch of macrourids, skates and rays is strictly controlled through vessel move-on rules when by-catch exceeds a given threshold. The fishery was conditionally certified as sustainable and well managed by the Marine Stewardship Council (MSC) in 2004. It was recertified without conditions in 2009 and again in 2014[3]. Though

[2] Seventy six white-chinned petrels were caught in 2014, just as this paper was being printed. The vessel caught birds on a single line set in breach of the night time setting requirements.

[3] See http://www.msc.org/track-a-fishery/fisheries-in-the-program/certified/south-atlantic-indian-ocean/south-georgia-patagonian-toothfish-longline (accessed 25 April 2014).

eco-labels such as the MSC have been criticised (Christian et al., 2013), they still provide confidence for consumers and can result in a price premium for fishers in exchange for more robust ecological management.

Fishing for mackerel icefish began in South Georgia waters in the late 1970s, with large catches taken by east European (then eastern bloc) vessels. Catches peaked in 1981/1982 with a reported catch of 178,000 tonnes, although there is some doubt about the accuracy of this figure. Following concerns about the depletion of stocks, CCAMLR closed the fishery in 1989. The fishery was later reopened, but with a highly conservative catch limit and was restricted to pelagic trawling to avoid impacts on non-target species. In recent years, catch limits have been set at less than 5000 tonnes. Icefish fishing activity is usually focussed on an area to the north-west of South Georgia. The pelagic trawls, with a minimum mesh size of 90 mm, have little by-catch and so have little impact on non-target species.

The fishery for Antarctic krill began in the early 1970s with vessels from Japan and the former Soviet Union catching krill in the Scotia Sea; the catches were mainly processed for human consumption being sold in tinned, frozen or paste form. The fishery focussed on three main areas: the South Shetland Islands (CCAMLR Subarea 48.1), the South Orkney Islands (Subarea 48.2) and South Georgia (Subarea 48.3). The early fishing fleet was later joined by vessels from Poland, Chile and Korea. It peaked in 1981/1982 with catches of 528,000 tonnes. Concern about the rapid expansion of the fishery and the potential impact on non-target species was one of the factors that led to the establishment of CCAMLR in 1982. Catches decreased from 1983 to 1985, but remained above 200,000 tonnes between 1985 and 1992. This reduction in catch is attributed to the discovery of high levels of fluoride in the exoskeleton of krill and with associated processing problems. The rapid reduction in catches in the early 1990s was due to the break-up of the former Soviet Union and reduction in effort from east European fleets. Catches remained stable at around 100,000–150,000 tonnes per year from 1992 to 2009, but have increased to over 200,000 tonnes in the past few years.

In the early years of the krill fishery, vessels employed large conventional pelagic mid-water trawls to catch krill; similar trawls are still in use today. However, in the last few years, a continuous trawling method has also been developed; these vessels use mid-water trawls, with krill pumped continuously from the cod-end to the factory processing deck. Modern conventional mid-water trawlers are now capable of catching and processing up to 400 tonnes per day, while continuous fishing vessels are capable of catching and processing up to 800 tonnes per day. Production varies greatly between vessels, but krill oil is now a major output. Some vessels now utilise the whole krill to produce krill oil, dried meat pellets and dried carapaces. These products are sold for health food supplements, for pet foods and for aquaculture.

The precautionary catch limit for krill in CCAMLR Area 48 has been set at 5.61 million tonnes, but with a lower interim catch limit of 620,000 tonnes (the 'trigger' level). This trigger level is intended to restrict the rapid development of the fishery and ensure that impacts of the fishery on krill-dependent predators are minimised. The full 5.61 million tonnes cannot be taken until such time as CCAMLR reaches an agreement upon the spatial distribution and management of catches so that impacts on krill-dependent predators are minimised. The trigger level has been spatially subdivided with separate limits agreed for each part of Area 48. These limits restrict catches to 155,000 tonnes around the South Shetland Islands, 279,000 tonnes around the South Orkney Islands, 279,000 tonnes around South Georgia and 93,000 tonnes around the South Sandwich Islands. The overall limit for Area 48 is set at 620,000 tonnes, but in setting the subdivided spatial limits, it was recognised that there needed to be operational flexibility for the fishery.

Each of the three fisheries has the potential to interact with the marine ecosystem in different ways. In accordance with the FAO Code of Conduct for Responsible Fisheries[4], their sustainable management should promote the protection of marine living resources and ensure that exploitation does not threaten the environment. The main threats from the three fisheries principally relate to ecosystem perturbation should key species be removed or depleted; incidental mortality of non-target species; or resource competition with predators that target the same species as fishermen. In the toothfish fishery, incidental mortality can happen: when seabirds are caught and drowned on baited hooks; when marine mammals become entangled in lines during hauling; when benthos is destroyed by the movement of lines over the seabed, or when it is caught on hooks; and when non-targeted fish species take baited hooks and are caught as by-catch (e.g. grenadiers or rattails (*Macrourus* spp.), or skates or rays). Data concerning dependent predators of adult toothfish are limited but are likely to include large, deep-diving vertebrates which have the capacity to dive to the depths which adult toothfish inhabit (Collins et al., 2010). Adult toothfish have been recorded in sperm

---

[4] See ftp://ftp.fao.org/docrep/fao/005/v9878e/v9878e00.pdf (accessed 25 April 2014).

whale stomachs (Clarke, 1980) and have occasionally been identified in the diet of elephant seals (Reid and Nevitt, 1998; Slip et al., 1994).

In the icefish fishery, seabirds are occasionally killed, usually as a consequence of warp-strike, or when diving through the large meshes of the net; however, management procedures now limit these problems. The fishery may also occasionally catch non-targeted fish species, including other icefishes and notothens; however, by-catch levels are generally very low and move-on rules apply. There are numerous known predators of icefish, including both marine mammals and seabirds (e.g. Everson et al., 1999; Hill et al., 2005); however, the competitive impacts on the population dietary requirements of these species are unknown, but are thought to be minor given the relatively low yield of icefish allowed for the fishery and the ability of icefish predators to switch diet to krill or to other species (e.g. Waluda et al., 2012).

The main by-catch species in the krill fishery are larval fish, although the numbers of these are relatively few (Agnew et al., 2010). Fur seals have occasionally been caught in the nets of krill trawlers, but since the mandatory requirement for seal exclusion devices in 2003, there have not been any killed[5]. The main threat from the krill fishery comes from the potential for resource competition with krill-eating predators, including with marine mammals and seabirds. The potential for competition is most likely to occur when land-based predators are constrained to feed within a limited distance of their breeding site and when they must also provision their offspring as well as feed themselves. Should the krill fishery become aggregated close to these breeding sites, the potential for resource competition may be very significant (Cury et al., 2011; Pikitch et al., 2012). Ensuring that this does not occur will require careful spatial, temporal and process management; this was a key objective for the SGSSI MPA.

Many of the main threats that arise from the three fisheries are already well managed, either through procedural processes and Conservation Measures agreed by CCAMLR or by fishing licence regulations agreed by GSGSSI. However, some threats are difficult to manage through procedural measures and are better managed by spatial and/or temporal measures such as MPAs.

1.3. Indigenous human populations

SGSSI is a UK Overseas Territory. The islands are governed by GSGSSI, with the Commissioner located in Stanley, Falkland Islands. The archipelago

---

[5] One fur seal was caught in 2014, just as this paper was being printed.

is uninhabited, apart from two scientific research stations operated by the British Antarctic Survey (BAS).

2. ESTABLISHMENT OF THE MPA

2.1. How it came about

The application of spatial and/or temporal protection as a management tool has a long history of use within CCAMLR. For example, CCAMLR has already introduced a ban on bottom trawling; has closed areas to allow for the recovery of previously depleted finfish stocks and has introduced a depth ban on toothfish fishing in exploratory fisheries. However, such measures do not constitute permanent management restrictions as they can potentially be changed at some future date. Therefore, in 2005, CCAMLR recognised the need to establish more permanent spatial protection measures for, *inter alia*, protection of (i) representative areas; (ii) scientific areas to assist with distinguishing between the effects of harvesting (or other activities) and natural ecosystem changes, as well as providing opportunities for understanding the Antarctic marine ecosystem without interference; (iii) areas potentially vulnerable to impacts by human activities, to mitigate those impacts and/or ensure the sustainable use of marine living resources; and, (iv) the protection or maintenance of important ecosystem processes in locations where those processes are amenable to spatial protection.

In 2008, CCAMLR identified 11 priority areas where it was agreed that there was a need to develop MPAs as part of a representative system across the Convention Area. These priority areas included both South Georgia and the South Sandwich Islands as separate areas for consideration. The development of the SGSSI MPA, along with those at some other Sub-Antarctic Islands (e.g. DeLord et al., 2014; Lombard et al., 2007), should therefore be viewed within this broader context.

Spatial and temporal management restrictions have also previously been used by GSGSSI as part of fishery licence regulations; these include the ban on bottom trawling; seasonal closures and depth restrictions in the toothfish fishery; as well as a ban on all fishing within 12 nautical miles of South Georgia and Shag Rocks. GSGSSI also introduced other spatial management restrictions as part of the MSC certification process for the toothfish fishery at South Georgia.

Recognition of the need for permanent spatial management measures by both CCAMLR and GSGSSI mirrors other international aspirations to develop comprehensive networks of MPAs. Such concerns have been voiced by the United Nations World Summit on Sustainable Development (UN WSSD) in 2002, the IUCN World Parks Congress in 2004, the Convention for the Protection of the Marine Environment of the North-East Atlantic in 2003 and CCAMLR in 2005. Indeed, the UN WSSD in 2002 set a target for governments to protect 20–30% of all marine habitats under their jurisdiction.

Consequently, in 2009, project proposals to support the creation and designation of a SGSSI MPA were submitted to the UK government; these led to the development of two pieces of funded work. The first project, funded through a UK Overseas Territories Environment Programme award (OTEP SGS701; 1 April 2010–31 March 2013), set out to identify important and vulnerable marine habitats that require conservation in order to better preserve the unique characteristics of the SGSSIMZ. Based on this informa-tion, the project sought to identify a representative and comprehensive network of MPAs. The project aimed to benefit GSGSSI in the sustainable management of the SGSSIMZ, through the conservation of habitats and species in the context of climate variability and change, as well as pressures from local sustainable fisheries and tourism. The second project, funded through a Darwin Initiative award (18-019; 1 April 2010–31 March 2012), aimed to establish baseline data on the macro- and mega-benthic bio-diversity of the South Georgia shelf and slope, identify key (endemic) species and biodiversity hotspots and utilise data to formulate management strategies for the conservation of biodiversity in the SGSSIMZ.

In February 2012, based on the initial outputs from these projects, coupled with expert scientific opinion, the GSGSSI declared the SGSSIMZ north of 60°S to be a sustainably managed MPA (IUCN Category VI). No fisheries activities are licensed in that part of the SGSSIMZ that falls south of 60°S. The MPA, approximately $1.07 \times 10^6 \text{ km}^2$, included a number of strict no-take (IUCN Category 1a) zones located within 12 nautical miles of South Georgia, Clerke Rocks, Shag Rocks and Black Rock, and within 3 nautical miles of each of the South Sandwich Islands, giving a total area of 20,431 km^2 that provided refugia for fish species (including for spawning fish), and that reduced competition between fisheries and land-based forag-ing seabirds and marine mammals. In June 2013, the provisions of the MPA were enhanced, following completion of both projects and further consul-tation with interested stakeholders.

In designating the MPA, GSGSSI sought to enhance spatial and tem-poral protection of the marine ecosystem. On advice from the scientific teams involved in the project, GSGSSI agreed high-level objectives to (i) better protect important biodiversity within the SGSSIMZ; (ii) protect representative benthic and pelagic environments; (iii) protect important or large-scale ecosystem processes; (iv) protect trophically important pelagic prey species, including Antarctic krill; (v) protect foraging areas used by spatially constrained krill-eating predators; (vi) protect localised areas of ecosystem importance; (vii) protect rare or vulnerable benthic habitats; (viii) protect areas important for key life cycle stages and processes of commercially important species such as Patagonian toothfish, mackerel icefish and Antarctic krill; (ix) promote recovery of the SGSSIMZ ecosystem following historical harvesting and (x) maintain ecosystem robustness and resilience to climate variability and change within the SGSSIMZ.

2.2. Rationale for its placement

In determining the spatial and temporal provisions of the MPA, a systematic conservation planning approach (Margules and Pressey, 2000) was used, as this can help balance competing demands for ecosystem services by setting clear and transparent objectives, the achievement of which should entail minimal cost or disruption to existing human activities. In determining how best to balance conservation against sustainable fishing, managers rely upon the infrastructure necessary to assess, regulate and police the ecosystem. The costs of supporting this infrastructure may be generated from either the ecosystem itself (e.g. from fishing licences or from tourism) or from external sources, such as governments or Non-Governmental Organisations (NGOs) interested in conserving biodiversity. Revenue from fisheries licences is the primary means by which GSGSSI can currently fund adequate control, monitoring and surveillance within the SGSSIMZ; this infrastructure is critical to the elimination of IUU fishing, which if allowed to flourish would almost certainly seriously degrade ecosystem values (cf. Edgar et al., 2014; Halpern, 2014). Currently, remotely sensed satellite monitoring methods of control and surveillance are difficult to use as the seas around South Georgia are often cloud obscured.

The development of the SGSSI MPA was guided by the scientific team working in close association with GSGSSI. This team had extensive experience and knowledge concerning the regional marine ecosystem, including expertise relating to benthic organisms, zooplankton (including krill), fish, squid, marine mammals, seabirds, commercial fisheries, oceanography, and climate variability and change. The team collated relevant and available geospatial scientific information and human use information into a Geographic Information System (ArcGIS Version 10: ESRI, Redlands, CA) so that candidate spatial and temporal management measures might be examined.

After collating the available spatial data, and despite the relative abundance of data for South Georgia compared with other oceanic islands, it was apparent that the data alone were inadequate for providing robust or comprehensive scientific advice about a number of important ecological issues (for example, *inter alia*; (i) inter- and intra-annual levels of environmental variability are both pronounced features of the ecosystem at South Georgia, yet the underlying causes and ecological responses are not well-understood; (ii) satellite tracking data for penguins, other seabirds and seals are only available from a limited number of sites, for a limited number of species and for a limited number of life-history categories, so considerable uncertainty remains about their preferred spatial habitat utilisation; (iii) benthic data are mostly only available over the shelf and not for the shelf slope or abyssal plain). Nevertheless, the use of the available spatial data, combined with the use of expert scientific knowledge, did allow progress to be made with the systematic conservation planning process, even in advance of collecting new field data. It was also recognised that expert knowledge was not a complete substitute for robust ecological data and that future research and monitoring would be necessary.

The scientific team reviewed all existing fishery licencing regulations and identified a number of additional spatial and temporal management measures that might be better implemented through spatial protection. The team then led an expert scientific workshop designed to consider potential new conservation objectives and candidate proposals for spatial and temporal management measures.

Following the workshop, the scientific team developed a series of specific conservation objectives which were subsequently agreed by GSGSSI; these are described below and summarised in Table 2.1. Each conservation objective addressed specific threats related to different components of the marine ecosystem. Appropriate management measures were then designed to mitigate these threats, while allowing other activities to take place where these were assessed to be compatible with the agreed conservation objectives. Following the scientific workshop and subsequent stakeholder consultations, the revised management measures were implemented under national legislation.

In designing the provisions of the MPA, different conservation objectives focussed on either the pelagic or benthic communities. Separating these two domains was a strategic decision; this separation also reflected the ecological expertise available within the two scientific projects advising GSGSSI. Better understanding of the threats associated with the different

Table 2.1 Protection objectives for the South Georgia and the South Sandwich Islands MPA

Protection objective	Target area description	Target area for protection (%)	Actual area protected (%)
Objective 1	**Representative pelagic environments**		
Protect representative examples of all pelagic habitats	1.1 All pelagic environment types	10–20	See Tables 2.3A and 2.3B 41.4% Via seasonal closure
Objective 2	**Large-scale ecological processes**		
Protect large-scale flux of Antarctic krill from the Antarctic to South Georgia	2.1 Pelagic off-shelf areas in the ACC	30–40	41.4% Via seasonal closure
Protect off-shelf to on-shelf transfer of Antarctic krill carried to South Georgia by the flow of the ACC	2.2 Pelagic areas over South Georgia shelf slope and submarine canyons to the south and east of the island	30–40	41.4% Via seasonal closure
Protect large-scale flux of marine snow to benthic domain	2.3 All environment types	30–40	41.4% Via seasonal closure
Objective 3	**Trophically important pelagic prey species**		
Protect Antarctic krill summer core distribution	3.1 Coastal and shelf areas	100	100%
Protect Antarctic krill winter core distribution	3.2 Coastal and shelf areas	30–40	33%
Objective 4	**Spatially constrained foraging areas of krill-eating predators**		
Protect Antarctic fur seal summer foraging distribution	4.1 Coastal, shelf, shelf slope and off-shelf foraging areas to 150 km	100	100%
Protect macaroni penguin summer foraging distribution	4.2 Coastal, shelf, shelf slope and off-shelf foraging areas to 80 km	100	100%

Continued

Table 2.1 Protection objectives for the South Georgia and the South Sandwich Islands MPA—cont'd

Protection objective	Target area description	Target area for protection (%)	Actual area protected (%)
Protect gentoo penguin summer foraging distribution	4.3 Coastal foraging areas to 22 km	100	100%
Protect chinstrap penguin summer foraging distribution, east South Georgia and South Sandwich Islands	4.4 Shelf, shelf slope and off-shelf foraging areas to 80 km	100	100%
Protect black-browed summer foraging distribution	4.5 Coastal and shelf foraging areas	100	100%
Protect seabird and marine mammal winter foraging areas	4.6 Coastal and shelf foraging areas	30–40	33%
Objective 5	**Localised areas of ecological importance**		
Protect pelagic larval fish from the krill fishery, to aid recovery of demersal fish stocks and recovery of previously depleted stocks	5.1 Shag Rocks	100	100%
Protect pelagic larval fish from the krill fishery, to aid recovery of demersal fish stocks and recovery of previously depleted stocks	5.2 Coastal areas and fjords	100	100%
Objective 6	**Representative benthic environments**		
Protect representative examples of benthic habitats	6.1 All shelf areas	10–20	100%
	6.2 Biodiverse areas on the shelf slopes in the depth zones used by the toothfish fishery	10–20	15%
	6.3 All abyssal areas	10–20	100%

Table 2.1 Protection objectives for the South Georgia and the South Sandwich Islands MPA—cont'd

Protection objective	Target area description	Target area for protection (%)	Actual area protected (%)
Objective 7	**Areas important for *Dissostichus eleginoides* life cycle processes**		
Protect juvenile *Dissostichus eleginoides* recruitment and feeding	7.1 Shag Rocks	100	100%
Protect *Patagonotothen guntheri* (prey for juvenile *Dissostichus eleginoides*) from the ice fish fishery	7.2 Shag Rocks	100	100%
Objective 8	**Rare or vulnerable benthic habitats**		
Protect benthos on South Georgia shelf	8.1 Benthic shelf communities on the South Georgia shelf	100	100%
Protect unique species, communities and habitats on black smokers and white smokers	8.2 Volcanic geothermal fracture zone	100	100%
Protect unique species, communities and habitats living in the deepest trench in the Southern Ocean	8.3 Scotia tectonic plate subduction zone and hadal communities	100	100%
Protect communities on isolated seamounts	8.4 Southern seamounts south of South Georgia	100	100%
	8.5 Northern seamounts north of South Georgia	100	100%
Protect the unique reef structure formed by polychaete worms	8.6 Serpulid reef structure	100	100%
Protect benthic abyssal communities	8.7 Benthic deep abyssal communities	100	100%

fisheries operating in a particular domain allowed these to be considered, while ensuring comprehensive protection for other parts of the ecosystem. Such a rationale provides for the conservation of biodiversity, while also allowing for sustainable harvesting, a pragmatic strategy critical for the economic underpinning of South Georgia and vital for effective fisheries management and successful enforcement (see also Dunne et al., 2014 (Chapter 3)). The agreed set of conservation objectives and management measures enabled a precautionary approach to management to be maintained.

In designing the management measures, each conservation objective was assessed and a target percentage for protection was agreed in terms of the proportion of the area type or feature to be included in the MPA (Table 2.1). Where conservation objectives were directed at protecting representative ecological features or communities, a target percent of 10–20% was considered appropriate; this follows recommendations from the Convention on Biological Diversity (CBD). Where conservation objectives were directed towards the protection of unique or vulnerable ecological processes, features or community components, a target percent of 100% was thought necessary; where ecological information was uncertain, or where threats were currently perceived to be minor (but which may increase in the future), a target percent of 30–40% was considered to be adequate. The scientific team recognised that these three categories of target percentages should be reviewed in the future, based on the outcomes of any additional research and monitoring.

In summary, the steps for the SGSSI MPA planning process were:

1. Collate available physical, environmental, biodiversity and human use data;
2. Review existing management measures and identify those that would be better implemented through spatial protection;
3. Convene an expert workshop to identify additional conservation objectives and candidate areas for spatial and/or temporal protection;
4. Determine target percentages for protection, based on agreed conservation objectives;
5. Refine the new management measures to be implemented;
6. Consult stakeholders and other interest groups;
7. Assess the performance of the new management measures against the protection target percentages;
8. Identify future work that may be required to achieve further protection targets or to revise aspects of the MPA where data were insufficient.

The specific conservation objectives, agreed between the scientific teams and GSGSSI, were as follows:

2.2.1 Objective 1: Protect representative pelagic environments

In order to determine how best to structure representative protection for the pelagic domain, a bioregionalisation analysis was undertaken following the methods of Grant et al. (2006) as revised by Raymond (2011)[6]. Defining bioregions allows the SGSSIMZ to be divided into 'proxy ecological regions' until better data become available which provide increased ecological understanding. Spatial protection can be developed to provide representative coverage of the different bioregions.

A geographic grid (with each cell of 0.1° resolution) for the planning area was set up that extended beyond the limits of the SGSSIMZ, thereby avoiding edge effects. The number of clusters was determined by the default settings in the software. Analyses were conducted in Matlab Version 7 (Mathworks, Natick, MA) and R Version 2 (R Foundation for Statistical Computing, Vienna).

Four variables were used for the pelagic bioregionalisation: (1) sea surface temperature (SST), comprising a summer climatology spanning the 2002/2003 to 2009/2010 Austral summer seasons, compiled from satellite-derived data from MODIS Aqua (Thomas and May, 2005)[7]; (2) mean sea surface height (SSH), comprising the mean dynamic topography spanning the 1993–1999 period, compiled from satellite-derived data integrated with *in situ* data (Rio et al., 2011)[8]; (3) satellite-derived chlorophyll *a*, comprising the mean summer surface concentrations in mg m^{-3} spanning the 2002/2003 to 2009/2010 Austral summer seasons[9]; and (4) ocean depth, comprising seafloor topography from satellite altimetry integrated with *in situ* ship depth soundings (Smith and Sandwell, 1997)[10]. Depth data were sub-sampled and log-transformed to moderate the effects of deep waters. Each variable was interpolated from its original spatial resolution to the 0.1° analysis grid using bilinear interpolation.

SST was used as a general indicator of water mass designation; however, satellite altimetry (SSH) generally gives frontal positions that better match those from subsurface hydrography than does SST. Both SST and SSH were therefore included in the regionalisation. Ocean depth and productivity (Chl-*a*) were used to provide general indicators of different habitats and

---

[6] Source code is available at https://github.com/jiho/atlasr/zipball/master (accessed 25 April 2014).

[7] Data obtained from http://oceancolor.gsfc.nasa.gov/ (accessed 25 April 2014).

[8] Data available from http://www.aviso.oceanobs.com/ (accessed 25 April 2014).

[9] Data obtained from http://oceancolor.gsfc.nasa.gov/ (accessed 25 April 2014).

[10] Data obtained from http://topex.ucsd.edu/WWW_html/mar_topo.html (accessed 25 April 2014).

bioregions. The results showed that 10 bioregions (ecological habitats) were evident in the gridded data covering the SGSSIMZ (Figure 2.3 and Table 2.2).

The derived habitat types (proxy ecological regions) show close similarities with the known oceanographic characteristics (compare Figures 2.1 and 2.3) of the Scotia Sea (see also Orsi et al., 1995). The habitat types also reflect the ecological regions described by Longhurst (1998), which align with the circumpolar structure of the ACC. The similarity of the derived habitat types to these published ecological regions suggests that representative protection for the full range of habitat types will adequately represent different ecological and oceanographic characteristics.

2.2.2 Objective 2: Protect large-scale ecological processes

There are a number of large-scale ecological processes that have been identified as critical to the maintenance of the SGSSIMZ. Potentially the most important are (a) the transport of krill in the ACC, (b) advection of krill from open ocean off-shelf habitats to neritic on-shelf areas and (c) the supply of

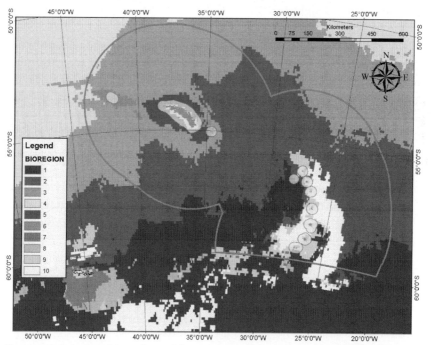

Figure 2.3 Pelagic habitats derived from the bioregionalisation analysis for the South Georgia and South Sandwich Islands MPA (see Table 2.2 for habitat descriptions).

Table 2.2 Pelagic habitats derived from the bioregionalisation analysis for the South Georgia and South Sandwich Islands MPA (see Figure 2.3)

Bioregion	Characteristics of the bioregion description	Bathymetry	Summer Chl-a	SST	SSH	Area (km)	Protected by the 12 nm closures (%)
1	South of SACCB	Mainly deep—4500 m	Low productivity	Coldest	Lowest	124,345.1	1.24
2	SACCB to SACCF	Mainly deep—3500 m	Mid productivity	Warmer	Mid level	471,525.7	0.39
3	SACCF to PF	Mainly deep—4000 m	Mid productivity	Warmer	High	328,471.6	0.14
4	North of PF	Mainly deep—4000 m	Low productivity	Warmest	Highest	8745.6	0.00
5	South Georgia shelf	Shelf	Productive	Warmer	High	33,026.6	31.20
6	Shag Rocks shelf	Deep shelf and shelf slope	Mid productivity	Warmer	Higher	12,005.2	26.49
7	South Scotia Arc shelf	Shelf	Productive	Cold	Low	2429.4	81.56
8	South Scotia Arc shelf slope	Shelf slope	Low productivity	Cold	Low	1156.9	22.07
9	Productive south of SACCF	Mid depth—2000 m	Productive	Cold	Low	29,348.2	20.23
10	Highly productive south of SACCF	Mainly deep—4500 m	Highly productive	Coldest	Lowest	54,324.7	9.81
	Total					1,070,266.6	

detritus and other organic matter to the benthic domain via benthic–pelagic coupling. While other processes are important, if these three were interrupted, for example, by fishing or by climate change (or other drivers), there is a high probability that both the pelagic ecosystem and the benthic community would be very different (e.g. Trathan et al., 2007).

2.2.2.1 Transport of krill in the ACC

Protection of pelagic communities is complex, as biological production in one area may be advected elsewhere by prevailing ocean currents (so-called allochthonous production that has been transported to a site, as opposed to autochthonous production that has been produced locally within a site). The advection of krill is particularly important in this respect (Murphy et al., 1998; Trathan et al., 1995). Indeed, the flux of krill to South Georgia is critical for sustaining the high biomass of krill-dependent predators such as whales, seals, penguins and other seabirds, as well as fish, squid and predatory zooplankton that feed upon juvenile or larval krill (Heywood et al., 1985). Maintaining the flux of krill to South Georgia is therefore critical. Protection within the SGSSIMZ is important; however, protecting the source of krill for South Georgia also requires coordinated international action to protect krill stocks at a circumpolar scale, in particular upstream of the SGSSIMZ where the commercial krill fishery operates at the South Orkney Islands, the South Shetland Islands and close to the Antarctic Peninsula (Everson and Goss, 1991; Murphy et al., 1997).

2.2.2.2 Advection of krill from open ocean off-shelf habitats to neritic on-shelf areas

Submarine canyons are important for the advection of krill and other zooplankton onto the shelf (Ward et al., 2007) where they may then become available to predators that are constrained by the location of their breeding sites or to predators with small home ranges. Flows of krill onto the shelf have been hypothesised for sites along the Antarctic Peninsula, with some submarine canyons being particularly important, especially those that open at the shelf edge to the pelagic off-shelf domain (Dinniman and Klink, 2004; Pinõnes et al., 2011, 2013); there is a high probability that such processes are also important at South Georgia.

2.2.2.3 Supply of detritus and organic matter to the benthic domain via benthic–pelagic coupling

Submarine canyons are also important sites for vertical advection as they accelerate and cause more efficient transfer of energy via particle flux to the deep-sea benthos. This advection is important when considering benthic–pelagic coupling. The faecal material of zooplankton is a major route of energy flow, while vertical migration of many zooplankton species, as well as the production of pelagic larvae by benthic organisms, represents different pathways to link the two domains (Schnack-Schiel and Isla, 2005). Particle fluxes in the Antarctic are highly variable in both volume and composition. There are also vertical differences throughout the water column. At depths shallower than 500 m, more than 90% of the annual flux commonly occurs during the spring–summer period. Faecal material usually reaches its maximum in February once the early phytoplankton bloom has developed. In contrast, at greater depths, particle fluxes close to the sea floor beyond the continental shelf break do not normally show such strong seasonal variation. In the Southern Ocean, these organic matter transfers are important since they allow the accumulation of highly nutritive material, which may fuel the benthos during the winter months. Faecal rain showers from krill during the summer months are potentially very important sources of material for benthic communities (Atkinson et al., 2012).

Ensuring that human activities do not disrupt large-scale ecological processes, such as krill flux and krill transfer from off-shelf to on-shelf as well as vertical transport, is important for the South Georgia ecosystem. Protection for these processes could be achieved by restricting fishing at key locations. However, any necessary protection is complicated by the very nature of these habitats and communities. Planktonic communities in the open ocean are ephemeral, develop seasonally and change with prevailing weather conditions (Tarling et al., 2012; Ward et al., 2012). These pelagic communities are also generally wide-ranging and extend beyond the SGSSIMZ (Ward et al., 2004).

Currently, direct threats from scientific activities, tourism activities or fishing do not threaten the pelagic off-shelf areas of the ACC, as in general most vessels only transit across the off-shelf areas. However, the consequences of climate change are known to be affecting the Southern Ocean, not just within the southwest Atlantic sector, but also more widely around the Antarctic continent. Currently, increases in ocean temperature (Gille, 2002, 2008; Whitehouse et al., 2008), changes in sea-ice extent and duration (Parkinson, 2002, 2004; Stammerjohn et al., 2008; Turner et al., 2009) and increases in ocean acidity levels (Bednaršek et al., 2012; Kawaguchi et al., 2013; Tarling et al., 2012) are known to be taking place. Protecting against these threats can only be addressed at a global level, as unilateral actions within the SGSSIMZ will do little to help mitigate these threats.

Also, management actions within the SGSSIMZ would not mitigate the impacts of direct threats upstream of the SGSSIMZ. However, implementing protection measures in pursuit of other objectives (e.g. Objective 1, Objective 3 and Objective 4) might also provide adequate precautionary protection for Objective 2, at least until such time as new more direct threats develop in off-shelf areas in the future.

2.2.3 Objective 3: Protect trophically important pelagic prey species

It has long been recognised that Antarctic krill populations at South Georgia show considerable spatial and temporal variability (e.g. Fielding et al., 2014; Reid et al., 2010; Trathan et al., 2003). Evidence is also accumulating that there is not only inter-annual variability but also intra-annual variability (Saunders et al., 2007; Reid et al., 2010), with a peak in krill biomass during the summer months. An increase in biomass might be expected at this time as krill feed and grow, laying down both somatic and reproductive tissues as a result of increases in seasonal primary production (Korb et al., 2004). However, increased advection into the South Georgia area may also be important as krill are released from the winter sea-ice and advected northwards (Murphy et al., 2004; Thorpe et al., 2004, 2007). Protecting this flow of krill towards South Georgia requires consideration of krill stocks at a circumpolar scale, also an understanding of how krill biomass is distributed both under sea-ice and in the open ocean (Brierley et al., 2002; see also Objective 2).

The total biomass of krill in open oceanic off-shelf habitats is greater than the biomass of krill in shelf habitats (Atkinson et al., 2008; Hewitt et al., 2004). However, locating predictable and accessible off-shelf aggregations of krill is uncertain. Hill et al. (2009) suggest that the probability of encountering high-density krill swarms (at a scale of 1 km) is a function of both the local depth (which is a proxy for distance from land) and the larger scale average krill density (at scales >100 km^2). However, once krill arrive at South Georgia and move to on-shelf habitats, aggregations are more predictable (see Trathan et al., 2003). Commercial fisheries (Trathan et al., 1998a) and krill-eating predators (e.g. Trathan et al., 1998b) both target these predictable aggregations. Central-place foragers rely critically upon accessible and predictable aggregations of krill, particularly during the breeding season, when many are constrained to return to land to provision their offspring. Pelagic predators may also prefer certain locations to feed; for example, during the early years of the whaling era at South Georgia, fin and blue whales were taken in very great numbers from a small number of sites where they may have been feeding preferentially (Everson, 1977). Protecting these areas is feasible if they can be identified through analyses of distribution and tracking data (e.g. Staniland et al., 2007; Trathan et al., 1998b, 2006b) or if robust habitat modelling can identify environmental correlates (Lascelles et al., 2012; Ratcliffe et al., 2014; Wakefield et al., 2011). Unfortunately, such evidence is not available for most species or from most breeding sites at South Georgia, and no evidence at all is available from the South Sandwich Islands.

2.2.4 Objective 4: Protect spatially constrained foraging areas of krill-eating predators

South Georgia is renowned for the diversity, abundance and high density of krill-eating predators (whales, seals, penguins, other seabirds, fish and squid). Populations of many of these species are changing, with some recovering from historical exploitation (Trathan et al., 2012). The relative biomass of these different species is also changing; nevertheless, their consumption of krill remains substantial. Towards the end of the 20th century, and assuming a diet mainly of krill, the annual consumption by Antarctic fur seals and macaroni penguins at South Georgia was estimated to be 3.84 and 8.08 million tonnes, respectively (Boyd, 2002). Their per capita food consumption varies depending upon sex and age but, overall, was 1.7 tonnes year^{-1} for Antarctic fur seals and 0.45 tonnes year^{-1} for macaroni penguins. Antarctic fur seals have their highest levels of energy demand during their breeding season (Boyd, 2002) when they forage around South Georgia and just beyond the edge of the shelf (Staniland et al., 2011). Outside the breeding season, much of the fur seal population leaves the island (Staniland et al., 2012). In contrast, macaroni penguins have high energetic demands during winter as well as during breeding and prior to moult (Boyd, 2002), but in winter they have usually left the immediate vicinity of South Georgia (Ratcliffe et al., 2014). Post-breeding macaronis are widely distributed with most foraging happening off-shelf over deep water, hundreds of kilometres distant from summer breeding sites.

Many other krill-eating marine mammals and seabirds also follow the same pattern and leave the vicinity of South Georgia in the winter after their breeding cycle is complete. This includes black-browed albatrosses (Phillips et al., 2005), grey-headed albatrosses (Croxall et al., 2005) and white-chinned petrels (Phillips et al., 2006). Southern right whales also leave the area in winter (Leaper et al., 2006), but this time in anticipation of calving in warmer, more sheltered waters.

In contrast, other krill-eating species are known to remain close to the island after breeding is complete, including gentoo penguins (Tanton et al., 2004). Some, such as chinstrap penguins, probably show variable behaviour post-breeding. No reports on foraging behaviour exist for chinstraps breeding at SGSSI, but elsewhere long-range pre-moult (Biuw et al., 2010) and post-moult (Trivelpiece et al., 2007) winter migrations of chinstraps have been recorded. These journeys have been to feeding areas in the vicinity of other colonies elsewhere in the southwest Atlantic, some-times hundreds of kilometres distant. However, near-range winter migra-tion has also been recorded with birds staying within a hundred kilometres of their breeding sites (Trivelpiece et al., 2007), with time spent both over the shelf and off-shelf. Such differences in winter migratory behaviour among chinstrap penguins might reflect different ancestral epicentres of population (Trivelpiece et al., 2007).

Protecting central-place foragers, including albatrosses, penguins and seals, which are considered to have high conservation status, remains chal-lenging in the absence of adequate telemetry data to indicate the location of important bird areas (Lascelles et al., 2012; Pichegru et al., 2012) or ecolog-ically and biologically significant areas (see CBD COP 9 Decision IX/20). At present, most tracking data from South Georgia for seabirds and seals have been collected from a limited number of sites, yet many of the species are widespread and it is therefore uncertain as to how representative telemetry data are from one site (Ronconi et al., 2012).

Protecting previously depleted populations of krill-eating predators that are now recovering is important, including populations of Antarctic fur seals (Payne, 1977), humpback whales and southern right whales that calve in the coastal waters off Brazil and Argentina (Leaper et al., 2006; Zerbini et al., 2004). These whale stocks feed around South Georgia and/or the South Sandwich Islands during the summer months and both are thought to feed on krill while in the Scotia Sea (Reilly et al., 2004; Tormosov et al., 1998).

Potentially, the most important direct threat to krill-eating species at South Georgia is from competition with the commercial fishery for krill, though currently the catch from the fishery remains low (Trathan et al., 2012). At the current catch level, the competitive impacts from the fishery are undetectable (SC-CAMLR-XXII/Annex 4), unless all the catch were to be spatially concentrated and taken close to monitored penguin and seal breeding sites. In the future, if the fishery increases as predicted (Nicol et al., 2011), care must be taken to ensure that it does not become concen-trated into small areas, or take such large catches, that it becomes a major competitor to krill-dependent species (Cury et al., 2011).

At present, there is no evidence to suggest that krill abundance at South Georgia has changed over the past 25 years, despite high levels of inter-annual and intra-annual variability in krill biomass (e.g. Fielding et al., 2014; Reid et al., 2010; Saunders et al., 2007; Trathan et al., 2003). Further, the continued recovery of marine mammal populations suggests that krill biomass remains adequate for most species.

2.2.5 Objective 5: Protect localised areas of ecological importance

In addition to protecting the large-scale ecological processes critical to the maintenance of the South Georgia and South Sandwich Islands marine eco-system (Objective 2), other smaller-scale ecological processes and localised areas of importance are also worthy of protection. These include localised protection of pelagic larval fish to reduce or eliminate their by-catch in the krill fishery, which may then aid recovery of previously depleted demersal finfish stocks.

At South Georgia, ichthyoplankton species form characteristic oceanic, neritic and near-shore assemblages (Loeb et al., 1993). The oceanic group (Loeb et al., 1993) is generally composed of mesopelagic species, which as adults are most abundant at depths >200 m; this assemblage typically includes larval *Electrona antarctica*, *Electrona carlsbergi* and *Notolepis coatsi*, and it also commonly includes late larval and early juvenile stages of some notothenioids, including *Notothenia rossii*, *Notothenia coriiceps* and *D. eleginoides*. The neritic group (Loeb et al., 1993) is dominated by notothenioids and includes several species that have previously been commercially exploited over the shelf as demersal adult or juvenile stages; it typically includes larval *C. gunnari*, *Lepidonotothen larseni* and *Gobionotothen gibberifrons*. The near-shore group (Loeb et al., 1993) contains species that are most commonly found in shallow water as adults, including *Gobionotothen marionensis* and *Harpagifer georgianus*.

Post-hatching and yolk sac stages of many species of fish are found within the fjords of the north coast of South Georgia, confirming the widespread use of these features as spawning grounds (Belchier and Lawson, 2013). Spawning of the majority of notothenioid fish species occurs during the Austral autumn and winter (Kock and Kellerman, 1991) with overwintering eggs hatching during early spring. Consequently, peak larval fish abundance at South Georgia is observed between September and November (Belchier and Lawson, 2013). As the majority of spawning occurs in the coastal and neritic regions, the density of fish larvae is highest inshore and decreases significantly with distance from the coast (Everson et al., 2001).

The abundance of numerous fish larvae in shallow waters and in near-shore areas is probably related to the extensive neritic spawning areas. The spawning locations of many demersal and bentho-pelagic fish species are found within the coastal fjords and gullies that are located within 12 nm of the coast of South Georgia. Spawning of *C. gunnari* is known to occur in the fjords and bays on the north coast of the island (Everson et al., 2001), while high densities of adult *N. rossii* frequently found to the east may represent spawning aggregations (Kock et al., 2004). In addition, restricted offshore transport and elevated food abundance are probably also important (Loeb et al., 1993). Distribution patterns of zooplankton at South Georgia suggest a slow rate of water exchange across the shelf break and circulation patterns that maintain neritic populations over the shelf (Atkinson and Peck, 1990). In near-shore areas, there are elevated levels of zooplankton species that constitute an important initial food source for fish larvae (North and Ward, 1989, 1990).

Protecting near-shore larval fish communities is not only potentially important for aiding the recovery of previously depleted demersal finfish stocks, but probably also for maintaining the food supplies of predators that consume finfish, including seals and penguins, and some of the smaller volant seabirds including e.g. storm petrels (Croxall and North, 1988). Protecting larval communities may also provide greater ecosystem resilience as some species show a higher average growth rate in certain summers, possibly related to large-scale patterns of environmental variability, indicated by sea temperature (Hill et al., 2005; North et al., 1998).

Currently, the krill fishery operates over the outer shelf and at the shelf break; therefore, protection for oceanic larval fish communities is probably not necessary, particularly if they are carried away from South Georgia in the flow of the ACC. Protection for these communities may become important in the future, should the krill fishery move off-shelf into oceanic regions.

2.2.6 Objective 6: Protect representative benthic environments
Generally, Antarctic and Sub-Antarctic benthic assemblages are poorly described, but in many locations are very rich (Arntz et al., 1994). The benthic fauna across the Scotia Arc shows high levels of species richness; for example, of the 1224 species recorded from the South Orkney Islands, 1026 are marine and 821 benthic (Barnes et al., 2009). There are very few non–Antarctic locations where the benthos constitutes such a high proportion of biodiversity. As elsewhere in the world, there are also major changes in faunal composition with depth (e.g. Griffiths et al., 2008), but this is difficult to assess as it is strongly confounded by sampling effort.

There are only a few studies on the benthic communities within the SGSSIMZ, including for the areas around Shag Rocks (e.g. Barnes,

2008), around South Georgia itself and around the South Sandwich Islands (e.g. Kaiser et al., 2008). Benthic communities at Shag Rocks and at South Georgia show a major biogeographic boundary, with Antarctic species reaching their northern limit and Magellanic species reaching their southern limit in this region (Barnes et al., 2009, 2011; Hogg et al., 2011). The benthic community on the South Georgia shelf is extremely rich and represents an anomaly compared with low regional endemism south of the Polar Front (PF) (Hogg et al., 2011). The South Sandwich Islands are thought to have a benthic fauna that is predominantly Antarctic, but most of the seabed around the archipelago has been too little sampled to provide meaningful analyses.

The slow growth and recovery times of these assemblages mean they are susceptible to disturbance; for example, from iceberg scour which is a common feature in many coastal and shelf assemblages (cf. Barnes and Souster, 2011). Benthic species in cross–shelf canyons and on deeper shelf slopes are seldom exposed to such processes, while assemblages in the abyssal depths are thought to be little disturbed.

The extent of the seabed in the SGSSIMZ is large, comprising some 1.07 million km^2. The diversity of benthic habitats and communities across the SGSSIMZ is considerable and includes a number that are unique; protection for the benthic domain is therefore important. As a consequence of planning for the MPA, bottom fishing is now limited to long–line fishing for Patagonian toothfish and Antarctic toothfish in depth zones between 700 and 2250 m. The aerial extent of the SGSSIMZ between depths of 700 and 2250 m is 85,928 km^2 or approximately 8.01% of the SGSSIMZ.

Despite considerable effort to collate biodiversity information for the seabed within the SGSSIMZ, it remains poorly described, particularly for the shelf slope, abyssal plains and at all depths around the South Sandwich Islands. A number of samples from the northern continental slope off South Georgia and Shag Rocks indicate that it is highly biodiverse (e.g. Barnes, 2008; Griffiths et al., 2008). The data compiled by Hogg et al. (2011) suggest that the shelf above 700 m is one of the most species-rich regions of the Southern Ocean with 1445 species known to date, the vast majority of which are benthic. Furthermore, most of this benthos is both rare (35% of species have been recorded just once) and endemic (more than 44% of bryozoans, cnidarians and mollusc species which occur there are only known from the region; see Hogg et al., 2011). For this reason, the benthos around the South Georgia archipelago has been highlighted as of high priority with respect to conservation and implementation of protocols agreed by the CBD (Barnes et al., 2011).

As shelf areas above 700 m and the abyssal depths below 2250 m are closed, representative protection is not needed apart from within those depths used by the toothfish fishery. Hence, and in order to provide protection for representative habitats and communities within these depths, the outputs from an analysis of historical data from the fishery documenting by-catch of both benthic organisms and demersal non-target fish species (see Agnew et al., 2007; Martin et al., 2012) were reviewed. Martin et al. (2012) used two separate analysis approaches. The first involved phylogenetic analysis of samples collected by fisheries observers during 2005 to explore areas of high species richness. The second used by-catch data from the long-line fishery to map the distribution of benthic by-catch and model the impacts on benthic communities and toothfish catches. The combined analysis identified a number of candidate areas that were deemed to be of sufficient biodiversity and importance that they should be closed to commercial fishing. Importantly, as most available species-habitat data are for depths shallower than 700 m, the abundance of habitat-forming organisms, such as octocorals, was generally used as a proxy for biodiversity in the selection of these candidate areas.

Martin et al. (2012) examined the effects of different candidate areas on catches of toothfish and on by-catch species (including macrourids, rays and a variety of benthic taxa) in order to select areas in which the highest percentage of octocorals would be protected with the least impact on toothfish catches. Proposals from Martin et al. (2012) led to the establishment of a number of Benthic-Closed Areas (BCAs) within which commercial fishing has been prohibited since 2008, and only very limited fishing occurs for research purposes (the impact of which should, however, be monitored).

In order that these BCAs provide effective protection to the benthic community within their boundaries, only a reduced level of experimental fishing is now allowed. The effort level for experimental fishing is a compromise between complete protection and sufficient effort to determine whether fish stocks increase within the boundaries of each area and to determine if the benthic fauna change after the cessation of commercial harvesting.

2.2.7 Objective 7: Protect areas important for D. eleginoides life cycle processes

The operational governance of South Georgia, including both terrestrial and marine management, relies upon the Patagonian toothfish fishery; approximately 52% of GSGSSI gross income is currently derived from the toothfish fishery[11]. Consequently, it is critical that this stock is harvested sustainably and that an adequate spawning stock biomass is maintained to ensure the continued recruitment of juvenile fish. Key factors for ensuring the sustainable exploitation of the stock include a robust stock assessment model, and protection for important dietary items, in particular, those for larval and juvenile fish.

At South Georgia, spawning in Patagonian toothfish generally occurs during the Austral winter with a large spawning event in late July/August and a smaller event in April/May (Agnew et al., 1999). Pre-spawning fish are thought to be distributed all around the shelf slopes of South Georgia and Shag Rocks and spawning is not localised in particular regions of the shelf. Patagonian toothfish attain maturity (length at 50% maturity) at an age of between 6 and 10 in males and between 10 and 13 years in females (Collins et al., 2010); Patagonian toothfish are the most fecund nototheniids, with absolute fecundities ranging from 56,940 to 567,490 at South Georgia (Nevinsky and Kozlov, 2002). Currently, the commercial fishery operates between 6 April and 31 August, but with an earlier start for vessels that have a consistent track record of meeting all existing fishery regulations. Fishing therefore occurs during the spawning season and protection of an adequate spawning stock biomass is crucial.

GSGSSI follows the CCAMLR ecosystem approach for managing the toothfish stock, which advocates assessments that are based upon a precautionary approach where scientific data are limited. Using the CCAMLR approach, the historic stock dynamics are projected 35 years into the future, with variability included in some of the input parameters. A constant catch projection allows for the calculation of a long-term yield that satisfies the CCAMLR decision rules. These rules stipulate that the yield is chosen such that the probability of the spawning stock biomass dropping below 20% of its median pre-exploitation level during the 35-year projection is not greater than 10% and that the median escapement in the spawning stock biomass at the end of the projection is not less than 50% (Agnew, 2004). Though the GSGSSI follows the CCAMLR approach, it generally implements an even more stringent set of management criteria than does CCAMLR, setting a lower Total Allowable Catch.

Designating closed areas that protect a proportion of the spawning stock biomass, particularly older, more fecund individuals, is likely to ensure that the fishery is more robust against unforeseen perturbations, such as changes in environmental relationships between recruitment and body size in

---

[11] See http://www.sgisland.gs (latest financial accounts for the period ending December 2012, accessed 25 April 2014).

relation to temperature (Belchier and Collins, 2008). Also, closing areas where juveniles may feed and grow and where they are protected from the fishery is also likely to help ensure that the stock is more robust to changing environmental conditions. Testing the utility of the BCAs to these assumptions should therefore be seen as a priority for the future.

Such closed areas are likely to be important, as at South Georgia toothfish move by only about 10 km year^{-1} on average (Agnew et al., 2006a), indicating little large-scale movement. Indeed, even when the very small numbers of individuals that may move more widely are included, tagging results show that animals move, on average, by only 27 km between tagging and recapture (Agnew et al., 2006b).

Though toothfish may not move very far horizontally, they show clear ontogenetic changes in their depth range and diet which are associated with down-slope migration as body size increases (Collins et al., 2010). Larval toothfish are pelagic and occur in the upper water column (North, 2002). The diet of larval toothfish has not been studied, but the diet of juvenile toothfish (<750 mm TL) has been determined from trawl caught specimens, which are primarily piscivorous (Collins et al., 2007). Collins et al. (2007) found that the diet of juveniles at Shag Rocks was dominated by the abundant yellow-finned nototothen (*Patagonotothen guntheri*), which comprised 90% of the diet, whereas at South Georgia, where the yellow-finned nototothen is absent, other nototheniids were the primary prey. Some pelagic prey species have also been reported to be taken, including myctophids and Antarctic krill (Collins et al., 2007). The diet of large, adult toothfish has been studied from both long-line and pot-caught fish; in general, adult toothfish are opportunistic scavengers or carnivores (Collins et al., 2010).

Closed areas are likely to be valuable for protecting toothfish, particularly juvenile animals that have specific prey requirements. However, because of their low mobility, such closed areas are unlikely to provide high levels of biomass export that may facilitate enhanced catches at other locations. For example, the toothfish population found in the north of the South Sandwich Islands may be an extension of the South Georgia stock (Roberts and Agnew, 2008), but it remains relatively small and overspill from South Georgia remains limited.

The MSC considers that the toothfish stock at South Georgia is in good condition, and that the fishery is well monitored and well managed[12].

[12] See http://www.msc.org/track-a-fishery/fisheries-in-the-program/certified/south-atlantic-indian-ocean/south-georgia-patagonian-toothfish-longline/2ndreassessment-documentation/20140708_PCDR_TOO111.pdf (accessed 31 July 2014).

However, the addition of no-take zones to protect benthos and key life-history stages, and/or key dietary items for juvenile fish, may further enhance both ecosystem and stock resilience.

2.2.8 Objective 8: Protect rare or vulnerable benthic habitats

Less than 5% of the area of the SGSSIMZ lies above 700 m depth. However, the South Georgia shelf is the most species-rich region of the Southern Ocean recorded to date (Hogg et al., 2011). The benthic fauna is marked by the cumulative dominance of endemic and range-edge species, potentially at their thermal tolerance limits. Consequently, the ecological implications of environmental change for the South Georgia benthic ecosystem could be severe. If ocean temperatures continue to rise (Whitehouse et al., 2008), changes could include depth profile shifts by some fauna towards cooler Antarctic Winter Water (90–150 m), the loss of some range-edge species from regional waters and the partial or complete loss of some of South Georgia's endemic species. This means that the South Georgia shelf warrants a very high level of protection.

Within the SGSSIMZ along the East Scotia Ridge, to the west of the South Sandwich Islands, a unique hydrothermal vent community is known to exist which is thought to represent a new vent biogeographic province (Rogers et al., 2012). The presence of black smokers, diffuse venting and associated chemosynthetically driven ecosystems along the geographically isolated back-arc spreading centre means this area should also be protected.

To the east of the South Sandwich Islands, a hadal trench exists which is the deepest trench in the Southern Ocean. Such trenches account for the deepest 45% of the oceanic depth range and host active and diverse biological communities though these are generally very poorly described (Jamieson et al., 2010). Approximately 24% of the SGSSIMZ lies below 6000 m. Hence, the uniqueness of this feature and of any hadal community that may exist in the trench means that this area should be protected.

There are a number of isolated seamounts in the SGSSIMZ that are likely to support important benthic communities, although most of these have not been scientifically sampled and remain undescribed. Seamounts are also often important for foraging pelagic predators (Morato et al., 2010). Therefore, protection of communities on isolated seamounts is generally thought to be important. Indeed, a recent review of seamounts (Rowden et al., 2010) supports the assertion that seamount communities are vulnerable to fishing, and that these communities have high sensitivity and low resilience to bottom fishing disturbance. Rowden et al. (2010) also found plausible evidence to support the suggestion that seamounts are stepping stones for dispersal, oases of abundance and biomass, and hotspots of species richness. This suggests that these undescribed seamounts in the SGSSIMZ should be protected.

A reef of the serpulid polychaete *Serpula narconensis* has been reported from the South Georgia shelf (Ramos and San Martin, 1999). This occurrence was the second serpulid reef reported for the Antarctic. It was reported at a depth of between 91 and 105 m depth and so is very different from the serpulid reefs previously reported which are generally situated in the intertidal zone or at shallow depths. Ramos and San Martin (1999) suggested that this reef may be one of the largest, if not the largest, serpulid reef in the world. Protection of this unique reef structure formed by polychaete worms is therefore warranted.

As with the seamounts in the SGSSIMZ, the benthic abyssal communities are poorly studied. Elsewhere, however, deep-sea biodiversity is among the highest on the planet (Ramirez-Llodra et al., 2010). Two large-scale diversity patterns are thought to exist for deep-sea benthic communities. Firstly, a relationship between diversity and depth is thought to exist, with a peak at intermediate depths of 2000–3000 m (Ramirez-Llodra et al., 2010). Secondly, a poleward trend of decreasing diversity has been suggested, although this remains controversial. The range of depths found across the SGSSIMZ suggests that much of the seabed falls within this depth zone; indeed, approximately 17.75% of the SGSSIMZ lies between these depths making it an important candidate for protection.

2.3. Who drove it

Overall responsibility for the development of the SGSSI MPA naturally lay with GSGSSI, with the support of the UK government, including for the development of the necessary legislation. Collation of scientific evidence was through the OTEP project and the Darwin Initiative project (see earlier). The OTEP award also included resources to advise GSGSSI about conservation and policy objectives.

2.4. Stakeholder involvement

GSGSSI hosts annual meetings in the UK to which a broad range of stakeholders are invited (scientists, the tourism industry, fishing industry, conservation organisations, etc.). Scientific results, conservation and policy objectives, and MPA proposals were presented from the two science projects at these meetings and stakeholders were invited to comment. GSGSSI also holds an annual meeting with fishing industry representatives where it generally discusses stock assessments, catch projections and other related issues; at these meetings, scientific results, conservation and policy objectives, and MPA proposals were also presented and discussed. Following the scientific workshop, a final consultation exercise elicited detailed responses from a broad range of interested parties. Revised conservation objectives and new spatial and temporal management proposals were developed in the light of advice from these various stakeholder groups. The MPA design procedure was thus lengthy and comprehensive.

2.5. Local legislation

The SGSSI (Territorial Sea) Order (1989) designates the boundaries at 12 nm of the Territories of SGSSI. The SGSSIMZ is defined in Proclamation No. 1 (1993) at 200 nm from the baselines. The Fisheries (Conservation and Management) Ordinance (2000, as amended) provides the legal basis for the regulation of fishing activities within the SGSSIMZ. The MPA was initially designated by the SGSSI MPA Order (2012), which was repealed and replaced by the SGSSI MPA Order (2013) which incorporated the additional protection outlined above. The Wildlife and Protected Areas Ordnance (2011, as amended) is the primary legislation which allows for the designation of MPAs.

2.6. International context

SGSSI lie within the management area of CCAMLR, so fisheries assessments are carried out within a multilateral, international context. The UK, through the GSGSSI, implements all CCAMLR Conservation Measures applicable to SGSSI waters and in some cases imposes even more stringent regulations and lower fishery catch limits in order to ensure the sustainable exploitation of marine resources. Further, in addition to the toothfish fishery, MSC certification has also been obtained for fishing vessels operating in the mackerel icefish and krill fisheries. UK Sovereignty over the islands is disputed by Argentina.

3. IMPACTS OF THE MPA
3.1. Social

There is no indigenous human population; therefore, there are no social impacts of the MPA. However, the management provisions of the MPA will potentially provide valuable lessons for managing the fishery for Antarctic krill under CCAMLR.

3.2. Ecological

3.2.1 Objective 1: Protect representative pelagic environments

The results of the pelagic bioregionalisation analysis (Grant et al., 2006; Raymond, 2011) are shown in Table 2.2 and Figure 2.3; the analysis identified 10 pelagic bioregions that fall into two categories: (i) those unique or having low representation outside the SGSSIMZ and (ii) those having high representation outside the SGSSIMZ.

For the unique pelagic bioregions, the waters of the South Georgia shelf (Bioregion 5), the Shag Rocks shelf (Bioregion 6), the South Scotia Arc shelf (Bioregion 7) and the South Scotia Arc shelf slope (Bioregion 8) are all protected within the coastal 12 nm pelagic no-take zones (management measures E1, E2, E3 and P3; Table 2.3A and Figure 2.4A) which afford between 20% and 80% protection for these bioregions (Table 2.2). Although the 12 nm pelagic no-take zones protect just 3.4% of the SGSSIMZ, they protect important habitats that are not well-represented elsewhere. In addition, the seasonal closure of the krill fishery (management measure P1; Table 2.3A) provides an equivalent of 41.4% space/time closure (151 out of 365 days); moreover, this is operative during the most critical time of year for the majority of krill predators. The pelagic space/time protection afforded by these five measures is greater than the target of 20–30% for representative protection set by the UN WSSD and greater than the 10% target set under Decision X/2 by the CBD.

For the non-unique pelagic bioregions, two (Bioregions 9 and 10) that occur in the productive waters to the south of the Southern ACC Front (SACCF) are also well protected within the 12 nm pelagic protection zones (management measure P3; Table 2.3A), which affords between 10% and 20% protection (Table 2.2). In contrast, the offshore waters to the south of the Southern ACC Boundary (SACCB) (Bioregion 1), between the SACCB and the SACCF (Bioregion 2), between the SACCF and the Antarctic PF (Bioregion 3), and to the north of the PF (Bioregion 4), are not well protected by the coastal no-take zones (management measures E1, E2, E3 and P3; Table 2.3A) which afford less than 1.25% protection for each of these bioregions (Table 2.2). Nevertheless, all these habitats are totally protected by the summer seasonal closure of the krill fishery for 5 months of the year (management measure P1; Table 2.3A), equivalent to 41.4% space/time closure (see earlier).

Table 2.3A Agreed protection measures for the pelagic domain for the South Georgia and South Sandwich Islands MPA (see Figure 2.4A)

	Measure	Area	Rationale	Specific considerations
Pelagic: implemented in 2012	E1	12 nm no-take zone around South Georgia. The seabed, overlying waters and associated organisms in an area of 13,899 km^2	Protects all marine life within 12 nm of South Georgia	The shallow marine environment around South Georgia including: 1. The spawning grounds of many fish species, including mackerel icefish; particularly protects juvenile/larval fish 2. The inshore foraging areas of marine predators such as gentoo penguins, cormorants, petrels and prions
	E2	12 nm no-take zone around Clerke Rocks. The seabed, overlying waters and associated organisms in an area of 1923 km^2	Protects all marine life within 12 nm of Clerke Rocks	The shallow marine environment to the southeast of South Georgia including: 1. The spawning grounds of many fish species, including mackerel icefish; particularly protects juvenile/larval fish 2. The inshore foraging areas of marine predators such as penguins, cormorants, petrels and prions
	E3	12 nm no-take zone around Shag and Black Rocks. The seabed, overlying waters and associated organisms in an area of 2337 km^2	Protects all marine life within 12 nm of Shag Rocks	The shallow marine environment of the Shag Rocks shelf incorporating: 1. A principal recruitment area for juvenile Patagonian toothfish 2. Spawning grounds of mackerel icefish; particularly protects juvenile/larval fish 3. A key foraging area for black-browed albatross, Antarctic fur seals and baleen whales

Continued

Table 2.3A Agreed protection measures for the pelagic domain for the South Georgia and South Sandwich Islands MPA (see Figure 2.4A)—cont'd

Measure	Area	Rationale	Specific considerations
E4	3 nm no-take zone around each of the South Sandwich Islands. The seabed, overlying waters and associated organisms in areas that total 2114 km^2	Protects all marine life within 3 nm of the South Sandwich Islands	The shallow marine environment around each of the South Sandwich Islands including: 1. The inshore foraging grounds of marine predators 2. The spawning grounds of fish species
Pelagic: implemented in 2013	P1	Seasonal summer closure of the krill fishery; 1 November to 31 March	Avoids potential competition between the krill fishery and krill predators	The pelagic environment across the whole SGSSIMZ: 1. Protects land-based krill predators (seals and seabirds) during the time when adults are feeding their young and are constrained in their foraging range during November–March which is the critical breeding period 2. Protects krill-eating fish and squid during the time when many are preparing to spawn 3. Protects recovering populations of krill-eating whales while on their summer feeding grounds and before they migrate north to calve

| P2 | Move-on rules in the mackerel icefish fishery and krill fishery around Shag and Black Rocks | Protects juvenile toothfish and *Patagonotothen guntheri* from by-catch in krill or icefish fisheries | The area is used by both krill and icefish fisheries. Seasonal closure of the krill fishery (P1) would provide significant protection, but some species forage extensively in this area during the winter, so the use of move-on rules and by-catch limits in these pelagic fisheries will provide protection for non-target fish species and limit potential resource competition |
| P3 | 12 nm pelagic no-take zone around each of the South Sandwich Islands. The seabed, overlying water and associated organisms in areas that total 18,042 km^2 | Protect near-shore foraging ranges of chinstrap penguins and recovering whale stocks | The area is very rarely fished by the krill fleet; in the longer term, there may be a desire to see the krill fishery move further offshore in order to distribute effort and catches |

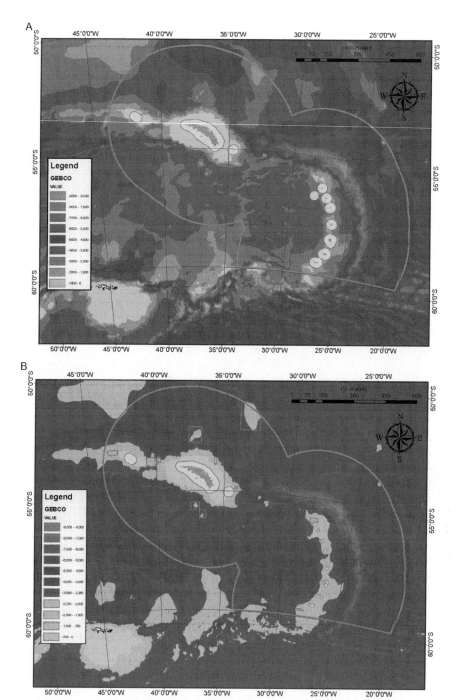

Figure 2.4 (A) Designated protection measures for the pelagic domain for the South Georgia and South Sandwich Islands MPA (see Table 2.3A). (B) Designated protection measures for the benthic domain for the South Georgia and South Sandwich Islands MPA (see Table 2.3B). Depths in yellow (light grey in the print version) are the only depths where bottom fishing gear is allowed to come into contact with the sea bed.

3.2.2 Objective 2: Protect large-scale ecological processes

Management measure P1 (Table 2.3A) provides considerable protection for some of the large-scale ecological processes critical to the maintenance of the marine ecosystem in the SGSSIMZ. During the summer months, the transport of krill in the ACC (Thorpe et al., 2007) is protected by the summer seasonal closure of the krill fishery, as is the advection of krill from open ocean off-shelf habitats to on-shelf areas (Ward et al., 2007).

The advection of krill (Thorpe et al., 2007) means that krill will be continually replenished throughout the year, not just during the summer. Thus, any depletion of krill, whether due to consumption by krill predators or by the fishery (in winter), will be replenished by advection from the southern Scotia Sea.

The summer closure of the krill fishery also helps maintain the supply of detritus and other organic matter to the benthic domain through benthic–pelagic coupling. The faecal material of zooplankton is a major route of energy flow (Schnack-Schiel and Isla, 2005) and the prohibition on krill fishing in summer means that this is not interrupted during the period when more than 90% of the annual flux occurs, generally during the spring–summer period, following the spring phytoplankton bloom.

3.2.3 Objective 3: Protect trophically important pelagic prey species

The summer seasonal closure of the krill fishery (management measure P1; Table 2.3A) means that adult krill remain directly available to higher trophic level predators, and also that adult krill are protected during their summer spawning season (Tarling et al., 2007). Calyptopis and early stage furcilia are found in the vicinity of South Georgia, though numbers are mostly lower than models predict (Tarling et al., 2007). Nevertheless, though reproduction and early stage development can be successful in the region, little is known about the fate of krill eggs and larvae spawned in the waters around South Georgia. It is generally assumed that krill spawned at South Georgia are carried away from the island in the flow of the ACC (Tarling et al., 2007). Such advection might be important for other downstream ecosystems, but this remains a topic for future research.

Though advection is probably one of the main causes of local recruitment failure at South Georgia, predation of krill larvae remains important (Tarling et al., 2007) and the summer seasonal closure of the krill fishery means that larvae (from spawning adults) will continue to be available to predators such as myctophid fish, chaetognaths and amphipods (Tarling et al., 2007), which in turn are important prey for higher trophic level predators.

3.2.4 Objective 4: Protect spatially constrained foraging areas of krill-eating predators

The average phenology for seabirds breeding at South Georgia (Figure 2.5) indicates that the summer seasonal closure of the krill fishery (management measure P1; Table 2.3A) protects the dietary requirements of most species during the time when they are constrained as central-place foragers, whether they are direct consumers of krill or indirect consumers. However, some species begin breeding before 1 November or complete after 31 March, meaning that they are not entirely protected by the seasonal closure. In practice, however, most of these species are not in direct competition with the krill fishery, which operates mostly over the northern outer shelf edge (Trathan et al., 1998a). These species are also not in direct competition as a high percentage of predator dietary demands are met off-shelf over oceanic habitats. Indeed, depending upon species, between 47% and 70% of consumption occurs in off-shelf habitats (Hill et al., 2012). Some of these long-range foragers, such as king penguins, are myctophid specialists (Trathan et al., 2008) and so are not vulnerable to direct competition with the krill fishery. Others, such as wandering albatross, feed on squid and fish and are also known to scavenge (Xavier et al., 2003a), so again are unlikely to be in competition with the krill fishery. However, some species do feed on krill, to a greater or lesser extent; these include Antarctic prions (Liddle, 1994), Wilson's storm petrels (Croxall and North, 1988), white-chinned petrels (Berrow and Croxall, 1999), black-browed albatrosses (Xavier et al., 2003b), grey-headed albatrosses (Xavier et al., 2003b), light-mantled sooty albatrosses and southern giant petrels (Croxall et al., 1985). Outside the summer, these species often forage much further afield than the South Georgia shelf, particularly during their pre-incubation exodus, incubation or during late chick rearing (e.g. Croxall et al., 2005; González-Solís et al., 2000; Navarro et al., 2013; Phillips et al., 2005, 2006, 2009; Quillfeldt et al., 2013; Wakefield et al., 2011); thus, they often obtain dietary items outside the SGSSIMZ, where they will not be in direct competition with the krill fishery.

Antarctic fur seals, like seabirds, are spatially constrained while provisioning their offspring (Staniland et al., 2007), so will be protected from competition with the krill fishery by the summer seasonal closure (management measure P1; Table 2.3A). During the winter, when fur seals are not central-place foragers, they might face resource competition with the commercial fishery. However, fur seals show distinct patterns of dispersal during the winter, with many leaving the SGSSIMZ and foraging close to the Patagonian Shelf or close to the edge of the winter seasonal sea-ice (Staniland et al., 2012).

South Georgia Breeding Birds Schedule

Species	September	October	November	December	January	February	March	April	May	June	July	August
King Penguin 1	C C C	C C C C	A A A / C	L L L F F	H H H L L F F	L L L H	L L/H H	L/H L/H CB CB	C C C	H CB CB	CB CB CB CB	C C C C
King Penguin 2	C C C	C C C C	C C	C C C C	H H H	C C C	H H H	C C C	C C C	C C C	CB CB CB	C C C C
Gentoo Penguin		L L	L/H L/H H	A A	L L F	C C	H H H	C				
Macaroni Penguin	A A	A A	L L	A A L	L F F	C C	H H					
Wandering Albatross	C C C	C C C	C C C	A A A F F	F F F H H	I I I	H H H	C C C	C C C	C C C	C C C	C C C C
Black-browed Albatross		A A A	L L	I I	H CB CB	CB CB C	C C	C C C	F F	F		
Grey-headed Albatross	A A	A A A	L L	I I	CB CB CB	CB C C	C C	C C C	F F	F F		
Light-mantled Sooty Albatross	A A	A A	L L	I I	I I	CB CB CB	C C C	C C C	C F F	F F		
Southern Giant Petrel	???		L L	I I	H H H	C C C	C C C	C C	F			
Northern Giant Petrel	???	L L		CB CB CB	H H	C C C	C C C	F F	F			
Cape Petrel*	A A	A A		L L	H H	C C C	C C C	F				
Snow Petrel					L L	C C C	C C	C C	F F			
Antarctic Prion	A A	A A	L L	I	L L	C C	C C	F F	F			
Blue Petrel	A A	A A L L	I	L L	O O	O F F	O C	C O F F	F			
White-chinned Petrel	A A	A A		L L	I I	H H H	H H	O F F	F F	F		
Wilson's Storm-Petrel	A A	A A		A A A	L L	I	H H	F F	F			
Black-bellied Storm-Petrel			???	A A A	L L L	I						
South Georgia Diving Petrel			???	H H	L L	H O	F F					
Common Diving Petrel			???	H H H	H H	O O/F C/F	F F					
South Georgia Shag			L L	H H H	H H	O O	C C C					
South Georgia Pintail			L L	L L	L L	O O	C C C	F F				
Greater Sheathbill*	A A	A A A	L L/H	L L	O O	C C	C F F	F				
Brown Skua	A A	A A	L L	L/H L/H	H H H	F F	F F F	F				
Kelp Gull		A A	L L	H H H	H H H	O O	O F F					
Antarctic Tern*			L L L	H H H	C C C	F F F						
South Georgia Pipit												

* No data from S. Georgia
A = Arrival
L = Laying
I = Incubating
H = Hatching
CB = Chick brooded (where a significant period)
C = Chick (not brooded)
F = Fledging
??? = Uncertain arrival time
—— = Exact dates and periods uncertain. Horizontal line runs from laying to fledging dates. Vertical bar indicated mean hatch date
Bold type indicates mean dates (where available)

Figure 2.5 Average phenology for birds breeding at South Georgia. The summer seasonal closure for the krill fishery (1 November to 31 March) is shown in green (grey in the print version).

Whales, unlike seabirds and fur seals, are not central-place foragers; however, their feeding grounds may still be spatially constrained. Humpback
whales (Breeding Stock A) show peak abundance at their wintering grounds
off north-eastern Brazil between July and October (Zerbini et al., 2004),
departing for their summer feeding grounds between October and
December and arriving there some 40–60 days later in February (Zerbini
et al., 2006). Assuming their return migration also takes 40–60 days, they
must depart the southern feeding grounds off SGSSI in April or May. Consequently, humpback whales, during much of their time in the SGSSIMZ,
are also protected by the summer closure of the krill fishery (management
measure P1; Table 2.3A). Southern right whales are at their wintering
grounds from mid-May until early December; although stragglers may be
present throughout the year, the main group of whales begins to arrive in
May or June, reaches peak numbers by the end of September or early October
and most depart for the feeding grounds by December (Payne, 1986). Thus,
they too are protected by the summer closure of the krill fishery while feeding
in the SGSSIMZ.

3.2.5 Objective 5: Protect localised areas of ecological importance
Management measures E1, E2, E3 and P3 (Table 2.3A) protect larval and
juvenile fish, thus aiding recovery of demersal fish stocks and recovery of
previously depleted stocks. These measures also protect larval and juvenile
toothfish from by-catch in krill or icefish fisheries.

3.2.6 Objective 6: Protect representative benthic environments
The only depth zone open for bottom fishing lies between 700 and 2250 m
(management measures E1–E6, B8, B10 and B11; Table 2.3B and
Figure 2.4A); however, the BCAs and seamount closures (management
measures B1–B7 and B9; Table 2.3B) ensure that 12,970 km^2 is protected,
equivalent to 15% of this particular depth zone.

Martin et al. (2012) highlight the West Shag Rocks and Gully areas
(management measure B1; Table 2.3B) as having high gorgonian abundance; they also highlight the area to the east of South Georgia as an area
of relatively high gorgonian abundance (management measure B2;
Table 2.3B). These areas, particularly West Shag Rocks and West Gully, also
give significant protection to other benthic taxa, including Ascidiacea,
Bryozoa, Cnidaria (Alcyonacea, Antipatharia, Actiniaria, Stylasteridae,
Anthoathecatae, Scleractinia and Zoanthidea), Echinodermata (such as
Euryalida) and Porifera. Analysis of toothfish maturity data shows that the

Table 2.3B Agreed protection measures for the benthic domain for the South Georgia and South Sandwich Islands MPA (see Figure 2.4B)

	Measure	Area	Rationale	Specific considerations
Benthic: implemented in 2012	E1	12 nm no-take zone around South Georgia. The seabed, overlying waters and associated organisms in an area of 13,899 km^2	Protects all marine life within 12 nm of South Georgia	The shallow marine environment around South Georgia including: 1. The spawning grounds of many fish species, including mackerel icefish; particularly protects juvenile/larval fish 2. The inshore foraging areas of marine predators such as cormorants
	E2	12 nm no-take zone around Clerke Rocks. The seabed, overlying water and associated organisms in an area of 1923 km^2	Protects all marine life within 12 nm of Clerke Rocks	The shallow marine environment to the southeast of South Georgia including: 1. The spawning grounds of many fish species, including mackerel icefish; particularly protects juvenile/larval fish 2. The inshore foraging areas of marine predators such as cormorants 3. The 'spirulid reef' at approximately 55°00'S, 34°31'W
	E3	12 nm no-take zone around Shag and Black Rocks. The seabed, overlying water and associated organisms in an area of 2337 km^2	Protects all marine life within 12 nm of Shag Rocks	The shallow marine environment of the Shag Rocks shelf incorporating: 1. A principal recruitment area for juvenile Patagonian toothfish 2. Spawning grounds of mackerel icefish; particularly protects juvenile/larval fish 3. A key foraging area for cormorants

Continued

Table 2.3B Agreed protection measures for the benthic domain for the South Georgia and South Sandwich Islands MPA (see Figure 2.4B)—cont'd

Measure	Area	Rationale	Specific considerations
E4	3 nm no-take zone around each of the South Sandwich Islands. The seabed, overlying water and associated organisms in areas that total 2114 km^2	Protects all marine life within 3 nm of the South Sandwich Islands	The shallow marine environment around each of the South Sandwich Islands including: 1. The inshore foraging grounds of marine predators 2. The spawning grounds of fish species
E5	Ban on bottom trawling	Protects the benthic fauna from the destructive practices of bottom trawling	Protects the benthic habitat from the most destructive form of bottom fishing
E6	700 m minimum depth for long-line bottom fishing	Protects the benthic fauna from any form of bottom fishing for depths shallower than 700 m	Protects juvenile toothfish at these depths
B1	Establish existing Benthic Closed Areas (BCA) as areas closed to commercial long-line fishing in areas that total 3715.5 km^2	Protects vulnerable marine fauna such as corals, gorgonians and sponges in locations within the depth zone used by the toothfish long-line fishery	The BCAs have been closed since 2008 and there is evidence that in addition to protecting the benthos these areas are refugia for toothfish
B2	Closure of the East South Georgia candidate BCA bounded by the box 54°48′ to 54°54′S and 34°00′ to 34°12′W; 142.9 km^2	Protects an area that is particularly rich in gorgonians	This was one of the original candidate BCAs, but was not included in the initial selection

Benthic: implemented in 2013

B3	Closure of the seamounts to the south of South Georgia; 2714.7 km^2	Protects the potentially (largely unknown) benthic fauna of these seamounts; provides refugia for large adult toothfish	The seamounts have been open to the toothfish fishery, but the licenced fleet has not been concentrated in this area in recent years; closed for the first time in 2012
B4	Close the North–East Georgia Rise within the box 51°12′ to 52°24′S and 32°36′ to 34°00′W; 9852.9 km^2	Protects the potentially (largely unknown) benthic fauna of this large seamount; provides refugia for large adult toothfish	This area was open to the toothfish fishery in the distant past, but the licenced fleet has not been fished in this area in recent years
B5	Close the North Georgia Rise bounded by the box 52°19.8′ to 53°00′S and 36°45′ to 37°39.6′W; 4590.1 km^2	Protects the potentially (largely unknown) benthic fauna of this large seamount; provides refugia for large adult toothfish	This area was open to the toothfish fishery, but licenced vessels only visit the area in some years
B6	Closure of the Protector Shoals within the box 55°49.5′ to 56°03′S and 27°39′ to 28°15′W; 756.8 km^2	Protects the potentially (largely unknown) benthic fauna of this large submarine plateau	This area was open to the toothfish fishery
B7	Closure of the Kemp Seamount within the box 59°40′ to 59°45′S and 28°00′ to 28°16′W, the Kemp caldera 59°40′ to 59°43.8′S and 28°16′ to 28°25′W, and the Adventure caldera 59°42′ to 59°43.8′S and 27°50.4′ to 27°51.6′W; 204.5 km^2	Protects the potentially (largely unknown) benthic fauna of this seamount and these calderas. Protects these chemosynthetic habitats, including unique white smoker vent fields	These areas are currently mostly closed to the toothfish fishery

Continued

Table 2.3B Agreed protection measures for the benthic domain for the South Georgia and South Sandwich Islands MPA (see Figure 2.4B)—cont'd

Measure	Area	Rationale	Specific considerations
B8	Closure of all areas deeper than 2250 m to any form of bottom fishing	Protects benthic fauna in deep-water areas. Represents a range of deep-water benthic habitats. Creates large benthic no-take zone	There is little or no toothfish fishing deeper than 2200 m so there are no threats in these depth strata at the moment
B9	Closure of representative benthic habitat on the South Georgia shelf slope	Precautionary protection covering representative examples of important benthic habitat types	Benthic Closed Areas contribute to representative protection
B10	Close the volcanic geothermal zone near the South Sandwich Islands, including sites E2 and E9	Protects the potentially unique fauna of this zone; this is a unique feature in the entire Southern Ocean and potentially in the World Ocean	Not an area commercially fished
B11	Close the Scotia tectonic plate subduction zone and associated hadal communities	Protects the potentially unique fauna of this zone; this is the deepest trench in the entire Southern Ocean and one of the deepest in the World Ocean	Not an area commercially fished

east South Georgia area may also be an important location for spawning too-thfish (Martin et al., 2012).

3.2.7 Objective 7: Protect areas important for D. eleginoides *life cycle processes*

Management measures P2, E6, B3, B4, B5 and B8 (Table 2.3B) provide considerable protection for the different life cycle processes of Patagonian toothfish. The 2250 m lower depth limit on commercial fishing (management measure B8; Table 2.3B) will protect the very largest fish, as toothfish are known to follow an ontogenetic down–slope migration, moving deeper as they grow (Collins et al., 2010). This is important as larger toothfish are known to be significantly more fecund than smaller fish (Chikov and Melnikov, 1990; Nevinsky and Kozlov, 2002). The closure of various sea-mounts (management measures B3, B4 and B5; Table 2.3B) will also allow fish to grow to a larger size where they are protected and hence increase the contribution to the spawning stock biomass. The ban on toothfish fishing at depths shallower than 700 m means that young toothfish are protected until they reach approximately 75 cm in length.

Management measure P2 (Table 2.3A), which includes move-on rules in the icefish fishery and krill fishery around Shag and Black Rocks, also protects juvenile toothfish and *P. guntheri* (prey for juvenile toothfish; Collins et al., 2007) from by-catch in krill or icefish fisheries; *P. guntheri* may still be caught, but the risk is much reduced.

3.2.8 Objective 8: Protect rare or vulnerable benthic habitats

Within the SGSSIMZ, there are a number of benthic habitats that are unique within the Southern Ocean, indeed within the global ocean. Under management measures E1–E6 and B1–B11 (Table 2.3B), the benthic domain in the SGSSIMZ receives protection. Benthic habitats shallower than 700 m over the South Georgia shelf (which are significant even globally, see Hogg et al., 2011) are completely protected (management measure E6; Table 2.3B), as are benthic habitats deeper than 2250 m (management measure B8; Table 2.3B). Management measure E5 (Table 2.3B) ensures that the destructive practice of bottom trawling is banned throughout the entire SGSSIMZ.

Unique species, communities and habitats on black smokers and white smokers along the East Scotia Ridge volcanic geothermal fracture zone (Rogers et al., 2012) are protected by management measures B6, B7, B8 and B10 (Table 2.3B); protection for the unique spirulid reef structure formed by polychaete worms on the South Georgia shelf (Ramos and

San Martin, 1999) is afforded by management measure E2 (Table 2.3B), while protection for the benthic abyssal communities and the as yet undescribed, but potentially unique species, communities and habitats in the deepest trench in the Southern Ocean in the Scotia tectonic plate subduction zone and hadal zone are given under management measures B8 and B11 (Table 2.3B). Protection is also afforded to communities on isolated seamounts to the south of South Georgia and to the north of South Georgia under management measures B3, B4 and B5 (Table 2.3B).

3.3. Economic

It is highly likely that the provisions of the MPA will cause significant restrictions to fisheries, particularly the krill fishery, should they expand in the future. However, the MPA is unlikely to curtail, to any significant degree, current fisheries or tourism activities. Maintaining existing revenue from fisheries licences was a policy objective for the GSGSSI which relies upon such revenue, and without which it would not be able to maintain adequate management, control and surveillance, including protection from IUU fishing.

In addition to using fisheries licence revenue for fisheries management, GSGSSI utilises licence revenue to finance other activities, including terrestrial habitat restoration projects such as the eradication of alien species (e.g. *Rangifer tarandus*), developing hydroelectric power to reduce carbon emissions and the restoration and cleanup of historical whaling stations. No revenues are raised for the UK Treasury.

The SGSSIMZ holds globally significant biodiversity, including species of high conservation value. The SGSSIMZ is also an area where fur seal stocks continue to increase after historical extirpation and where whales and finfish stocks are showing increasing signs of recovering (Trathan et al., 2012). As such, SGSSIMZ requires careful management to facilitate conservation of these species and communities.

Given existing levels of biodiversity, the net ecological benefits that might be derived by creating a no-take closed fisheries zone across the whole of the SGSSIMZ are uncertain. However, the loss of revenue to maintain necessary fisheries protection would be very likely to lead to a return of IUU fishing.

3.4. Fisheries extraction post-MPA

At present, the MPA has had no detectable impact upon the current operation of fisheries at South Georgia, even for the krill fishery. The krill fishery has historically operated mainly during the winter, and this period still remains open. The summer closure (management measure P1) limits the fishery so that it cannot expand into the summer when it would overlap with many krill predators during a critical period of their breeding cycle. This precautionary restriction is likely to become increasingly important should the krill fishery expand in the future, when it may wish to undertake operations in the summer period.

Outside the SGSSIMZ, krill fisheries operate during the summer at the Antarctic Peninsula and at the South Orkney Islands where many krill-dependent predators also breed. Consequently, CCAMLR will need to ensure that appropriate management restrictions are developed as the krill fishery increases so that it does not impact upon krill-eating predators.

3.5. Effects on local and foreign fisheries fleets

South Georgia does not maintain a register of fishing vessels; therefore, technically all fleets, including those of the UK, belong to foreign states. Since its designation, there have been no objections concerning the MPA from any fisheries operators, regardless of nationality.

3.6. Enforcement issues

The GSGSSI already maintains sophisticated fisheries management, control and surveillance. Existing resources have the capacity to deal with any additional burden caused by the designation of the MPA.

3.7. Ecosystem services

The spatial and temporal management measures introduced by the MPA should enhance both ecosystem and fish stock resilience.

3.8. Interface with RFMOs

The SGSSIMZ remains one of the most important sites for biodiversity within the Southern Ocean (Atkinson et al., 2001; Barnes et al., 2011; Clarke et al., 2012; Hogg et al., 2011; Murphy et al., 2007a; Rogers et al., 2012); this has led to calls by conservation organisations that a very high level of protection should be implemented[13]. The ecosystem within the SGSSIMZ has been severely degraded by commercial activities in the past (Bonner, 1984; Kock, 1992, 2000; Laws, 1953), through poorly regulated harvesting of seals, whales and finfish stocks; however, such poor stewardship is no longer acceptable under modern management practices. Some

---

[13] See http://www.marinereservescoalition.org (accessed 25 April 2014).

years ago, the GSGSSI adopted a highly precautionary approach and now manages fisheries in a conservative manner based on the best available scientific evidence. The MPA now provides additional legal protection for species, communities and habitats.

Despite this increased protection, the SGSSIMZ will continue to face threats from outside the geographic extent of the GSGSSI management framework, such that adequate protection for some ecosystem components may only be assured through international cooperation. Cooperation must include the mitigation of seabird by-catch (e.g. Ridley et al., 2010) in long-line fisheries and trawl fisheries outside the SGSSIMZ[14]. Other fisheries, often within the territorial waters of individual sovereign nations or in the high seas under the governance of Regional Fisheries Management Organisations, urgently need to regulate for mandatory use of proven mitigation measures, including, *inter alia*, closed seasons, limits on discharge of offal, appropriate net and line weighting, use of streamer lines, night-time setting and more effective and widespread monitoring of compliance by approved and accredited international scientific observers. Incidental mortality of seabirds remains a critical conservation issue for South Georgia seabirds, which requires concerted international action.

The management of the fishery for Antarctic krill also requires international action through CCAMLR. At present, the catch taken by the krill fishery remains relatively small (Trathan et al., 2012); however, should the fishery expand in the future (Nicol et al., 2011), depletion of krill stocks in the SGSSIMZ could have major impacts upon ecosystem stability. Such depletion might also happen if the level of krill harvesting upstream from the SGSSIMZ was poorly regulated, regardless of the actual fishing pressure within the SGSSIMZ. An expanded krill fishery therefore potentially represents one of the most important threats to the krill-based marine ecosystem within the SGSSIMZ.

Displacement of krill fishing effort so that it becomes spatially concentrated would have important ecological impacts, particularly if the fishery becomes aggregated close to land-based krill-eating predator colonies. This cannot now happen at South Georgia, because of the summer closure of the krill fishery (management measure P1; Table 2.3A) and the implementation of strict no-take zones around each of the islands (management measures E1, E2, E3 and P3; Table 2.3A). However, if the krill fishery were to move away from South Georgia (either because of new fishing opportunities in ice-free

[14] See http://www.acap.aq/bycatch-mitigation (accessed 25 April 2014).

areas further south, following regional temperature increases, or if GSGSSI were ever to impose more restrictive legislation within the SGSSIMZ in the future), harvesting could become more concentrated at the Antarctic Peninsula and thus have more significant ecological impacts at that location. Such impacts might not only affect local predator colonies at the Peninsula, but given the advection and flow of krill in the ACC (Thorpe et al., 2007), also potentially at remote sites downstream, including at South Georgia. Similar unforeseen ecological consequences have been reported elsewhere (e.g. Hilborn, 2013). Therefore, an important goal for CCAMLR must be to ensure that the fishery does not become spatially concentrated, either within the SGSSIMZ or in upstream areas.

Recent technological advances in both catching methods and in the development of new markets suggest that the krill fishery will expand beyond its current level (Nicol et al., 2011). At present, Norwegian and Korean companies dominate the catch, though vessels from China have recently entered the fishery, and Ukrainian vessels are once again fishing for krill (Nicol et al., 2011). As China is the world's largest aquaculture producer, and given the recent emphasis on using krill as an aquaculture feed, the continued expansion of the fishery will require that CCAMLR continues to develop a robust management framework as a matter of urgency. The successful implementation of this will be critical for the SGSSIMZ and thus the success of the MPA. The UK, on behalf of GSGSSI, must therefore engage with future management developments for the krill fishery within CCAMLR.

Although the commercial catch of Antarctic krill in the southwest Atlantic is low in relation to the potential catch limit (Nicol et al., 2011; Trathan et al., 2012), once an appropriate management procedure has been agreed, the allowable catch could rise to 5.61 million tonnes or approximately 9% of the regional standing stock. At this level, the fishery would be equivalent to approximately 11% (by mass) of global fishery landings currently reported to FAO (Grant et al., 2013). Thus, in a world where the human population already stands above 7 billion, Antarctic krill potentially offers one of the few remaining major under-exploited sources of marine protein. Consequently, it is critical that CCAMLR and GSGSSI ensure that robust ecologically based management, including the use of appropriately scaled MPAs, remains a key objective.

3.9. Measures of success

Having proposed the management measures described in Tables 2.3A and 2.3B and shown in Figure 2.4A and B, it is important to assess how well they performed against the target percentage levels of protection identified in Table 2.1; Tables 2.1 and 2.2 show the actual levels of protection achieved. In almost all cases, the target percentage levels of protection were achieved in full; the only areas where the targets were not completely reached are shown in Table 2.2. Four representative pelagic bioregions (Bioregions 1, 2, 3 and 4) were not protected to the target extent desired (Objective 1). This also has implications for some of the large-scale ecological processes (Objective 2). Nonetheless, over 41.4% protection was finally achieved through spatial/temporal closure (management measure P1), thereby exceeding the original target for representative protection. Protection for open ocean pelagic communities is difficult, as biological production that has been transported to a site may also potentially move beyond that site. A mix of procedural management and spatial and temporal protection is therefore potentially the best conservation approach for large-scale pelagic systems. Management measure P1 ensures that protection is achieved during the most critical times of the year, albeit not all year round.

One of the key tools for managing MPAs is a clearly articulated management plan which must also include a research and monitoring plan that allows for the collection of data that will facilitate the future refinement of MPA management provisions. Enhanced scientific understanding will facilitate further understanding of existing threats or other threats that may arise in the future. The SGSSI MPA Management Plan will therefore be reviewed every 5 years.

Additional conservation objectives should also be developed to ensure that any future development of non-living resource exploitation (e.g. seabed mining and mineral extraction) does not damage the marine system at South Georgia.

4. LOOKING AHEAD

4.1. What has been achieved

The existing SGSSI MPA balances conservation with sustainable harvesting. In the future, other ways of underpinning the economics of GSGSSI may become available; however, as it exists at present, the MPA provides an instrument to help improve conservation and management.

4.2. Resistance to global change

Climate change requires concerted international action to minimise regional temperature increases and increased levels of ocean acidification (Solomon et al., 2007). Such action is particularly important for South Georgia, where warming is already apparent (Whitehouse et al., 2008) and where species at their range-edge are potentially susceptible (Barnes et al., 2009). CCAMLR also needs to introduce more robust management methods that take into account climate change impacts for all Southern Ocean fisheries, including for krill (Trathan and Agnew, 2010). The efficacy of the MPA will need to be closely monitored to ensure that in the future it continues to meet the stated conservation objectives.

4.3. What needs to be done

The SGSSI MPA will be reviewed every 5 years, with the first review due in 2018. At this and every subsequent review, it will be necessary to determine whether the provisions of the MPA are adequate, or whether enhanced management measures are needed. Consequently, an appropriate scientific research and monitoring plan remains to be developed. Some of the authors are therefore in the process of designing such a plan, including seeking to secure the necessary sustainable funding. The NGO community could also work with GSGSSI to help prioritise and enhance this plan.

4.4. Could it have been achieved more effectively

The development of the SGSSI MPA represents the outcome of a complex project reflecting close cooperation between the UK government, GSGSSI and scientists with an interest in advising policy makers.

Three fundamental principles were incorporated into the SGSSI MPA planning process. Without incorporating these principles, it is doubtful whether the MPA could have been designated. Perhaps the most important of these principles was the recognition of the policy need to balance conservation against sustainable fishing and tourism. SGSSI includes some of the most biodiverse areas in the Southern Ocean and as such warrants a very high level of conservation and protection. However, high levels of IUU fishing have occurred in the SGSSIMZ in the past (Agnew, 2004) and would undoubtedly resume without adequate control, monitoring and surveillance. To maintain a high level of fisheries protection, GSGSSI currently relies upon the revenue from fisheries licences; funds for these functions are not readily available from other sources. Consequently, maintaining strongly regulated fisheries is critical for the security of the SGSSIMZ.

The second fundamental principle was that where feasible, spatial and temporal protection measures should be based upon scientific evidence.

However, despite a long history of scientific endeavour at South Georgia (since the *Discovery Expeditions* in the early 20th century; Kemp, 1929), scientific data are not available to address a number of important and relevant ecological questions. This means that a precautionary approach was a necessary condition. As part of this approach, different habitats (bioregions) were identified (Table 2.2 and Figure 2.3), to inform the designation of representative protection. In time, as more information becomes available about both habitats (bioregions) and communities, and about specific threats to different parts of the marine system, representative protection could be augmented by new directed and specific management measures.

The third principle incorporated into the SGSSI MPA planning process was that the pelagic and benthic domains should be considered separately; this was based upon two important premises. (a) The first of these was that the harvesting of benthic species (such as toothfish), if carried out with appropriate mitigation procedures, will not impact upon the pelagic community; similarly, the harvesting of pelagic species (such as krill or icefish) will not impact upon the benthic community, particularly if benthic–pelagic coupling is protected over the summer months. Separating the benthic and pelagic communities allows for the coexistence of conservation and fishing and potentially increases flexibility in the designation of conservation and management measures. (b) The second premise was that benthic species are less mobile or move only relatively short distances (though many may be mobile for the larval phase of their lifecycle—meroplankton). Spatial protection is thus an appropriate tool for conserving animals in these life-history stages. In contrast, pelagic species that are mobile, or which form part of the planktonic community, drifting with the ocean currents, may not be properly protected by spatial conservation measures. Providing adequate protection for pelagic species may therefore require a mixture of procedural management as well as spatial and temporal measures.

We suggest that many of the threats affecting the SGSSIMZ are representative of those present in marine systems elsewhere and that managing these threats so that biodiversity is conserved and maintained is a general need. Thus, the tools and concepts used to help implement the SGSSI MPA have considerable relevance to the development of such protection elsewhere, particularly for oceanic islands and for exclusive economic zones across the world's oceans.

Mankind has utilised marine products for over 200,000 years and will surely continue to do so into the foreseeable future. Over the past two centuries, the biggest impacts on marine systems have been from anthropogenic activities, both directly and indirectly. For example, in 2011, over 78.9 million tonnes of marine fish and other seafood were landed from marine capture fisheries, with a further 19.3 million tonnes obtained from marine aquaculture (FAO, 2012). Since the catches actually landed do not include estimates of discard, incidental mortality and illegal fishing, this figure is a gross underestimate of the impacts of marine fisheries. Meeting current (and future) demands for fishery products has the potential for continued massive impacts on ecosystem structure and stability, including loss of biodiversity and productivity. At present, almost 30% of stocks are overexploited, about 57% are fully exploited (that is, at or very close to their maximum sustainable production) and only about 13% are not yet fully exploited (FAO, 2012). Thus, balancing conservation with sustainable use, including through the use of spatial and temporal management and zoning, remains a universal requirement across the world's oceans.

ACKNOWLEDGEMENTS

We thank the participants of the scientific workshop for their valuable input and contribution. We also thank colleagues from the UK Foreign and Commonwealth Office, from the South Georgia fishing industry and from numerous NGOs for their thoughtful and considered advice. B. Raymond provided software and advice that facilitated the pelagic bioregionalisation. This project is a contribution to the Ecosystems programme at BAS, part of the NERC. Two anonymous referees and Magnus Johnson provided valuable criticism of an earlier draft of this chapter.

REFERENCES

Agnew, D.J., 2004. Fishing South: The History and Management of South Georgia Fisheries. Penna Press, London, UK, p. 123.

Agnew, D.J., Heaps, L., Jones, C., et al., 1999. Depth distribution and spawning pattern of *Dissostichus eleginoides* at South Georgia. CCAMLR Sci. 6, 19–36.

Agnew, D.J., Clark, J.M., McCarthy, P.A., et al., 2006a. A study of Patagonian toothfish (*Dissostichus eleginoides*) post tagging survivorship in Subarea 48.3. CCAMLR Sci. 13, 279–289.

Agnew, D.J., Kirkwood, G.P., Pearce, J., Clark, J., 2006b. Investigation of bias in the mark-recapture estimate of toothfish population size at South Georgia. CCAMLR Sci. 13, 47–63.

Agnew, D.J., Roberts, J., Moir-Clark, J., et al., 2007. Options for Restricted Impact Areas to Protect Areas of High Coral Biodiversity at South Georgia and Shag Rocks. A Report for the Government of South Georgia and the South Sandwich Islands. .

Agnew, D.J., Grove, P., Peatman, T., Burn, R., Edwards, C.T.T., 2010. Estimating optimal observer coverage in the Antarctic krill fishery. CCAMLR Sci. 17, 139–154.

Arntz, W.E., Brey, T., Gallardo, V.A., 1994. Antarctic zoobenthos. Oceanogr. Mar. Biol. Annu. Rev. 32, 241–304.

Atkinson, A., Peck, J.M., 1990. The distribution of zooplankton in relation to the South Georgia shelf in summer and winter. In: Kerry, K., Hempel, G. (Eds.), Antarctic

Ecosystems: Ecological Change and Conservation. Springer-Verlag, Berlin/Heidelberg, pp. 159–165.

Atkinson, A., Whitehouse, M.J., Priddle, J., et al., 2001. South Georgia, Antarctica: a productive, cold water, pelagic ecosystem. Mar. Ecol. Prog. Ser. 216, 279–308.

Atkinson, A., Siegel, V., Pakhomov, E.A., et al., 2008. Oceanic circumpolar habitats of Antarctic krill. Mar. Ecol. Prog. Ser. 362, 1–23.

Atkinson, A., Schmidt, K., Fielding, S., et al., 2012. Variable food absorption by Antarctic krill: relationships between diet, egestion rate and the composition and sinking rates of their fecal pellets. Deep Sea Res. II 59–60, 147–158.

Barnes, D.K.A., 2008. A benthic richness hotspot in the Southern Ocean: slope and shelf cryptic benthos of Shag Rocks. Antarct. Sci. 20, 263–270.

Barnes, D.K.A., Souster, T., 2011. Reduced survival of Antarctic benthos linked to climate-induced iceberg scouring. Nat. Clim. Chang. 1 (7), 365–368.

Barnes, D.K.A., Griffiths, H.J., Kaiser, S., 2009. Geographic range shift response to climate change by Antarctic benthos: where we should look. Mar. Ecol. Prog. Ser. 393, 13–36.

Barnes, D.K.A., Collins, M.A., Brickle, P., et al., 2011. The need to implement the convention on biological diversity at the high latitude site, South Georgia. Antarct. Sci. 23, 323–331.

Bednaršek, N., Tarling, G.A., Bakker, D.C.E., et al., 2012. Extensive dissolution of live pteropods in the Southern Ocean. Nat. Geosci. 5, 881–885.

Belchier, M., Collins, M.A., 2008. Recruitment and body size in relation to temperature in juvenile Patagonian toothfish (Dissostichus eleginoides) at South Georgia. Mar. Biol. 155, 493–503.

Belchier, M., Lawson, J., 2013. An analysis of temporal variability in abundance, diversity and growth rates in the coastal ichthyoplankton assemblage of South Georgia (sub-Antarctic). Polar Biol. 36, 969–983.

Berrow, S., Croxall, J.P., 1999. The diet of white-chinned petrels Procellaria aequinoctialis, Linnaeus 1758, in years of contrasting prey availability at South Georgia. Antarct. Sci. 11, 283–292.

Biuw, M., Lydersen, C., De Bruyn, P.J.N., et al., 2010. Long-range migration of a chinstrap penguin from Bouvetoya to Montagu Island, South Sandwich Islands. Antarct. Sci. 22, 157–162.

Bonner, W.N., 1984. Conservation in the Antarctic. In: Laws, R.M. (Ed.), In: Antarctic Ecology, vol. II. Academic Press, London, UK, pp. 821–847.

Boyd, I.L., 2002. Estimating food consumption of marine predators: Antarctic fur seals and macaroni penguins. J. Appl. Ecol. 39, 103–119.

Brierley, A.S., Fernandes, P.G., Brandon, M.A., et al., 2002. Antarctic krill under sea ice: elevated abundance in a narrow band just south of ice edge. Science 295, 1890–1892.

Chikov, V.N., Melnikov, Y.S., 1990. On the question of the fecundity of the Patagonian toothfish Dissostichus eleginoides in the region of Kerguelen Islands. J. Ichthyol. 30, 122–125.

Christian, C., Ainley, D., Bailey, M., et al., 2013. A review of formal objections to Marine Stewardship Council fisheries certifications. Biol. Conserv. 161, 10–17.

Clarke, M.R., 1980. Cephalopods in the diet of sperm whales of the southern hemisphere and their bearing on sperm whale biology. Discov. Rep. 37, 1–324.

Clarke, A., Croxall, J.P., Poncet, S., Martin, A.R., Burton, R., 2012. Important bird areas: South Georgia. Br. Birds 105, 118–144.

Collins, M.A., Ross, K.A., Belchier, M., Reid, K., 2007. Distribution and diet of juvenile Patagonian toothfish on the South Georgia and Shag Rocks shelves (Southern Ocean). Mar. Biol. 152, 135–147.

Collins, M.A., Brickle, P., Brown, J., Belchier, M., 2010. The Patagonian toothfish: biology, ecology and fishery. In: Lesser, M. (Ed.), In: Advances in Marine Biology, vol. 58. Academic Press, Burlington, USA, pp. 227–300.

Croxall, J.P., North, A.W., 1988. Fish prey of Wilson storm petrel *Oceanites oceanicus* at South Georgia. Br. Antarct. Surv. B 78, 37–42.

Croxall, J.P., Prince, P.A., Ricketts, C., 1985. Relationships between prey life-cycles and the extent nature and timing of seal and seabird predation in the Scotia Sea Antarctica. In: Siegfried, W.R., Condy, P.R., Laws, R.M. (Eds.), Antarctic Nutrient Cycles and Food Webs. Springer-Verlag, New York, NY, USA/Berlin, Germany, pp. 516–533.

Croxall, J.P., Silk, J.R.D., Phillips, R.A., Afanasyev, V., Briggs, D.R., 2005. Global circum-navigations: tracking year-round ranges of non-breeding albatrosses. Science 307, 249–250.

Cury, P.M., Boyd, I.L., Bonhommeau, S., et al., 2011. Global seabird response to forage fish depletion—one-third for the birds. Science 334, 1703–1706.

Delord, K., Barbraud, C., Bost, C.-A., et al., 2014. Areas of importance for seabirds tracked from French southern territories, and recommendations for conservation. Mar. Policy 48, 1–13.

Dinniman, M.S., Klink, J.M., 2004. A model study of circulation and cross-shelf exchange on the west Antarctic Peninsula continental shelf. Deep Sea Res. II 51, 2003–2022.

Dunne, R.P., Polunin, N.V.C., Sand, P.H., Johnson, M.L., 2014. The creation of the Chagos marine protected area: a fisheries perspective. Adv. Mar. Biol. 69, 79–127.

Edgar, G.J., Stuart-Smith, R.D., Willis, T.J., et al., 2014. Global conservation outcomes depend on marine protected areas with five key features. Nature 506, 216–220.

Everson, I., 1977. The Living Resources of the Southern Ocean. FAO GLO/S0/77/1, Rome, Italy, 156 pp.

Everson, I., Goss, C., 1991. Krill fishing activity in the southwest Atlantic. Antarct. Sci. 3, 351–358.

Everson, I., Parkes, G., Kock, K.-H., Boyd, I.L., 1999. Variation in standing stock of the mackerel icefish *Champsocephalus gunnari* at South Georgia. J. Appl. Ecol. 36, 591–603.

Everson, I., North, A.W., Cooper, A.P.R., McWilliam, N.C., Kock, K.-H., 2001. Spawning location of mackerel icefish at South Georgia. CCAMLR Sci. 8, 107–118.

FAO, 2012. World Review of Fisheries and Aquaculture. Food and Agriculture Organisation of the United Nations, Rome. 148 pp. http://www.fao.org/docrep/016/i2727e/i2727e01.pdf.

Fielding, S., Watkins, J.L., Trathan, P.N., et al., 2014. Interannual variability in Antarctic krill (*Euphausia superba*) density at South Georgia, Southern Ocean: 1997–2013. ICES J. Mar. Sci. http://dx.doi.org/10.1093/icesjms/fsu104.

Gille, S.T., 2002. Warming of the Southern Ocean since the 1950s. Science 295, 1275–1277.

Gille, S.T., 2008. Decadal-scale temperature trends in the Southern Hemisphere Ocean. J. Climate 21, 4749–4765.

González-Solís, J., Croxall, J.P., Wood, A.G., 2000. Foraging partitioning between giant petrels *Macronectes* spp. and its relationship with breeding population changes at Bird Island, South Georgia. Mar. Ecol. Prog. Ser. 204, 279–288.

Grant, S., Constable, A., Raymond, B., Doust, S., 2006. Bioregionalisation of the Southern Ocean: Report of Experts Workshop, Hobart, September 2006. WWF-Australia and ACE CRC, 45 pp.

Grant, S.M., Hill, S.L., Trathan, P.N., Murphy, E.J., 2013. Ecosystem services of the Southern Ocean: trade-offs in decision-making. Antarct. Sci. 25 (5), 603–617.

Griffiths, H.J., Linse, K., Barnes, D.K.A., 2008. Distribution of macrobenthic taxa across the Scotia Arc, Antarctica. Antarct. Sci. 20 (3), 213–226.

Halpern, B.S., 2014. Making marine protected areas work. Nature 506, 167–168.

Hewitt, R.P., Watkins, J.L., Naganobu, M., et al., 2004. Biomass of Antarctic krill in the Scotia sea in January/February 2000 and its use in revising an estimate of precautionary yield. Deep Sea Res. II 51, 1215–1236.

Heywood, R.B., Everson, I., Priddle, J., 1985. The absence of krill from the South Georgia zone, winter 1983. Deep Sea Res. 32 (3), 369–378.

Hilborn, R., 2013. Environmental cost of conservation victories. Proc. Natl. Acad. Sci. U.S.A. 110, 9187.

Hill, S.L., Reid, K., North, A.W., 2005. Recruitment of mackerel icefish (*Champsocephalus gunnari*) at South Georgia indicated by predator diets and its relationship with sea surface temperature. Can. J. Fish. Aquat. Sci. 62, 2530–2537.

Hill, S.L., Trathan, P.N., Agnew, D.J., 2009. The risk to fishery performance associated with spatially resolved management of Antarctic krill (*Euphausia superba*) harvesting. ICES J. Mar. Sci. 66, 2148–2154.

Hill, S.L., Keeble, K., Atkinson, A., Murphy, E.J., 2012. A foodweb model to explore uncertainties in the South Georgia shelf pelagic ecosystem. Deep Sea Res. II 59–60, 237–252.

Hogg, O.T., Barnes, D.K.A., Griffiths, H.J., 2011. Highly diverse, poorly studied and uniquely threatened by climate change: an assessment of marine biodiversity on South Georgia's continental shelf. PLoS One 6 (5), e19795.

Jamieson, A.J., Fujii, T., Mayor, D.J., Solan, M., Priede, I.G., 2010. Hadal trenches: the ecology of the deepest places on earth. Trends Ecol. Evol. 25 (3), 190–197.

Kaiser, S., Barnes, D.K.A., Linse, K., Brandt, A., 2008. Epibenthic macrofauna associated with the shelf and slope of a young and isolated Southern Ocean island. Antarct. Sci. 20 (3), 281–290.

Kawaguchi, S., Ishida, A., King, R., et al., 2013. Risk maps for Antarctic krill under projected Southern Ocean acidification. Nat. Clim. Chang. 3, 843–847. http://dx.doi.org/10.1038/nclimate1937.

Kemp, S., 1929. The objects of the investigations. Discov. Rep. 1, 141–150.

Kock, K.-H., 1992. Antarctic Fish and Fisheries. Cambridge University Press, Cambridge, UK, 359 pp.

Kock, K.-H., 2000. Understanding CCAMLR's Approach to Management. http://www.ccamlr.org/en/system/files/am-all.pdf.

Kock, K.-H., Kellerman, A., 1991. Review: reproduction in Antarctic notothenioid fish. Antarct. Sci. 3, 125–150.

Kock, K.-H., Belchier, M., Jones, C.D., 2004. Is the attempt to estimate the biomass of Antarctic fish from a multi-species survey appropriate for all targeted species? *Notothenia rossii* in the Atlantic Ocean sector—revisited. CCAMLR Sci. 11, 141–153.

Korb, R.E., Whitehouse, M.J., Ward, P., 2004. SeaWiFS in the southern ocean: spatial and temporal variability in phytoplankton biomass around South Georgia. Deep Sea Res. II 51, 99–116.

Lascelles, B.G., Langham, G.M., Ronconi, R.A., Reid, J.B., 2012. From hotspots to site protection: identifying marine protected areas for seabirds around the globe. Biol. Conserv. 156, 5–14.

Laws, R.M., 1953. The elephant seal industry at South Georgia. Polar Rec. 6, 746–754.

Leaper, R., Cooke, J., Trathan, P.N., et al., 2006. Global climate drives southern right whale (*Eubalaena australis*) population dynamics. Biol. Lett. 2, 289–292.

Liddle, G.M., 1994. Interannual variation in the breeding biology of the Antarctic prion *Pachyptila desolata* at Bird Island, South Georgia. J. Zool. (Lond.) 234, 125–139.

Loeb, V.J., Kellermann, A.K., Koubbi, P., North, A.W., White, M.G., 1993. Antarctic larval fish assemblages—a review. Bull. Mar. Sci. 53 (2), 416–449.

Lombard, A.T., Reyers, B., Schonegevel, L.Y., et al., 2007. Conserving pattern and process in the Southern Ocean: designing a marine protected area for the Prince Edward Islands. Antarct. Sci. 19 (1), 39–54.

Longhurst, A., 1998. Ecological Geography of the Sea. Academic Press, San Diego, USA, 398 pp.

Margules, C.R., Pressey, R.L., 2000. Systematic conservation planning. Nature 405, 243–253.

Martin, S.M., Peatman, T.P., Pearce, J., Mitchell, R.E., 2012. Approach to the Establishment of Reduced Impact Areas at South Georgia. A Report for the Government of
South Georgia and the South Sandwich Islands.

Morato, T., Hoyle, S.D., Allain, V., Nicol, S.J., 2010. Seamounts are hotspots of pelagic biodiversity in the open ocean. Proc. Natl. Acad. Sci. U.S.A. 107, 9707–9711.

Murphy, E.J., Trathan, P.N., Everson, I., et al., 1997. Detailed distribution of krill fishing
around South Georgia. CCAMLR Sci. 4, 1–17.

Murphy, E.J., Watkins, J.L., Reid, K., et al., 1998. Interannual variability of the South
Georgia marine ecosystem: biological and physical sources of variation in the abundance
of krill. Fish. Oceanogr. 7, 381–390.

Murphy, E.J., Thorpe, S.E., Watkins, J.L., Hewitt, R., 2004. Modeling the krill transport
pathways in the Scotia sea: spatial and environmental connections generating the seasonal
distribution of krill. Deep Sea Res. II 51, 1435–1456.

Murphy, E.J., Trathan, P.N., Watkins, J.L., et al., 2007a. Climatically driven fluctuations in
Southern Ocean ecosystems. Proc. R. Soc. Ser. B 274, 3057–3067.

Murphy, E.J., Watkins, J.L., Trathan, P.N., et al., 2007b. Spatial and temporal operation of
the Scotia Sea ecosystem: a review of large-scale links in a krill centred food web. Philos.
Trans. R. Soc. B 362, 113–148.

Navarro, J., Votier, S.C., Aguzzi, J., et al., 2013. Ecological segregation in space, time and
trophic niche of sympatric planktivorous petrels. PLoS One 8 (4), e62897. http://dx.doi.
org/10.1371/journal.pone.0062897.

Nevinsky, M.M., Kozlov, A.N., 2002. On fecundity of Patagonian toothfish *Dissostichus
eleginoides* in the South Georgia region (South Atlantic). J. Ichthyol. 42 (4), 571–573.

Nicol, S., Foster, J., Kawaguchi, S., 2011. The fishery for Antarctic krill—recent developments. Fish Fish. 13, 30–40.

North, A.W., 2002. Larval and juvenile distribution and growth of Patagonian toothfish
around South Georgia. Antarct. Sci. 14, 25–31.

North, A.W., Ward, P., 1989. Initial feeding by Antarctic fish larvae during winter at South
Georgia South Atlantic Ocean. Cybium 13, 357–364.

North, A.W., Ward, P., 1990. The feeding ecology of larval fish in an Antarctic fjord, with
emphasis on *Champsocephalus gunnari*. In: Kerry, K., Hempel, G. (Eds.), Antarctic Ecosystems: Ecological Change and Conservation. Springer-Verlag, Berlin/Heidelberg,
pp. 299–307.

North, A.W., White, M.G., Trathan, P.N., 1998. Interannual variability in the early growth
rate and size of the Antarctic fish *Gobionotothen gibberifrons* (Lonnberg). Antarct. Sci.
10, 416–422.

Orsi, A.H., Whitworth III, T., Nowlin, W.D., 1995. On the meridional extent and fronts of
the Antarctic circumpolar current. Deep Sea Res. I 42, 641–673.

Parkinson, C.L., 2002. Trends in the length of the Southern Ocean sea-ice season, 1979–99.
Ann. Glaciol. 34, 435–440.

Parkinson, C.L., 2004. Southern Ocean sea ice and its wider linkages: insights revealed from
models and observations. Antarct. Sci. 16, 387–400.

Payne, M.R., 1977. Growth of a fur seal population. Philos. Trans. R. Soc. B 279, 67–79.

Payne, R., 1986. Long term behavioural studies of the southern right whale, *Eubalaena australis*. J. Cetacean Res. Manag. 10, 161–167.

Phillips, R.A., Silk, J.R.D., Croxall, J.P., Afanasyev, V., Bennett, V.J., 2005. Summer distribution and migration of non-breeding albatrosses: individual consistencies and implications for conservation. Ecology 86 (9), 2386–2396.

Phillips, R.A., Silk, J.R.D., Croxall, J.P., Afanasyev, V., 2006. Year-round distribution of
white-chinned petrels from South Georgia: relationships with oceanography and fisheries. Biol. Conserv. 129, 336–347.

Phillips, R.A., Bearhop, S., Mcgill, R.A.R., Dawson, D.A., 2009. Stable isotopes reveal individual variation in migration strategies and habitat preferences in a suite of seabirds during the non-breeding period. Oecologia 160, 795–806.

Pichegru, L., Ryan, P.G., van Eeden, R., et al., 2012. Industrial fishing, no-take zones and endangered penguins. Biol. Conserv. 156, 117–125.

Pikitch, E., Boersma, P.D., Boyd, I.L., et al., 2012. Little Fish, Big Impact: Managing a Crucial Link in Ocean Food Webs. Lenfest Ocean Program, Washington, DC, USA, 108 pp.

Pinõnes, A., Hofmann, E.E., Dinniman, M.S., Klinck, J.M., 2011. Lagrangian simulation of transport pathways and residence times along the western Antarctic Peninsula. Deep Sea Res. II 58, 1524–1539.

Pinõnes, A., Hofmann, E.E., Daly, K.L., et al., 2013. Modeling the remote and local connectivity of Antarctic krill populations along the western Antarctic Peninsula. Mar. Ecol. Prog. Ser. 481, 69–92.

Quillfeldt, P., Masello, J.F., Navarro, J., Phillips, R.A., 2013. Year-round distribution suggests spatial segregation of two small petrel species in the South Atlantic. J. Biogeogr. 40, 430–441.

Ramirez-Llodra, E., Brandt, A., Danovaro, R., et al., 2010. Deep, diverse and definitely different: unique attributes of the world's largest ecosystem. Biogeosciences 7, 2851–2899.

Ramos, A., San Martin, G., 1999. On the finding of a mass occurrence of *Serpula narconensis* Baird, 1885 (Polychaeta, Serpulidae) in South Georgia (Antarctica). Polar Biol. 22 (6), 379–383.

Ratcliffe, N., Crofts, S., Brown, R., et al., 2014. Love thy neighbour or opposites attract? Patterns of spatial segregation and association among crested penguin populations during winter. J. Biogeogr. 41, 1183–1192.

Raymond, B., 2011. A Circumpolar Pelagic Regionalisation of the Southern Ocean. CCAMLR, Hobart, Australia, Paper WS-MPA-11/6.

Reid, K., Nevitt, G.A., 1998. Observation of southern elephant seal, *Mirounga leonina*, feeding at sea near South Georgia. Mar. Mamm. Sci. 14, 637–640.

Reid, K., Watkins, J.L., Murphy, E.J., et al., 2010. Krill population dynamics at South Georgia: implications for ecosystem-based fisheries management. Mar. Ecol. Prog. Ser. 399, 243–252.

Reilly, S., Hedley, S., Borberg, J., et al., 2004. Biomass and energy transfer to baleen whales in the South Atlantic sector of the Southern Ocean. Deep Sea Res. II 51, 1397–1409.

Ridley, C., Harrison, N.M., Phillips, R.A., Pugh, P.J.A., 2010. Identifying the origins of fishing gear ingested by seabirds: a novel multivariate approach. Aquat. Conserv. 20, 621–631.

Rio, M.H., Guinehut, S., Lamicol, S., 2011. New CNES-CLS09 global mean dynamic topography computed from the combination of GRACE data, altimetry, and *in situ* measurements. J. Geophys. Res. 116, C07018. http://dx.doi.org/10.1029/2010JC006505.

Roberts, J., Agnew, D.J., 2008. Proposal for an Extension to the Mark Recapture Experiment to Estimate Toothfish Population Size in Subarea 48.4. CCAMLR WG-FSA-08/46.

Rogers, A.D., Tyler, P.A., Connelly, D.P., et al., 2012. The discovery of new deep-sea hydrothermal vent communities in the Southern Ocean and implications for biogeography. PLoS Biol. 10 (1), e1001234. http://dx.doi.org/10.1371/journal. pbio.1001234.

Ronconi, R.A., Lascelles, B.G., Langham, G.M., Reid, J.B., Oro, D., 2012. The role of seabirds in Marine Protected Area identification, delineation, and monitoring: introduction and synthesis. Biol. Conserv. 156, 1–4.

Rowden, A.A., Dower, J.F., Schlacher, T.A., Consalvey, M., Clark, M.R., 2010. Paradigms in seamount ecology: fact, fiction and future. Mar. Ecol. Evol. Perspect. 31, 226–241.

Saunders, R.A., Brierley, A.S., Watkins, J.L., et al., 2007. Intra-annual variability in the density of Antarctic krill (*Euphausia superba*) at South Georgia, 2002–2005: within-year variation provides a new framework for interpreting previous 'annual' estimates of krill density. CCAMLR Sci. 14, 27–41.

SC-CAMLR-XXII, 2003. Report of the Twenty Second Meeting of the Scientific Committee. CCAMLR, Hobart, Australia, p. 577.

Schnack-Schiel, S., Isla, R., 2005. The role of zooplankton in the pelagic-benthic coupling of the Southern Ocean. Sci. Mar. 69, 39–55.

Slip, D.J., Hindell, M.A., Burton, H.R., 1994. Diving behaviour of southern elephant seals from Macquarie Island: an overview. In: Le Boeuf, B.G., Laws, R.M. (Eds.), Elephant Seals: Population Ecology, Behaviour and Physiology. University of California Press, Berkeley, USA, pp. 253–270.

Smith, W.H.F., Sandwell, D.T., 1997. Global seafloor topography from satellite altimetry and ship depth soundings. Science 277, 1957–1962.

Solomon, S., Qin, D., Manning, M., et al., 2007. Contribution of Working Group I to the Fourth Assessment Report of the Intergovernmental Panel on Climate Change. Cambridge University Press, Cambridge, UK/New York, NY, USA.

Stammerjohn, S.E., Martinson, D.G., Smith, R.C., Iannuzzi, R.A., 2008. Sea ice in the western Antarctic Peninsula region: spatio-temporal variability from ecological and climate change perspectives. Deep Sea Res. II 55, 2041–2058.

Staniland, I.J., Boyd, I.L., Reid, K., 2007. An energy-distance trade-off in a central-place forager, the Antarctic fur seal (*Arctocephalus gazella*). Mar. Biol. 152, 233–241.

Staniland, I.J., Morton, A., Robinson, S.L., Malone, D., Forcada, J., 2011. Foraging behaviour in two Antarctic fur seal colonies with differing population recoveries. Mar. Ecol. Prog. Ser. 434, 183–196.

Staniland, I.J., Robinson, S.L., Silk, J.R.D., Warren, N., Trathan, P.N., 2012. Winter distribution and haul-out behaviour of female Antarctic fur seals from South Georgia. Mar. Biol. 159, 291–301.

Tanton, J.L., Reid, K., Croxall, J.P., Trathan, P.N., 2004. Winter distribution and behaviour of gentoo penguins *Pygoscelis papua* at South Georgia. Polar Biol. 27, 299–303.

Tarling, G.A., Cuzin-Roudy, J., Thorpe, S.E., et al., 2007. Recruitment of Antarctic krill *Euphausia superba* in the South Georgia region: adult fecundity and the fate of larvae. Mar. Ecol. Prog. Ser. 331, 161–179.

Tarling, G.A., Ward, P., Atkinson, A., et al., 2012. DISCOVERY 2010: spatial and temporal variability in a dynamic polar ecosystem. Deep Sea Res. II 59–60, 1–13.

Thomas, D., May, B.F., 2005. Overview of MODIS Aqua Data Processing and Distribution. NASA Goddard Space Flight Center, USA.http://oceancolor.gsfc.nasa.gov/.

Thorpe, S.E., Heywood, K.J., Stevens, D.P., Brandon, M.A., 2004. Tracking passive drifters in a high resolution ocean model: implications for interannual variability of larval krill transport to South Georgia. Deep Sea Res. I 51, 909–920.

Thorpe, S.E., Murphy, E.J., Watkins, J.L., 2007. Circumpolar connections between Antarctic krill (*Euphausia superba* Dana) populations: investigating the roles of ocean and sea ice transport. Deep Sea Res. I 54, 792–810.

Tormosov, D.D., Mikhaliev, Y.A., Best, P.B., et al., 1998. Soviet catches of southern right whales *Eubalaena australis* 1951–1971. Biological data and conservation implications. Biol. Conserv. 86, 185–197.

Trathan, P.N., Agnew, D., 2010. Climate change and the Antarctic marine ecosystem: an essay on management implications. Antarct. Sci. 22, 387–398.

Trathan, P.N., Everson, I., Miller, D.G.M., et al., 1995. Krill biomass in the Atlantic. Nature 367, 201–202.

Trathan, P.N., Daunt, F.H.J., Murphy, E.J., 1996. South Georgia: An Ecological Atlas. British Antarctic Survey, Cambridge, UK, 80 pp.

Trathan, P.N., Everson, I., Murphy, E.J., Parkes, G., 1998a. Analysis of haul data from the South Georgia krill fishery. CCAMLR Sci. 5, 9–30.

Trathan, P.N., Murphy, E.J., Croxall, J.P., Everson, I., 1998b. Use of at-sea distribution data to derive potential foraging ranges of macaroni penguins during the breeding season. Mar. Ecol. Prog. Ser. 169, 263–275.

Trathan, P.N., Brierley, A.S., Brandon, M.A., et al., 2003. Oceanographic variability and changes in Antarctic krill (*Euphausia superba*) abundance at South Georgia. Fish. Oceanogr. 12, 569–583.

Trathan, P.N., Green, C., Tanton, J., et al., 2006a. Foraging dynamics of macaroni penguins *Eudyptes chryolophus* at South Georgia during brood-guard. Mar. Ecol. Prog. Ser. 323, 239–251.

Trathan, P.N., Murphy, E.J., Forcada, J., et al., 2006b. Physical forcing in the southwest Atlantic: ecosystem control. In: Boyd, I.L., Wanless, S., Camphuysen, C.J. (Eds.), Top Predators in Marine Ecosystems. Cambridge University Press, Cambridge, pp. 28–45.

Trathan, P.N., Forcada, J., Murphy, E.J., 2007. Environmental forcing and Southern Ocean marine predator populations: effects of climate change and variability. Philos. Trans. R. Soc. B 362, 2351–2365.

Trathan, P.N., Bishop, C., Maclean, G., et al., 2008. Linear tracks and restricted temperature ranges characterise penguin foraging pathways. Mar. Ecol. Prog. Ser. 370, 285–294.

Trathan, P.N., Ratcliffe, N., Masden, E.A., 2012. Ecological drivers of change at South Georgia: the krill surplus, or climate variability. Ecography 35, 983–993.

Trivelpiece, W.Z., Buckelew, S., Reiss, C., Trivelpiece, S.G., 2007. The winter distribution of chinstrap penguins from two breeding sites in the South Shetland Islands of Antarctica. Polar Biol. 30, 1231–1237.

Turner, J., Comiso, J.C., Marshall, G.J., et al., 2009. Non-annular atmospheric circulation change induced by stratospheric ozone depletion and its role in the recent increase of Antarctic sea ice extent. Geophys. Res. Lett. 36, L08502.

Wakefield, E.D., Phillips, R.A., Trathan, P.N., et al., 2011. Habitat preference, accessibility and competition limit the global distribution of breeding black-browed albatrosses. Ecol. Monogr. 81, 141–167.

Waluda, C.M., Hill, S.L., Peat, H.J., Trathan, P.N., 2012. Diet variability and reproductive performance of macaroni penguins *Eudyptes chrysolophus* at Bird Island, South Georgia. Mar. Ecol. Prog. Ser. 466, 261–274.

Ward, P., Grant, S., Brandon, M.A., et al., 2004. Mesozooplankton community structure in the Scotia sea during the CCAMLR 2000 survey: January–February 2000. Deep Sea Res. II 51, 1351–1367.

Ward, P., Whitehouse, M., Shreeve, R., et al., 2007. Plankton community structure south and west of South Georgia (Southern Ocean): links with production and physical forcing. Deep Sea Res. I 54, 1871–1889.

Ward, P., Atkinson, A., Venables, H.J., et al., 2012. Food web structure and bioregions in the Scotia sea: a seasonal synthesis. Deep Sea Res. II 59–60, 253–266.

Whitehouse, M.J., Meredith, M.P., Rothery, P., et al., 2008. Rapid warming of the ocean around South Georgia, Southern Ocean, during the 20th century: forcings, characteristics and implications for lower trophic levels. Deep Sea Res. I 55, 1218–1228.

Xavier, J.C., Croxall, J.P., Trathan, P.N., Rodhouse, P.G., 2003a. Inter-annual variation in the cephalopod component of the diet of the wandering albatross, *Diomedea exulans* breeding at Bird Island, South Georgia. Mar. Biol. 142, 611–622.

Xavier, J.C., Croxall, J.P., Trathan, P.N., Wood, A.G., 2003b. Feeding strategies and diets of breeding grey-headed and wandering albatrosses at South Georgia. Mar. Biol. 143, 221–232.

Zerbini, A.N., Andriolo, A., da Rocha, J.M., et al., 2004. Winter distribution and abundance of humpback whales (*Megaptera novaeangliae*) off Northeastern Brazil. J. Cetacean Res. Manag. 6 (1), 101–107.

Zerbini, A.N., Andriolo, A., Heide-Jørgensen, M.P., et al., 2006. Migration and Summer Destinations of Humpback Whales (*Megaptera novaeangliae*) in the Western South Atlantic Ocean. International Whaling Commission, Cambridge, UK, Paper SC/A06/HW46.

> CHAPTER THREE

The Creation of the Chagos Marine Protected Area: A Fisheries Perspective [*]

Richard P. Dunne[*,1], Nicholas V.C. Polunin[†], Peter H. Sand[‡], Magnus L. Johnson[§]
*West Briscoe, Barnard Castle, Durham, United Kingdom
[†]School of Marine Science and Technology, Newcastle University, Newcastle, England
[‡]Ludwig-Maximilians-Universität München, München, Germany
[§]Centre for Environmental and Marine Sciences, University of Hull, Scarborough, United Kingdom
[1]Corresponding author: e-mail address: richardpdunne@aol.com

Contents

[*]*Note*: In-depth assessments of the fisheries of the Chagos have been made in compiling this review. These and the data on which they are based are available as separate Appendices in Electronic Supplementary Material (ESM) at http://dx.doi.org/10.1016/B978-0-12-800214-8.00003-7. These are cross-referenced in this text.

Advances in Marine Biology, Volume 69
ISSN 0065-2881
http://dx.doi.org/10.1016/B978-0-12-800214-8.00003-7

79

Abstract

From a fisheries perspective, the declaration of a 640,000 km^2 "no-take" Marine Protected Area (MPA) in the Chagos Archipelago in 2010 was preceded by inadequate consideration of the scientific rationale for protection. The entire area was already a highly regulated zone which had been subject to a well-managed fisheries licensing system. The island of Diego Garcia, the only area where there is evidence of overfishing has, because of its military base, been excluded from the MPA. The no-take mandate removes the primary source of sustenance and economic sustainability of any inhabitants, thus effectively preventing the return of the original residents who were removed for political reasons in the 1960s and 1970s. The principles of natural resource conservation and use have been further distorted by forcing offshore fishing effort to other less well-managed areas where it will have a greater negative impact on the well-being of the species that were claimed to be one of the primary beneficiaries of the declaration. A failure to engage stakeholders has resulted in challenges in both the English courts and before an international tribunal.

Keywords: Chagos, British Indian Ocean Territory, Fisheries, Tuna, Reef fish, Marine Protected Area

1. INTRODUCTION

There is a viewpoint that ocean ecosystems are subject to unsustainable anthropogenic pressures (Halpern et al., 2008), particularly from commercial fishing. Marine Protected Areas (MPAs) have been used extensively to protect resident tropical reef fish populations (Graham et al., 2011; Halpern, 2003) but their effectiveness depends upon their size relative to the geographic range of the species that they are designed to protect (Palumbi, 2004) and is influenced by the absence of one or more key features (no-take, enforced, old, large, isolated) (Edgar et al., 2014). Proposed solutions to problems of size have included "scaling up" into networks of reserves (Gaines et al., 2010) or large-scale MPAs (LSMPAs) such as those advocated by the Pew Environment Group (Nelson and Bradner, 2010). In recent years, the number of LSMPAs has grown, and there are now 10, of which two-thirds have been designated no-take (Leenhardt et al., 2013) including the Chagos MPA. For highly migratory species, however, there is still little evidence of the effectiveness of fixed-area MPAs (Kaplan et al., 2014; Palumbi, 2004).

The Chagos MPA was first proposed by the Chagos Environment Network (CEN), an association of scientific and conservation groups led by the Chagos Conservation Trust and the Pew Environment Group, in 2007 (Chagos Conservation Trust, 2009). The CEN approached the UK Foreign and Commonwealth Office (FCO) which agreed to hold a public consultation in late 2009. The FCO Consultation Document championed the "growing scientific support for establishing large-scale marine reserves to protect fish stocks", claimed that a Chagos MPA would make a "significant contribution to the wider biological productivity of the Indian Ocean", and gave the impression that Mauritius, which claimed sovereignty over the archipelago, welcomed the proposal, and further that it would have "no direct immediate impact" on the inhabitants of the islands (FCO, 2009) who had been exiled in the 1960s and 1970s. At the same time, a report commissioned by Pew Environment Group argued that no-take MPAs provided proven and effective protection for large, migratory, pelagic fish species, and that a 200 nautical mile (nm) Chagos MPA would be a safe haven for tuna, billfish and sharks that could only facilitate their recovery in the Indian Ocean and end a "significant" tuna fisheries by-catch in the British Indian Ocean Territory (BIOT) (Koldewey et al., 2010a). Following its creation, the MPA was immediately hailed as a fisheries conservation success (Nelson, 2010; Sheppard, 2010).

In order to evaluate these justifications for the Chagos MPA and more generally to assess its necessity and value, we have adopted an evidence driven approach to address the question whether a no-take MPA was appropriate from a fisheries perspective, and is likely to be meaningful and/or necessary. In so doing, we have posed several key questions: (1) Is the UK claim and proclamation legal? (2) Does the MPA afford protection for highly migratory tuna species? (3) Can the MPA effectively protect pelagic by-catch species? (4) Is the MPA necessary for the protection of reef fish?

(5) Is the exclusion of Diego Garcia from the MPA justified? (6) Will the MPA help protect other Indian Ocean coral reefs? (7) Was the Chagos Archipelago already adequately protected?

Before addressing each of these questions in detail, it is first necessary to examine the background of the Chagos Archipelago together with what is known about the fisheries in the area prior to the MPA declaration.

2. BACKGROUND

2.1. Chagos Archipelago/British Indian Ocean Territory

The Chagos Archipelago lies almost in the middle of the Indian Ocean (Figure 3.1). It consists of about 60 low-lying coral islands which together with the 200 nm MPA cover approximately 640,000 km^2 of ocean. The land area is only about 60 km^2. The islands were discovered in the sixteenth century. France assumed sovereignty in the late 1700s and exploited the islands for coconut oil and copra, ceding them to Britain in 1814. The Chagos were then a dependency of the colony of Mauritius until they were detached to become the BIOT in 1965. The formation of the new colony was a consequence of the United States (US) identifying Diego Garcia as an excellent location for military facilities. The US required the islands to remain under British sovereignty for security reasons (Vine, 2009). The detachment was condemned by the United Nations (United Nations, 1965) and was effectively imposed on Mauritius as the price it had to pay for independence (PREM 13/3320, 1965). On the 30 December 1966, a UK/US treaty made the Chagos available for joint defence purposes for an indefinite period (Exchange of Notes, 1966).

Since 1982, Mauritius has repeatedly claimed the Chagos (Abraham, 2011), while the UK maintains that it has no doubts about its sovereignty (Baroness Warsi, 2013) but acknowledges that it will cede the islands to Mauritius when there is no longer a defence requirement (FCO, 2002).

The disputes over sovereignty (ESM Appendix 1) have resulted in competing international claims around the Chagos Archipelago. In 1977, Mauritius declared a 200 nm Exclusive Economic Zone (EEZ) (ESM Figure 1-1), followed by a claim by the UK in 2003 to a 200 nm Environment (Protection and Preservation) Zone (ESM Figure 1-2). In June 2008, Mauritius deposited coordinates of archipelagic baselines with the UN (United Nations, 2008a,b) and in May 2009 submitted a claim to an

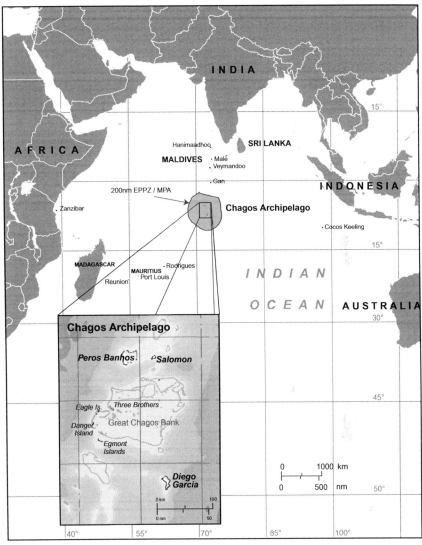

Figure 3.1 Location of the Chagos Archipelago and its 200 nm Marine Protected Area/Environment (Protection and Preservation) Zone (MPA/EPPZ). See also ESM Figure 8-1 for a more detailed map of the MPA.

extended Continental Shelf (Republic of Mauritius, 2009c), subsequently informing the UN that it proposed to complete the submission by June 2014 (Republic of Mauritius, 2013). The Maldives also dispute the median line between it and BIOT (Republic of Maldives, 2010; Sand, 2013).

2.2. The Former Islanders and traditional fishing rights (ESM Appendix 2)

The Chagos were first inhabited in the 1780s. Between 1965 and 1973, the British removed the population to Mauritius and Seychelles where they were left destitute (Gifford and Dunne, 2014). In 2000, the English High Court ruled the removal unlawful (Bancoult (1), 2000) but in 2004 the islanders were again prevented from returning when the British Government implemented a prerogative Order-in-Council, an undemocratic process which bypassed the UK Parliament and overturned the Court's decision (BIOT Constitution Order, 2004). The justification for this was claimed to be defence considerations and the unfeasibility of resettlement in the islands.

In 1965, when the islands were originally detached, Britain agreed to preserve traditional fishing rights for Mauritius to fish the internal waters and territorial sea. In 1971 and again in 1984, these rights were enshrined in BIOT Fishery Ordinances. When a 200 nm Fisheries Conservation and Management Zone (FCMZ) was introduced in the BIOT in 1991, licences were issued free to Mauritian registered vessels which fished, predominantly in the inshore fishery from 1977 until its closure on 1 November 2010 following the MPA declaration. In the early years, Mauritian purse seiners also operated in the pelagic tuna fishery.

2.3. Earlier proposals for MPAs (ESM Appendix 3)

The idea of MPAs in the Chagos was first raised in 1996 by the Marine Resources Assessment Group (MRAG), a company contracted since 1991 to administer the fisheries for the BIOT. Permanently protected areas were proposed, together with temporary closed areas but neither were progressed further (MRAG Ltd., 1998a). However, several islands (together with their territorial seas) (ESM Table 3-1) were made "Strict Nature Reserves" (BIOT Ordinance No. 1, 1970; BIOT Statutory Instrument No. 4, 1998), and part of Diego Garcia a "Restricted Area" (BIOT Ordinance No. 6, 1994). In 1999, the UK government extended the Ramsar Wetlands Convention (Ramsar, 1971) to BIOT and designated a 354.24 km^2 site on Diego Garcia in July 2001. This includes all of the lagoon waters and part of the territorial sea (3 nm) (ESM Figure 3-1).

The creation of the EPPZ in 2003, although not portrayed as a marine reserve, was nonetheless made with the intent of protecting the Great Chagos Bank (BIOT Administration, 2009), an area of submerged coral reefs for which BIOT could otherwise claim no outright sovereignty under international law. Although a draft management plan in 2003 recommended more extensive, fully protected areas to cover one-third of all terrestrial and marine habitats (Sheppard and Spalding, 2003), nothing further was implemented.

2.4. Declaration of the Marine Protected Area

On the 1 April 2010, the British Foreign Secretary established a MPA within the BIOT EPPZ (BIOT Proclamation, 2010). The limits of the EPPZ, and therefore the MPA, extend from the outer limit of a 3 nm territorial sea to 200 nm or the median line between the BIOT and the Maldives (United Nations, 2004). Within both EPPZ and MPA, Britain has claimed identical "sovereign rights and jurisdiction enjoyed under international law, including the United Nations Convention on the Law of the Sea, with regard to the protection and preservation of the environment". The UK has not claimed an EEZ for the Chagos (Simmonds, 2013). The definitive boundaries of the MPA have not yet been clarified, other than that it does not include "Diego Garcia and its waters out to 3 nautical miles" (Simmonds, 2013), an unnecessary statement since the MPA was only defined within the limits of the EPPZ which does not include *any* islands or their territorial sea (see ESM Figure 8-1). Prior to this, in 1991, the UK had promulgated a fisheries management zone, the FCMZ (BIOT Proclamation, 1991), with the same boundaries as the EPPZ/MPA. On the 20 December 2010, Mauritius challenged the legality of the MPA, and in April 2014, an international tribunal heard opposing arguments by the United Kingdom and Mauritius (Permanent Court of Arbitration, 2014). The adjudication of the tribunal is expected at the end of 2014.

3. BIOT FISHERIES MANAGEMENT REGIME 1965–2010 (ESM APPENDIX 4)

3.1. Commercial fisheries—The fisheries zones 1971–2010

At the time of the formation of BIOT, it was permissible to claim exclusive sovereignty over a 3 nm territorial sea. Seawards of this lay the High Seas to which all nations had free access. During the late 1960s and 1970s, fisheries zones became established practice and in 1969 a 9 nm zone was claimed by the UK (BIOT Proclamation, 1969). In 1971, the Fishery Limits Ordinance (BIOT Ordinance No. 2, 1971) applied within a "fisheries zone" (3 nm territorial sea + 9 nm contiguous zone) (ESM Figure 4-1). This was replaced in 1984 and in 1991, the latter Ordinance introduced a licensing scheme and applied to a new FCMZ contiguous to the territorial sea out to 200 nm (BIOT Proclamation, 1991) (Figure 3.2). The 1991, Ordinance was replaced in 1998 and again in 2007 (BIOT Ordinance No. 5, 2007). In 2003, an EPPZ was also claimed in addition to, and with the same geographical extent as the existing FCMZ but no further legislation was invoked.

3.2. Recreational fishing and fishing from yachts

"Recreational fishing" commenced with the arrival of US personnel on Diego Garcia in 1971. Permission was formally agreed between the US and the UK in 1976 (Exchange of Notes, 1976) and is governed by specific provisions in the Fisheries Ordinances. Persons lawfully present in BIOT are exempt from holding a fishing licence, subject to a number of conditions, including that fishing is for personal consumption, and sharks or other large game fish are released alive. There are also restrictions on the areas that can be fished at Diego Garcia (ESM Figure 4-3). The licence exemption applies throughout BIOT without distinction between personnel on Diego Garcia or elsewhere, including visiting yachts.

3.3. Implementation and enforcement

The BIOT Administration is in the FCO in London. On Diego Garcia there is a British Representative and a small number of UK military personnel who also perform BIOT functions. A Senior Fisheries Protection Officer is appointed to work in the territory. Since 1991, MRAG has administered the fisheries and provided research and guidance. Violations of the fisheries legislation are dealt with at Diego Garcia.

Illegal Unregulated Unreported (IUU) fishing in the FCMZ poses particular challenges because the area is on a transit route for fishing vessels and because the shallow banks and reefs provide a refuge for small vessels such as those that operate out of Sri Lanka (MRAG Ltd., 2004). Prior to 1994, there was no enforcement. Since 1997, the "Pacific Marlin", a converted tugboat operated under contract, has been the sole enforcement platform. Over the period 2006–2011, it has spent 54.5% of the year on fishery patrol, 18.7% on

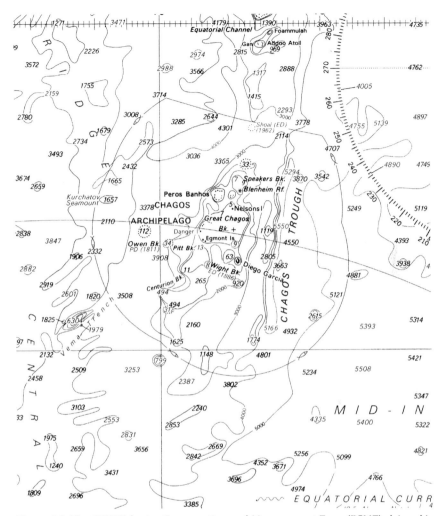

Figure 3.2 The BIOT Fisheries Conservation and Management Zone (FCMZ) claimed in 1991 shown by the line at 200 nm (fish symbols). Extract of Admiralty Chart 4073—Indian Ocean—Eastern Part, © Crown Copyright and/or database rights. *Reproduced by permission of the Controller of Her Majesty's Stationery Office and the UK Hydrographic Office (www.ukho.gov.uk).*

military patrol duties, 7.4% on other tasks, and 19.4% alongside in Diego Garcia (ESM Table 4-1).

In 1995, fishery administration costs were about £1 million per year (Figure 3.3, ESM Table 4-3). Between 1995 and 2010, costs increased by a further ~£65,000 per year and since the declaration of the MPA the annual

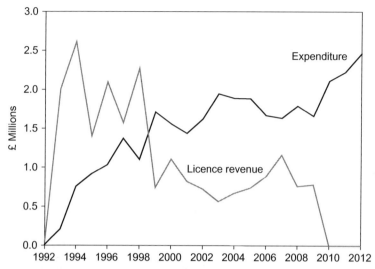

Figure 3.3 BIOT fisheries expenditure and licence revenue 1992–2012. *Source: see ESM Table 4-3.*

increase has been ~£250,000 per year. Between 1992 and 2010, revenue from fishing licences was variable since it depended on the annual catch, averaging £2 million from 1993 to 1998, thereafter falling to £770,000. Current fisheries management expenditure amounts to about £2.5 million per year. Prior to 2010, the net costs of managing the fishery have been £3.6 million over an 18-year period.

When the MPA was declared, the private Bertarelli Foundation offered to replace the lost licence revenues with a donation of £3.5 million for the period 2010–2015 (FCO Press Release, 2011). A further agreement for the US Government to provide £1.5 million worth of fuel arranged by the Pew Environment Group (The Independent, 2010) has not been honoured (Alistair Gammell, personal communication).

These costs can be compared with another British Overseas Territory, the Falkland Islands, a sea area slightly smaller than the BIOT FCMZ, where revenue is about £20 million per year and £6 million per year is allocated to fisheries protection and research (Falklands Islands Government, 2013; Chapter 2). The latter is more than double the BIOT expenditure. On the basis of the typical running costs of other MPAs worldwide (Balmford et al., 2004), the realistic running costs of the Chagos MPA might be expected to be around £11.8 million per year; the present expenditure is one-fifth of that.

4. REGIONAL FISHERY ORGANISATIONS (ESM APPENDIX 5)

The BIOT participated in two bilateral fisheries organisations to promote, facilitate and co-ordinate conservation and scientific research of fish stocks; the British/Mauritian Fisheries Commission (1994–1999), and since 1995, the British/Seychelles Fisheries Commission (BSFC). The UK on behalf of the BIOT is also a member of the Indian Ocean Tuna Commission (IOTC) which was established in November 1993 as the regional fisheries management organisation for the management of tuna and tuna-like species in the Indian Ocean. The Chagos lie on the edge of the Western IOTC area (Figure 3.4).

5. THE FISHERY AND ITS EXPLOITATION PRIOR TO 2010 (ESM APPENDIX 6)

5.1. Inshore/demersal

A commercial inshore fishery originated from the 1965 undertaking which recognised and preserved Chagossian/Mauritian traditional fishing rights. It targeted demersal species, principally lutjanids (snappers), lethrinids (emperors) and serranids (groupers), on a seasonal basis between 1 April and 31 October. Since 1991, up to 6×80 day licences per year were available, with a mean of 3.2 licences per year issued between 1992 and 2009 (ESM Table 6-2). Mauritian vessels were entitled to fish both within the territorial sea and the FCMZ. Most fishing was on the Great Chagos Bank (64% of the total BIOT catch), Speakers Bank (10.2%) and Pitt Bank (6.4%) (Figure 3.5; ESM Table 6-1) (MRAG Ltd., 1996). Only one island location, Peros Banhos, contributed substantially (5.1%) to the catch. The mean catch between 1977 and 2009 was about 200 t year^{-1} (Figure 3.6; ESM Table 6-2) but with considerable inter-annual variation (standard deviation ± 80.4 t). Over the licensing period (1991–2009), the total catch, number of days in BIOT, and the number of fishing licences exhibited a linear decline, suggesting that the catch decline was simply a result of less fishing. This is corroborated bycatch rate data which remained constant between 1991 and 2009 (ESM Appendix 6).

A conservatively estimated sustainable fishery yield for the shallow waters (<70 m) was 859 t year^{-1} (ESM Table 6-3 and 243 t year^{-1} in deeper water (70–150 m) (Mees et al., 1999b), five times greater than the average

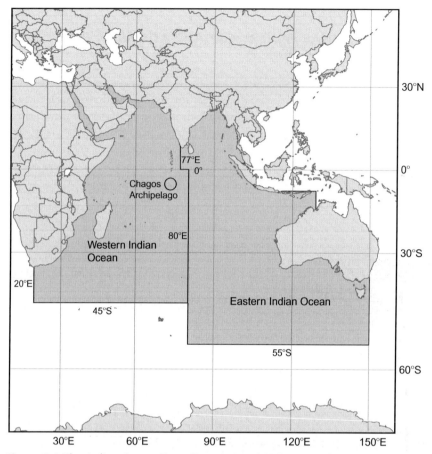

Figure 3.4 The Indian Ocean Tuna Commission (IOTC) areas of competence. The Chagos (BIOT) is situated in the Western Indian Ocean sub-area.

annual catch. In 2010, shallow reef fish biomass (kg ha^{-1}) surveyed using underwater visual census in the northern Chagos atolls was six times greater compared to even the most successful marine reserves elsewhere in the Indian Ocean (Graham et al., 2013). Much of this was due to abundance of higher trophic level fishes with large overall body size (Graham and McClanahan, 2013).

Limited information on discards and by-catch for the inshore fishery was available from an observer programme which operated intermittently after 1991. Discards included undersized fish, by-catch species such as sharks, and potentially ciguatoxic fish (Mees et al., 1999b) and represented a further 16% (32 t year^{-1}) of the initial catch by weight (ESM Table 6-4).

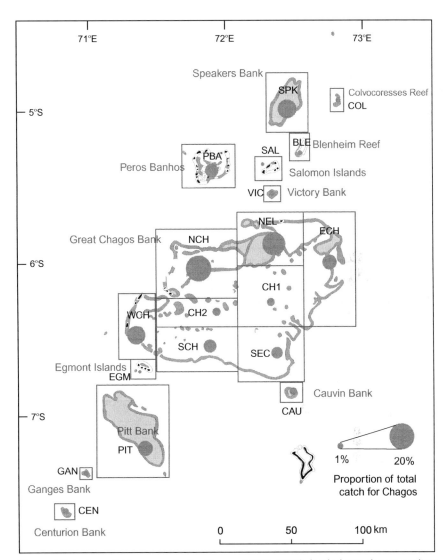

Figure 3.5 Inshore fishing areas in the Chagos (1991–2010) divided into data recording sectors. Circles indicate fishing effort in each sector (circle area proportional to catch). For a key to abbreviations see ESM Figure 6-1 and catch data Appendix 6.

IUU fishing was assessed to be the principal major threat to the inshore fishery and ecosystem in 2009 (Nugent et al., 2010) and this continues to be the case (Martin et al., 2013). Between 1996 and 2013, there have been a total of 76 arrests of illegal fishing vessels with no clear inter-annual pattern apart from an unusually large number in 2010 (Table 3.1). The majority of arrests took

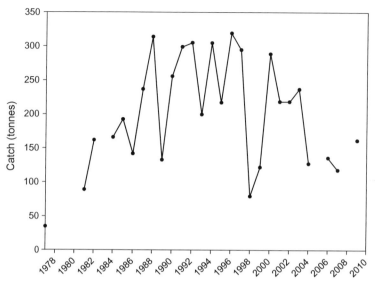

Figure 3.6 Annual catch data for the Chagos inshore fishery (1977–2009). *Source: ESM Table 6-2.*

place in inshore areas with a mean catch per vessel of 2030 kg. Sharks were found on 91% of the vessels and comprised an average 78% of the catch weight.

5.2. Indian Ocean tuna

The Indian Ocean yields 25% of the world's tuna production with about an equal split between industrial and artisanal fisheries. A high proportion of the industrial fishery catch comes from beyond national jurisdictions. The BIOT did not have an artisanal tuna fishery but between 1991 and November 2010, it licenced foreign purse seine and longline vessels to fish in its FCMZ. Tuna species of major commercial importance in the tropical Indian Ocean are skipjack (*Katsuwonus pelamis*), yellowfin (*Thunnus albacares*) and bigeye tuna (*Thunnus obesus*) with around 70% of the industrial catch from the western area (Figure 3.4) and with different species thought to occupy discrete spatial distributions as a function of latitude and depth (Figure 3.7) (Reygondeau et al., 2012).

The results of a large-scale Indian Ocean tagging programme illustrate the highly migratory nature of these tuna species (Figure 3.8). Distances travelled had modal values of 1482 km for yellowfin, and 1111 km for bigeye and skipjack tuna, with maximum distances up to 5370 km for all species, lending support to a hypothesis that there may be a single Indian Ocean

Table 3.1 Arrests by the BIOT Patrol Vessel for Illegal Unregulated Unreported (IUU) fishing in the BIOT FCMZ 1996–2013

Year	Arrests	Year	Arrests	Year	Arrests
1996	2	2002	3	2008	2
1997	2	2003	8	2009	6
1998	2	2004	5	2010	12
1999	6	2005	1	2011	8
2000	6	2006	2	2012	3
2001	2	2007	2	2013	4

Source: Martin et al. (2013).

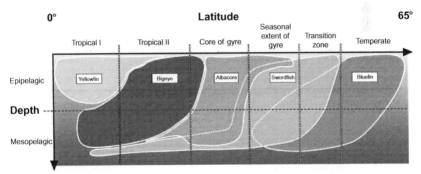

Figure 3.7 Conceptual scheme of global spatial distribution of dominant tuna species and "ecoregions" as a function of latitude and depth. Epipelagic = surface to mixed layer depth, mesopelagic = mixed layer to 1000 m. Data are from the longline fishery only and therefore exclude juveniles and skipjack tuna. The Chagos lie in Tropical I ecoregion. *Modified from Reygondeau et al. (2012).*

stock for each species (Hallier and Million, 2009). Mark–recapture distances are on average greater in the Indian Ocean than in other tropical oceans (Kaplan et al., 2014). The large range and the influence of oceanographic conditions were also illustrated by an Indian Ocean Dipole event in 1998 (Webster et al., 1999) when the purse seine fleet followed the tuna and moved thousands of kilometres into the eastern Indian Ocean (Marsac and Le Blanc, 1999). Other species which form part of the by-catch appear to have an even greater migratory spread. A population genetic study of the swordfish *Xiphias gladius* found a single Indian Ocean-wide panmictic population (Muths et al., 2013) with movements of up to 6670 km (Kadagi et al., 2011).

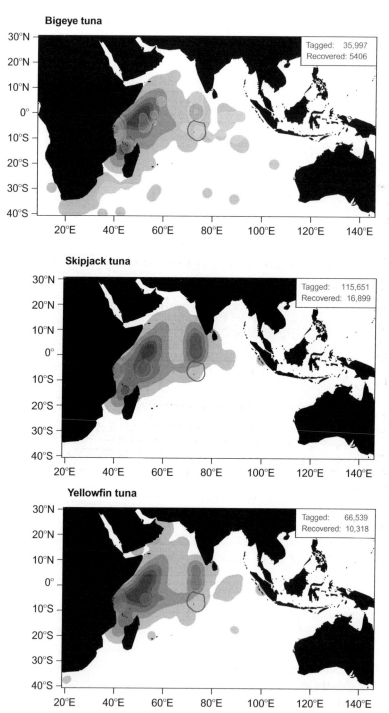

Figure 3.8 Density map of releases (dark grey) and recoveries (lighter grey) of tuna during the Regional Tuna Tagging Project-Indian Ocean. Position of BIOT FCMZ shown by outline (Hallier and Million, 2012). *Courtesy of Julien Million.*

The industrial fishery in the Indian Ocean consists predominantly of longline and purse seine fishing, with some gillnet fishing in areas beyond national jurisdiction. Longline primarily targets yellowfin (~30% of biomass caught) and bigeye tuna (~37%) in subsurface waters (typically 100 to >250 m depth) with ~93% of the catch consisting of large fully mature individuals. Purse-seining occurs in surface waters (maximum depth ~200 m) either on free-swimming schools or around fish aggregating devices. Free-swimming schools yield mainly large, mature yellowfin, skipjack and bigeye (~70%, ~25% and ~4% of biomass, respectively). FAD fishing targets mainly skipjack (~63% of biomass), but includes yellowfin (~28%) and bigeye (~9%), a large proportion of which are small juveniles (Kaplan et al., 2014).

The licenced fishing in the BIOT FCMZ had been seawards of 12 nm from islands and was seasonal and highly variable from year to year. Longliners typically operated from June to September and November to February, and the purse seine fishery from November to February. The Chagos lie centrally within the tropical longline fishing grounds (Figure 3.9) but were more lightly fished compared to adjacent areas to the west and east. For purse seine, fishing on FADs was relatively light, and fishing effort for free-swimming schools was of medium intensity (Figure 3.10).

A tuna migratory pattern is implied from the cyclical deployment of the European Union purse seine fishing vessels in the Western Indian Ocean (Figure 3.11) with the BIOT at its easternmost limit and is reflected in the proportion of the total Indian Ocean catch in the BIOT each month (Figure 3.12) (MRAG Ltd., 2010; Pearce, 1996). Between 1993 and 2008, the BIOT contributed a mean of 0.53% of the longline Indian Ocean tropical tuna catch and 2.73% of the purse seine (1.86% of the combined catches). Catches of bigeye, yellowfin and skipjack tuna in the Indian Ocean rose particularly rapidly between 1980 and 1990 (Figure 3.13) driven by the new purse seine fishery whose catch peaked in 2002 at 482,000 t before declining to around 300,000 t in the past few years (Figure 3.14). In the same period, the catch of the longline fishery doubled and stabilised at ~200,000 t from 1992 to 2007, reducing thereafter. The yellowfin tuna stock is currently not overfished while neither bigeye tuna or skipjack stocks are overfished or subject to overfishing (IOTC, 2013).

Unlike catch trends for the wider Indian Ocean (Figures 3.13 and 3.14), the catch in the BIOT was highly variable (Figure 3.15, ESM Table 6-7) with catches fluctuating between a maximum of 32,051 t in 1993–1994

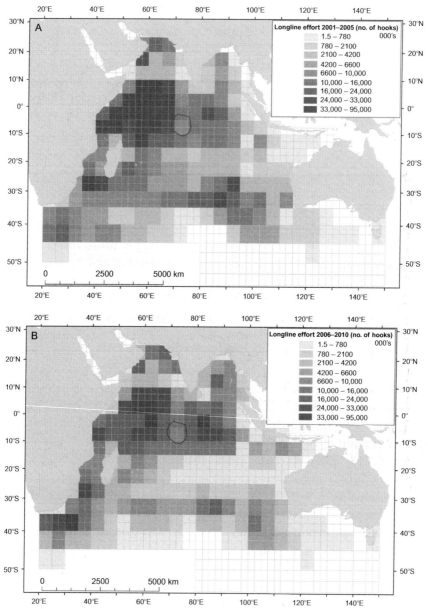

Figure 3.9 Indian Ocean longline fishery effort (A) 2001–2005, (B) 2006–2010. The position of the BIOT FCMZ is shown outlined. The footprint of the Western Indian Ocean fishery is ∼26 million km² (MRAG Ltd., 2010). *Modified from Nel et al. (2013).*

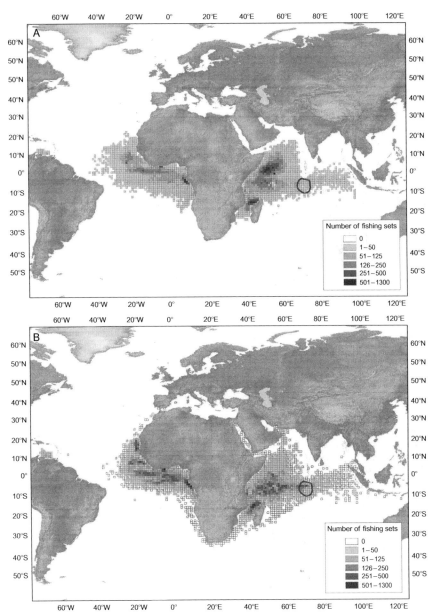

Figure 3.10 Purse seine fishing effort by French and Spanish fleets 1995–2011 in the Atlantic and Indian Oceans. Position of BIOT FCMZ is outlined. (A) Fish aggregating devices (FADs). (B) Free-swimming schools (FSC). The footprint of the Western Indian Ocean fishery is ~11.6 million km² (MRAG Ltd., 2010). *Modified from Clermont et al. (2012).*

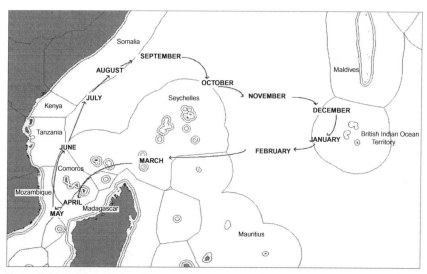

Figure 3.11 Western Indian Ocean purse seine tuna fishery. Cyclical pattern of activity by month based upon the European Union purse seine fleet movements. EEZ/Fishery Zones shown. After MRAG Ltd. (2010).

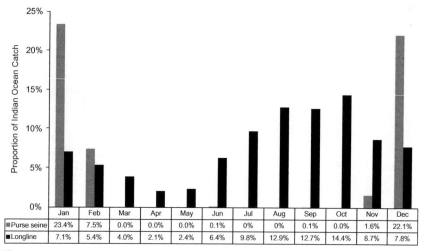

	Jan	Feb	Mar	Apr	May	Jun	Jul	Aug	Sep	Oct	Nov	Dec
Purse seine	23.4%	7.5%	0.0%	0.0%	0.0%	0.1%	0%	0%	0.1%	0.0%	1.6%	22.1%
Longline	7.1%	5.4%	4.0%	2.1%	2.4%	6.4%	9.8%	12.9%	12.7%	14.4%	8.7%	7.8%

Figure 3.12 Average proportion of tuna catch by vessel type for the period 1999–2008 taken within the BIOT FCMZ per month. *Data from MRAG Ltd. (2010).*

to a minimum of 685 t in 2006–2007. While the longline element has remained relatively constant at 1096 ± 688 t $year^{-1}$ (mean $\pm$ standard deviation), the much larger (up to 10 times) purse seine catches have shown dramatic inter-annual fluctuations (9571 ± 9537 t $year^{-1}$) and it is these that

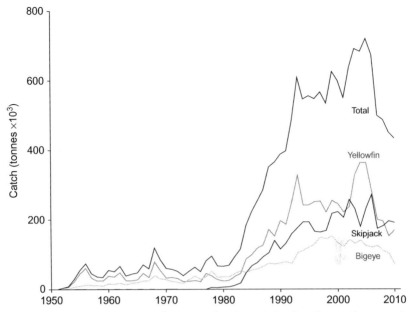

Figure 3.13 Indian Ocean tropical tuna catches by species (longline and purse seine fisheries) 1950–2012 (tonnes). *Data source: Working Party on Tropical Tunas (WPTT) http:/iotc.org/English/meetings/wp/wpttcurrent.php.*

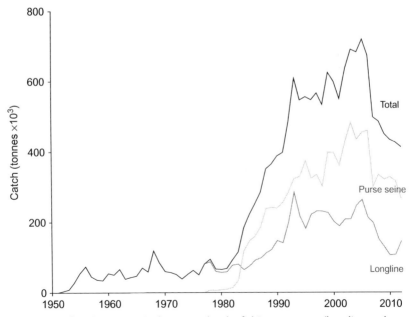

Figure 3.14 Indian Ocean tropical tuna catches by fishing gear type (longline and purse seine fisheries) 1950–2012 (tonnes). *Data source: Working Party on Tropical Tunas (WPTT) http:/iotc.org/English/meetings/wp/wpttcurrent.php.*

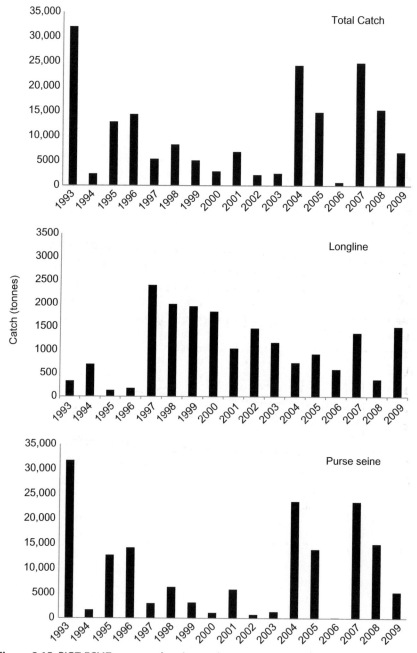

Figure 3.15 BIOT FCMZ tuna catches (tonnes) 1993–2010.

have driven the variability in the overall catch. Over the timescale of the fishery, catches were directly related to fishing effort (days fished) (OLS regression; longline: $R^2 = 0.80$, $p < 0.0001$; purse seine: $R^2 = 0.79$, $p < 0.0001$) indicating that catches were driven by fishing effort and not by changes in catch per unit effort (CPUE). However, between 1998 and 2009 there is evidence of a decrease in CPUE for the longline fishery of 3.0% per year ($R^2 = 0.47$, $p = 0.0136$), but it is not possible to deduce whether this represents a decrease in stock, changes in migration patterns, the fishing skills of individual vessel captains, or learned behaviour by the fish themselves. For both fisheries in the BIOT, fishing effort and therefore catch variability is likely to be primarily driven by changing oceanographic conditions that determine the extent of tuna migration inside the FCMZ (MRAG Ltd., 2010).

Fishing results in the removal of non-target species; for longliners this is typically approximates to an additional 20% of the total catch, and while purse seiners take less, the proportion is greater with FADs compared to free schools. The by-catch for free schools based on observer data is 0.2–1.0%, and for FADs 2.7–5.0% (MRAG Ltd., 2010; Pianet et al., 2009). Within BIOT, the purse seine by-catch rate is likely to be lower than for the wider Indian Ocean since 66% of sets are on free schools (MRAG Ltd., 2010). For silky sharks (*Carcharhinus falciformis*), a "near threatened" species which represents 85% of all sharks caught by the purse seine fishery, the majority are caught by the FAD fishery (92%) (Amande et al., 2011) with a by-catch "hot spot" in the north-western Indian Ocean.

A detailed record of longline by-catch in BIOT is available for three fishing seasons (ESM Table 6–8). Catches of numbers of shark amounted to 4.29% of the total catch and were dominated by blue shark (2.01% of total catches), with pelagic thresher sharks (0.66%), and silky sharks (0.56%) comprising the next largest groups. Because these data predate the introduction of legislation to limit shark catch it may be expected that the shark element of the by-catch reduced post-2006/2007 (MRAG Ltd., 2010).

5.3. Recreational fishery

In the northern atolls, visiting yachts have been allowed to fish in any location except the Strict Nature Reserves. The very high shallow-water reef fish biomass data from Salomon and Peros Banhos in 2000 and 2010 (Graham et al., 2013; Spalding, 2000) suggest that the fishery has not had any major impact.

At Diego Garcia, recreational fishing takes place from the shore, from boats (for pelagic species and for reef fish species) and from ships at anchor. The shore fishery and "reef" boat fishery target mainly top predators, while the "pelagic" boat fishery catches oceanic species and a small number of reef fish (Mees et al., 1999a; MRAG Ltd., 1999). In the first 20 years, the entire fishery was a "free for all" and catches were plentiful but no records exist. There has been only one comprehensive survey of the fishery which was conducted in 1997/1998 and included an assessment of the shore catch, when the total catch (reef and pelagic) was estimated to be 103.75 t year^{-1} and was thought to be within what were considered sustainable yield limits of 4–5 t km^{-2} year^{-1} (Mees et al., 1999a; MRAG Ltd., 1998b, 1999). This sustainable yield value is however much higher than for the comparable commercial inshore fishery (0.1 t m^{-2} year^{-1}) (Mees et al., 1999b) which is considered equally applicable to the recreational fishery (Sheppard and Spalding, 2003). Reviews of coral reef fisheries highlight the large regional variability in sustainable limits and the difficulty of estimating this for a particular location (Dalzell et al., 1996; McClanahan, 2006). For the pelagic fishery, similar concepts of sustainable yields are not applicable because of the migratory nature of the fish.

Attempts to introduce log-sheet records for the shore fishery have consistently failed and for the boat fishery were poorly completed for many years. Reliable boat catch statistics only exist after 2006 (ESM Tables 6-10, 6-11, 6-12) recording an average catch of 35.1 t year^{-1}, with 7.7 t year^{-1} of this coming from the reef and lagoon. The shore fishery would add around 40 t year^{-1} (based on 1997/1998) resulting in a total reef derived catch of about 48 t year^{-1}. Recent trends (Figure 3.16) show a sharp decline in the pelagic catch which is partly attributable to closure and subsequent restrictions following two US fatalities in September 2011 off the southern end of the atoll (US Navy Support Facility Diego Garcia, 2011a,b).

From 1971 until 2008, there was no restriction on catch but in 2008 all fishing was required to be for "a reasonable amount for personal consumption within 3 days" (Fisheries (Conservation and Management) Ordinance, 2007). Despite this, photographs of catches in 2008 and 2009, and data show that catches are far in excess of the restriction (ESM Appendix 6). It is interesting to note that the formal US Navy environmental guidance (CNRJ, 2011) omits any reference to the 3-day limit.

Hitherto, the "recreational" fishery has been largely unreported in earlier studies (Koldewey et al., 2010b; Sheppard et al., 2012). A recent (2012) reef fish survey at Diego Garcia found that reef fish biomass was less than

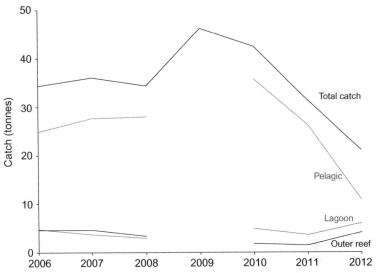

Figure 3.16 Annual catches from the boat-based recreational fishery at Diego Garcia 2006–2012. *Data from ESM Table 6-10.*

one-third that at Peros Banhos and the Great Chagos Bank, with smaller fish and a lower trophic level, despite including no lagoon sites which are the most heavily fished (Graham et al., 2013). The significant depression of biomass on the seaward reefs of Diego Garcia compared to elsewhere in BIOT suggests that the Diego Garcia fishery has depleted target species.

5.4. Sharks

Little fisheries or ecological information exists for sharks in the Indian Ocean as a whole (Dulvy et al., 2008). In the Chagos area, visual sightings of reef sharks during SCUBA dives in 1975/1979, 1996, 2001, 2005, 2006, 2010, and 2012 have led to semi-quantitative comparisons of shark numbers which show a decline of ~76% between 1975/1979 and 2012. This has been variously attributed to (a) the Mauritian inshore fishery (Anderson et al., 1998; Koldewey et al., 2010b) and (b) illegal fishing by Sri Lankan boats (Anderson et al., 1998; Graham et al., 2010), the latter hypothesis finding support in the predominance of reef sharks in IUU catches (Martin et al., 2013). However, the data and methodology are not without their problems (different observers/locations, very high variability) so, for example, in the 2010/2012 data no inter-atoll differences are detectable despite substantial differences in mean shark biomass (up to 28 times) (Graham et al., 2013), and whereas numbers declined by 90% between 1975 and 1996, there was no further change between 1996 and 2010 (Graham et al., 2010) despite continued poaching (ESM Table 6-5) and licenced inshore fishing (ESM Table 6-2). The "name change" from FCMZ to MPA affords no greater protection against IUU fishing for reef sharks, unless it is met with increased enforcement. Indeed evidence from the Great Barrier Reef suggests that even in one of the world's most well-managed reef ecosystems no-take zones alone do not provide effective protection for reef sharks (Robbins et al., 2006; Chapter 7).

Elasmobranch catches by the gillnet fishery (not present in BIOT) are a particular issue and for longliners also but to a lesser extent. Pelagic oceanic shark species however are highly migratory species and not normally considered resident within discrete locations. For the Chagos, preliminary results from satellite tagging of three silky and five silvertip sharks (Bertarelli Foundation, 2013) show displacements of up to 1500 km and 100 km, respectively (David Tickler, unpublished). Within the FCMZ, tuna vessels were not permitted to fish within 12 nm of land, lessening the potential for reef shark species to be caught, and indeed none have been observed in catches during observer programmes (MRAG Ltd., 2010). There have also been specific measures adopted to reduce shark by-catch, including bans since 2006 on the removal of fins (BIOT Statutory Instrument No. 1, 2006) and since 2007 on using wire trace.

For the shore-based recreational fishery at Diego Garcia, there is no record of the number of sharks caught each year. From the sparse data that exist, sharks form a not insubstantial part of the catch (MRAG Ltd., 1999) and it is unknown if the current regulations requiring their release are in fact complied with. For the boat fishery, sharks were no longer landed from about 2002. A diver survey in February 2012 recorded the lowest shark biomass in the Chagos at Diego Garcia; between 7- and 28-fold less than elsewhere (Graham et al., 2013). Given that there has been no commercial fishing or IUU in the vicinity of the island, it must be assumed that the recreational fishery (or some other activity or environmental factor) has had a significant local impact on the abundance of sharks.

5.5. Marine mammals

Although Chagos is at the heart of the Indian Ocean Whale Sanctuary established in 1979, very little is known about the abundance of cetaceans in the area, hampering any meaningful assessment of the effect of a no-take MPA.

A recent review makes no mention of the Chagos (de Boer et al., 2002). Sightings of whales (short-finned pilot whales, *Globicephala macrorhynchus*) and of dolphins (spinner dolphins, *Stenella longirostris*, bottlenose dolphins, *Tursiops* sp., striped dolphin, *Stenella coeruleoalba*, spotted dolphin, *Stenella attenuata*) include those by Anderson (1996), the BIOT inshore observer programme (Clark et al., 2002), and at Diego Garcia (Deslarzes et al., 2005; NAVFAC Pacific, 2005). Acoustic monitoring has also demonstrated that blue whales of at least three different populations visit Chagos waters (Stafford et al., 2004, 2011) and sperm whales, *Physeter macrocephalus* are also found (Ocean Alliance, 2009; Townsend, 1935; Wray and Martin, 1983).

In the Indian Ocean generally, there is little information on cetacean by-catch in the tuna fishery but it is thought to be small for the longline fishery (Huang and Liu, 2010). In the purse seine fishery although tuna schools are sometimes associated with cetaceans the scale of any interactions is not known and there is only one record of a young baleen whale entangled and killed (Romanov, 2001). There is concern over the impact of underwater noise pollution on marine mammals (DiMento, 2006), particularly the use of military sonar. The remains of Cuvier's beaked whales, *Ziphius cavirostris*, a species that is especially vulnerable to military sonar have been found on the Chagos islands (Mata'irea Blogspot, 2009; NAVFAC Pacific, 2005).

5.6. Turtles

Sea turtles in the Chagos have been reviewed by Mortimer and Day (1999) and the interaction of sea turtles and tuna fisheries in the wider IOTC region by Nel et al. (2013). Green turtles *Chelonia mydas* and hawksbills *Eretmochelys imbricata*, nest and forage in the Chagos with nesting populations crudely estimated in 1996 at 400–800 and 300–700, respectively (Mortimer and Day, 1999). Leatherbacks (*Dermochelys coriacea*) may forage there (Mortimer, 2006) and olive ridleys (*Lepidochelys olivacea*) may also pass through the area (Nel et al., 2013). The green turtle population forms part of the south-west Indian Ocean regional management unit (ESM Figure 6-16) although there are recent reports of individuals tracking to Somalia and the Maldives (Jeanne Mortimer, unpublished). Preliminary genetic studies indicate that the nesting hawksbill population is a distinct stock but some hawksbills foraging in the Chagos may originate from rookeries in Seychelles (Mortimer and Broderick, 1999; Mortimer et al., 2002).

The routes of migration for both hawksbill and green turtles to and from nesting and foraging grounds principally lie to the west of the Chagos and this is where they are most susceptible to the tuna fishery. In the FCMZ, the only records of turtle by-catch for both longline and purse seine fleets are of catches by longliners of a leatherback in 2001–2002 (Sheppard and Spalding, 2003) and a female green turtle in 2002–2003 (MRAG Ltd., 2003). The dangers posed by the longline fishery within the FCMZ to nesting populations would seem to have been insignificant. If there is an impact of the fishery, it is likely spread across the wider Indian Ocean and the Chagos MPA would therefore not represent a cost-effective conservation measure.

The capture, killing and sale of green turtles has been prohibited in BIOT since 1968 and since 1984 all species of turtles and their eggs have been protected. Although there are no baseline studies from this time, the turtle populations are thought to be increasing from earlier exploited levels (Mortimer, 2007). There is also no known IUU fishery for turtles in the Chagos although anecdotal accounts exist of turtle egg collection by fishers camping on the islands (MRAG Ltd., 2011).

A potential negative impact on the nesting populations has been identified as artificial night-time lighting from the US base on Diego Garcia (Mees and Polunin, 1994; Mortimer, 2000), particularly since the island supports approximately one-third of all turtle nesting in the Chagos (Mortimer and Day, 1999). The MPA will have no impact on this threat.

5.7. Holothurians

Surveys of holothurians (sea cucumbers) in the Chagos were undertaken in 2002 (Posford Haskoning Ltd., 2002) and in 2006 and 2010 (Price et al., 2013; Price et al., 2010). While the 2002 surveys only covered Peros Banhos and Salomon (the outer atolls), those in 2006 and 2010 extended to Diego Garcia and the Great Chagos Bank. Abundance in the outer atolls decreased dramatically between 2002 and 2006 (between 82- and 8-fold depending on location and species). Abundances at Diego Garcia in 2006 were between 5.3- and 10.6-fold higher than in the outer atolls (Price et al., 2010). The decline seen in the outer atolls and differences between those areas and Diego Garcia were attributed to poaching. In the later re-survey in 2010, however, while abundances at Salomon, Peros Banhos and on the Great Chagos Bank were unchanged from 2006, the population at Diego Garcia plummeted by at least sevenfold (Price et al., 2013). Since the ban on collection of holothurians at Diego Garcia is strictly enforced, the authors were unable to attribute the decline to fishing pressure and could only conclude that the explanation lay in other ecological factors. Collectively from the holothurian survey results, it is difficult to reach any conclusion on the magnitude of the effect of holothurian poaching other than that it must deplete stocks.

6. CREATION OF THE MPA—1 APRIL 2010 (ESM APPENDIX 7)

6.1. Background, scientific and conservation rationale

The catalyst for the MPA came from a proposal in July 2007 by the Pew Environment Group for a no-fishing zone that was claimed necessary for environmental protection. Simultaneously, the CEN was created to campaign for the MPA. On 9 March 2009, the CEN, in publicity material financed by Pew, highlighted a claim of the "destructive impact(s) of … unsustainable fishing", the need for "a no-take fishing zone" which would "increase Indian Ocean fish stocks", and how no-take zones can allow fish stocks to recover "remarkably quickly", together with a page entitled "The massacre of Indian Ocean sharks" (Chagos Conservation Trust, 2009). However, there was a notable absence of facts or scientific evidence about the Chagos to underpin what was written and the evidence from the fisheries perspective was contrary to the basic claims.

On 5 May 2009, the Foreign Secretary, David Miliband, was briefed on the potential benefits of declaring the entire FCMZ "a no-take MPA" and to "bring an end to fishing and legislate for the protection of the seas and atolls" (Roberts, 2009). The briefing advocated either "a Big Bang approach" of a full reserve or "a gradualist approach", a need for "a strong scientific basis to support the creation of an MPA" which the note acknowledged only partially existed, and that a full reserve would be "a political as well as a scientific decision". Risks included (1) Mauritian claims to sovereignty, fishing rights, and potential opposition; (2) Chagossian opposition, adding that the creation of the reserve "could create a raft of measures designed to weaken the movement" for Chagossian rights which could be achieved by "activating the environmental lobby"; and (3) US military opposition which could be overcome by assurances. The potential benefits which had been claimed by those briefing the Foreign Secretary, although largely unsubstantiated, caused him to be understandably "fired up" and "enthusiastic" and FCO officials were told to "ensure that the creation/announcement of the reserve is scheduled within a reasonable timeframe" (Gould, 2009).

On 12 May 2009, US Embassy officials were assured that an MPA would have no impact on the US military use of BIOT and Diego Garcia, in particular (Bancoult (3), 2012; US Embassy (London), 2009), and were informed that the timeline was tight and the announcement had to be before the next General Election which "must occur no later than May 2010". FCO officials told them that the MPA would help further restrict access to BIOT, and "put paid to resettlement claims" by Chagossians. In achieving the latter goal, the BIOT Commissioner would use the environmental lobby which he stated was "far more powerful than the Chagossians' advocates" (US Embassy (London), 2009).

In June 2009, MRAG advised that while the proposal had broad conservation aims it would have little value for the tuna fisheries and might be invalid under international law. It proposed that the reserve only extend to the islands, their territorial seas and an area out to 12 nm from the 200 m depth contour round the Great Chagos Bank, totalling 53,270 km^2 (MRAG Ltd., 2009). MRAG Ltd. (2009) also pointed out it was likely to increase IUU fishing activity and result in significantly increased enforcement costs and that the historical fishing rights of Mauritius needed to be considered.

On 21 July 2009, FCO officials held formal talks with the Mauritian Government. The UK pointed out that for a full no-take MPA "the scientific basis had not yet been fully established but the idea merited consideration" and they also presented the alternative gradual route which included the plan proposed by MRAG. The delegations issued a joint communiqué welcoming the proposal in principle and stating that officials and scientists should meet to examine the implications with a view to informing the next round of talks (Republic of Mauritius, 2009a). However, on 10 November 2009, the FCO unilaterally launched a public consultation which resulted in the breakdown of further bilateral talks (Republic of Mauritius, 2009b).

In August 2009, a meeting examined the science issues (Williamson, 2009). MRAG presented arguments for only partial no-take fishing areas but failed to persuade the other 22 invited workshop participants several of whose interests lay in imposing a no-take fisheries regime throughout the 200 nm MPA. The report concluded that any decision to impose a no-take area and also to extinguish Mauritian fishing rights would be a political one, and cautiously stated that any phasing out of tuna fishing would need further research to avoid unintended consequences.

6.2. Public consultation

A public consultation commenced on 10 November 2009. A confidential memorandum recommended only "interim measures" until the continued inadequacy of scientific information was addressed (Yeadon, 2009). In contrast, the public consultation (FCO, 2009) gave the overwhelming impression that there was a strong scientific case for the MPA and portrayed the BIOT fisheries generally in negative terms as a loss-making business. The Consultation Document was largely devoid of any useful factual information about fisheries in BIOT.

On 7 January 2010, a meeting convened by those anxious about the lack of any socio-economic considerations in the FCO consultation document, came to a majority view that none of the MPA options put forward by the FCO were appropriate and that a 4th option involving sustainable use of natural resources alongside conservation was required. From a tuna fisheries perspective, participants agreed that the FCMZ was insufficiently large to protect any of the highly migratory fish stocks during their life cycle, but could not agree if a no-take MPA would nonetheless provide some net positive benefit. For the demersal fishery, the benefits of a no-take MPA were recognised but difficulties were foreseen concerning Mauritian historical fishing rights leading to suggestions that a zoned approach be adopted to accommodate this and the possible return of the Chagossians. The greatest problem was identified as being IUU fishing and its effect on reef sharks and holothurians (Mangi et al., 2010).

The consultation, which included about 450 written responses, produced a variety of viewpoints (Stevenson, 2010). In support of the full no-take option, there was a tendency simply to use "campaign-type" letters, including statements such as "tuna fishing should be banned as tuna stocks are declining" and that by-catch was a major concern, rather than providing reasoned argument. The CEN's own submission for a total no-take MPA (CEN, 2010) made sweeping claims for the predicted fisheries benefits which drew on a "comprehensive review" commissioned by the Pew Environment Group and undertaken by the Zoological Society of London (Koldewey et al., 2010a,b).

On 30 March 2010, FCO officials recommended to the Foreign Secretary that the results of the consultation be placed in the public domain and that a "positive, but not definitive, announcement" be made, but that before an MPA was declared further consultation with stakeholders was required (Yeadon, 2010a). The Foreign Secretary's immediate response was that he wished to be "bolder" and that "we should actually decide to go ahead" (Clayton, 2010). The rejection of the official advice caused a flurry of concerned activity and a memorandum the next day advised that this proposed path was fraught with danger, and that "the best defence against legal challenges which are likely to be forthcoming... is to demonstrate a conscientious and careful decision-making process. A rapid decision now would undermine that" (Yeadon, 2010b). Despite this advice, the Foreign Secretary instructed the BIOT Commissioner that day to declare a full no-take MPA (BIOT Proclamation, 2010).

6.3. Diego Garcia—A military exclusion zone

Diego Garcia and its 3 nm territorial sea (470 km^2) have now been formally exempted from the Chagos MPA (Simmonds, 2013). The exemption had been a preferred option in the consultation (FCO, 2009) leading to questions concerning the environmental impact of the military base (Stevenson, 2010). It is currently the only inhabited island in the Chagos, with a resident population of approximately 3000 (Lunn and Mills, 2013) and houses the "US Naval Support Facility Diego Garcia" which was established under the 1966 UK/US treaty (Sand, 2009b). Accumulated infrastructure build costs over the years are estimated at over US$3 billion (US Department of Defense, 2013) with a further US$200 million upgrade programme underway (Erickson et al., 2010). A major part of the construction work involved dredging 30.8 km^2 of the island's lagoon (Sheppard, 1980) and removal of the 6.4 km of seaward coral pavement for "coral landfill" for the airport and a highway (EG&G Environmental Consultants, 1980; Tucker and Doughty, 1988). Some of the resulting damage to the island's coral resources and marine ecosystem is now considered irreparable (Sheppard and Spalding, 2003; Vine et al., 2012).

7. ANALYSIS (ESM APPENDIX 8)

In the 1960s as former colonial countries around the world achieved independence, Great Britain at the behest of the US created a new colony, the BIOT to further the defence interests of the UK and US (BIOT Constitution Order, 2004). The expulsion of the native population between 1965 and 1973 (Gifford and Dunne, 2014) and the building of a US military base on the island of Diego Garcia in 1971 created a territory under exclusive British control, free of democratic scrutiny, parts of which are beyond the reach of international environmental agreements (Sand, 2009a). It is against this background that the present day Chagos MPA was created on 1 April 2010 by the then UK Foreign Secretary, David Miliband.

Access to information concerning Chagos is severely hampered due to the lack of open government and the BIOT Administration's claim that freedom of information laws do not apply (First Tier Tribunal (Information Rights), 2013; Sand, 2013). It is inevitable therefore that access to factual information remains incomplete, and it is this, together perhaps with a lack of diligence that has led to inaccurate or incomplete assessments, particularly by those on behalf of the Pew Foundation, of the territory's fisheries and unjustified projection of the benefits and necessity of fishing closures (Koldewey et al., 2010a,b).

Detailed analyses of the balance between the services that ecosystems can or should provide to society and the necessity to preserve and protect those ecosystems are beyond the scope of this review, nonetheless the underlying issues require mention. Kareiva et al. (2007) concluded that the reality of the human footprint on this planet meant that discussion about what areas of the world to set aside as protected areas was irrelevant and that in reality we should consider what trade-offs we are willing to accept as a result of the exploitation and control of nature. Such an approach is equally applicable to the marine environment, however, the overwhelming perception is that overfishing is the primary threat to the world's oceans and as a result they are suffering biodiversity loss and declining or collapsing fishery resources (Worm et al., 2006). Therefore, it is argued that urgent and widespread conservation in the form of areas closed to all fishing is needed.

Direct evidence to support the creation of MPAs for fisheries benefits is limited and many arguments are in fact supported by nothing more than normative assumptions. "Advocacy coalitions" comprised of non-governmental organisations (NGOs), journalists, and the public have been particularly effective, added to by the voice of scientists who have aligned themselves with their favoured faction or interest group (Caveen et al., 2013). The creation of the Chagos MPA is a classic example with prominent players including NGOs such as the CEN and the Pew Environment Group all lobbying in favour of the creation of a full no-take MPA. Armed with access to officials, financial resources, and supported by scientists keen to preserve a "pristine" natural laboratory to which they have preferential or exclusive access (and perhaps with an eye on potential funding from rich environmental groups), the pro-advocacy coalition produced what appeared to be persuasive arguments (CEN, 2010; Chagos Conservation Trust, 2009; Williamson, 2009). These arguments were welcomed by officials not exclusively for their conservation benefits but also for the purpose of weakening further challenges by the former inhabitants in restoring their right of return, and for giving greater security and control over the territory (Bancoult (3), 2012; Roberts, 2009; US Embassy (London), 2009). Pitted against this coalition was a number of small groups who advocated a more inclusive approach, supported by a number of scientists, lawyers and academics who had also begun to question the purported science and assumptions (Dunne and Johnson, 2011; Sand, 2012) some going so far as to argue that the Chagos MPA designation might be considered as representing a neo-colonial return to fortress conservation (De Santo et al., 2011). These groups however were significantly hampered by a politically driven timescale and the refusal of officials to allow access to documents required for a challenge to the advocacy claims, necessitating applications to the courts (First Tier Tribunal (Information Rights), 2012) and over 3 years of litigation (First Tier Tribunal (Information Rights), 2013, 2014). As a consequence, there was never any real public debate of the scientific merits behind the Chagos MPA, at least not prior to the declaration on 1 April 2010, although officials at the time were clearly worried by the weakness of the scientific justification, together with the confidential advice of their own fisheries experts that as regards the pelagic fishery, the major fishery in the area, there was no justification for an outright closure (MRAG Ltd., 2009, 2010).

Pro-advocacy scientists applied the perception of declining or collapsing fishery resources in their arguments for a complete no-take MPA, painting an Indian Ocean-wide picture of an overexploited fishery with ineffectual management. Their data and analyses of the BIOT commercial tuna fishery were restricted to a 5-year period (2004–2009) and no clear picture of patterns or trends emerged from this. There was a general pessimistic overview of Indian Ocean tuna stocks, and no attempt was made to evaluate the recreational fishery at Diego Garcia or the Mauritian inshore fishery. While acknowledging that the latter might be within sustainable limits, the overwhelming impression offered was that even here all was not well (Koldewey et al., 2010b). Their view of shark by-catch from the BIOT longline fishery with their estimated catch of 31069 blue sharks (59,749 all sharks) over a 5-year period (2003–2008) if correct would represent 20% of the entire catch (all shark species >40%). On this basis alone, their figures seem highly improbable. More realistic approaches show the blue shark by-catch to be 3.1% of the total catch weight (5.9% all shark species) which is 6.6-fold less than that of (Koldewey et al., 2010b) (for details, see ESM Appendix 8). Independent BIOT observer data of the proportion of the total catch (2.01% for blue shark by numbers and 4.29% for all shark species) over a 3-year period (2000–2003) (ESM Table 6-8) together with observer by-catch records from the Taiwanese bigeye tuna fleet (2004–2008) (2.0% for blue shark and 3.9% for all sharks by number) (Huang and Liu,

2010) cast further doubt on the Koldewey et al. (2010b) computations. As a result of the shark by-catch computations, the by-catch of the tuna fishery has been portrayed as substantial (Koldewey et al., 2010b) if not immense (Sheppard, 2011) and therefore in need of further regulation.

Rather than examining fisheries data for the FCMZ and considering whether or not a Chagos MPA might help preserve and protect the fisheries of the archipelago, Koldewey et al. (2010b) restated opinion on the global utility of pelagic MPAs and the usefulness of protected areas for reef fish communities. FCO/BIOT officials recognised the weakness of this evidence being presented by the CEN and its scientists as a problem (Yeadon, 2009) even though they may not have appreciated the full extent of the flaws.

There is also now an increasing consensus that when it comes to the use of MPAs to protect pelagic ecosystems what is needed are "well-selected pelagic MPAs" (Game et al., 2009) together with a science-driven analysis to achieve this (Game et al., 2010; Kaplan et al., 2010). Indeed, an evaluation of fixed-area closures on the High Seas for Pacific bigeye tuna stocks concluded that "spatial fishery conservation measures need to be rigorously evaluated before implementation" (Sibert et al., 2012) and that in contrast to popular scientific press (Pala, 2010) those closures were actually ineffectual (Sibert et al., 2012).

7.1. Is the UK claim and proclamation of the Chagos MPA legal?

The proclamation of the 200 nm Chagos EPPZ in 2003 was at the forefront of claims to environmental regulatory powers (as opposed to fishery regulation) in sea areas beyond the coastal state's territorial sea. In contrast, other claims at the time had been only to a distance of 60 nm (French, Croatian, and Italian claims in the Mediterranean) and in 2006 to 50 nm in the Papahānaumokuākea Marine National Monument (Sand, 2007). While fishery control afforded by the 1991 FCMZ was firmly embedded in customary international law, the legal basis of the EPPZ lay in Part V of UNCLOS 1982 which had created the new concept of the EEZ as an area *sui generis* (a legal term meaning of "its own kind") where powers and obligations derive exclusively under the Convention. The Chagos MPA which now co-exists with the EPPZ neither extends the area nor the powers first claimed in 2003. To date, however, neither the rights claimed under the EPPZ nor the MPA have been exercised; the only management act has been the cessation of fishing licences under existing domestic legislation.

The Chagos MPA/EPPZ, as all-encompassing nature protection areas purporting to occupy the full extent of the EEZ and exercising degrees of exclusive control, would therefore appear to require some form of multilateral mandate for legitimacy, as was done in the case of the Great Barrier Reef in 1990, the Florida Keys in 2002, the Baltic in 2005 (Sand, 2007) and Papahānaumokuākea Marine National Monument in 2008 (US Department of Commerce, 2008). Problems of legality concerned the FCO and BIOT Administration when first receiving advice from MRAG (MRAG Ltd., 2009), and again when this was raised in the public consultation (Stevenson, 2010). When the MPA was proclaimed on 1 April 2010, these were still unresolved (Yeadon, 2010a). Similar questions of international legality have now been raised by the Republic of Mauritius in its case against the UK before an international arbitral tribunal under the UN Convention on the Law of the Sea (Papanicolopulu, 2011); the tribunal's adjudication on this is expected at the end of 2014.

7.2. Does the Chagos MPA afford protection for highly migratory tuna species?

The scientific facts do not support the presumption that a no-take Chagos MPA would confer benefits on migratory fish species based upon the islands and seamounts acting as natural aggregation devices, and that accordingly the MPA could provide "an excellent area for the recovery of shark, tuna and other large predators" (Koldewey et al., 2010b). Firstly, all three tuna species fished in the Chagos are not overfished and they all roam widely within the Western Indian Ocean (Figures 3.8–3.11) with average linear distances of 1289–1315 km. Secondly, the purse seine fishery, which was responsible for 89% of the total catch in the BIOT from 1993 to 2010 (ESM Table 6-7), only fished for 2 months (December and January) of the year inside the FCMZ (Figures 3.11 and 3.12). For the remainder of the year, a Chagos MPA would thus afford no protection, even if protection were necessary, nor is it certain that during those two months effort would not be redirected elsewhere. Thirdly, the extreme variability in the purse seine catch means that for 10 of the 17 years the catch was particularly small (average 2832 t) (ESM Table 6-7) and the protection of a no-take MPA therefore minimal. Fourthly, the BIOT only contributed on average ~1.86% of the total Indian Ocean tuna catch, and the area covered by the Chagos MPA is only about 2.5% of the footprint of the Western Indian Ocean longline fishery (Figure 3.9) and 5.5% of the purse seine (Figure 3.10).

Subsequent arguments in support of the MPA superimposed the median lifetime displacement of Pacific tuna species on maps of the Chagos in order to imply that a high degree of protection would be afforded (Sheppard et al., 2012, 2013), but this ignores the evidence that exists for the Indian Ocean and implies that there is a resident tuna population whose range is located in the middle of the Chagos. The reality is that the Chagos is on the eastern extremity of a highly mobile stock which only spends a very limited time within the FCMZ. Nor is there evidence that the Chagos might be a nursery area for tuna and although the peak purse seine fishery was during the spawning season, spawning is known to take place very widely across the rest of the Indian Ocean (Mees et al., 2009; MRAG Ltd., 2010).

For adult bigeye and yellowfin tuna, although longline fishing was in every month of the year (Figure 3.12) and catches less variable (Figure 3.15, ESM Table 6-7), nevertheless, this fishery only contributed 11% of the BIOT tuna catch and 0.53% of the total Indian Ocean catch (Figures 3.14 and 3.15; ESM Table 6-7) so that the net benefits could at best be very limited.

The overall conclusion reached by the BIOT Fishery Advisers both in June 2009 and again in February 2010 was that "making the whole of the BIOT FCMZ a fully no-take area will provide no conservation benefit to tuna" (MRAG Ltd., 2010). Modelling demonstrates that the Chagos MPA has little impact on yellowfin stocks whether or not fishing effort was eliminated or redistributed (Martin et al., 2011) and an evaluation of the closure of several areas in the Indian Ocean came to a similar conclusion (Kaplan et al., 2014). The overwhelming evidence therefore is that while the net-benefit of fixed-area pelagic MPAs is contentious for the conservation of tuna stocks elsewhere in the world, in the specific instance of the Chagos MPA benefit from the closures is even less likely. The current available evidence does not support the supposition of a benefit.

7.3. Can the Chagos MPA effectively protect pelagic by-catch species?

An area where fishing results in an unusually high by-catch in particular of protected, endangered or threatened (PET) species may be of some value if enclosed in an MPA. However, a recent research programme into PET by-catch, areas of high juvenile retention, and biodiversity has not identified Chagos as a hotspot (MRAG Ltd., 2010). Similarly, our review suggests that neither the inshore or pelagic fisheries posed a significant threat to turtles, cetaceans or other marine mammals. For pelagic sharks, the purse seine fishery is likely to have had minor impact given that the majority (66%) of the fishing in BIOT was on free schools where the by-catch is very small (0.2–1.0%). In the longline fishery, the by-catch for all shark species (4%) and low fishing effort (0.53% of the Indian Ocean catch) mean that the numbers caught were relatively insignificant. Similar arguments concerning migratory fish species also apply, with silky sharks which are regularly caught in purse seines, travelling up to 3700 km in 100 days (Filmalter et al., 2010). For blue shark which is the predominant shark by-catch species in the longline fishery, studies from the North Atlantic indicate mean ranges of 857 km and a maximum range of 6926 km (Kohler et al., 2002). Displacement of longline effort outside the MPA also actually reduces the total level of protection of sharks because displaced vessels will still fish but they will not be affected by the BIOT ban on using wire trace.

7.4. Is the Chagos MPA necessary for the protection of reef fish?

Fixed-area MPAs can under appropriate conditions be an effective conservation tool to protect tropical reef fish populations from some threats (Graham et al., 2011; Halpern, 2003) and this logic has been uncritically extended to apply to the Chagos MPA (Koldewey et al., 2010b; MRAG Ltd., 2010). In the Chagos, however, reef fish biomass is currently remarkably high at most locations apart from Diego Garcia (Graham et al., 2013). While Graham et al.'s surveys need to be extended to mirror the fishing locations of the Mauritian inshore fishery, there is no evidence of the latter causing deleterious depletion of stocks. Reported annual average catches (200 t) were well within the sustainable yield (1102 t year^{-1}), and unchanging CPUE over 19 years corroborate sustainable practices. Although in the past 37 years, there appears to have been a substantial reduction (76%) in Chagos wide reef shark species, apparently stable shark numbers between 1996 and 2010/2012 do not implicate the licenced inshore fishery. Observers attribute the decline to IUU fishing by boats from Sri Lanka (Graham et al., 2010) which continues to be a problem in the Chagos MPA. It would seem therefore that in the absence of more effective enforcement, improvements in reef shark numbers are unlikely to be seen.

7.5. Is the exclusion of Diego Garcia from the MPA justified?

Diego Garcia has been described as a "legal black hole" because the UK has chosen not to apply certain international environmental agreements and for the past environmental degradation which the US have caused to the both the land and marine environment (Sand, 2009a). This criticism received a vigorous riposte from others who claim that the military presence has resulted in pristine reefs and no resource depletion, referring both to Diego Garcia and the Chagos (Sheppard et al., 2009). Diego Garcia is now a legislative anomaly within the Chagos MPA and its marine environment is neither pristine nor untouched. Its reef fish biomass and trophic structure have been significantly impacted, not because fishing is a necessary food source but because fishing is allowed for recreational pleasure. Damage to the fish community may well have had indirect effects on the nearby atolls and reefs, which lie between 60 km (Great Chagos Bank) and 123 km (Egmont) to the north and west, given the likelihood of biological connectivity over these distances (Harrison et al., 2012; Roberts, 2012).

The depletion of resources on the reefs of Diego Garcia which is the second largest atoll island (by area of submerged reef habitat <70 m deep) in the Chagos after Peros Banhos (Dumbraveanu and Sheppard, 1999) cannot simply be viewed in isolation as an acceptable or unavoidable consequence of a military treaty between the UK and US. The financial cost of some of the documented environmental degradation may be $3–18 million per year (Vine et al., 2012) but this does not as yet include the depletion of the fishery resources. Neither would it seem that the existing fishery legislation is being enforced adequately or the shore fishery properly documented or controlled, or that steps are being taken to minimise the impact of the US base on the nesting green turtle population. The simplest and most effective conservation measures that could have been taken at practically no cost were the very ones that were ruled out of consideration by the UK Government (FCO, 2009) for fear, it seems of upsetting its US allies (US Embassy (London), 2009).

7.6. Will the Chagos MPA help protect other Indian Ocean coral reefs?

Arguments have been advanced that the potential benefits from a Chagos MPA extend far wider than the 200 nm limit alone, but the claims to Indian Ocean-wide biological connectivity (Sheppard, 2011; Sheppard et al., 2012, 2013) are inadequately documented and have been disputed (Dunne and Johnson, 2011; MRAG Ltd., 2010) and are poorly supported by what little is known about the ocean circulation in the Chagos which might transport organisms and larvae, and whether larvae are in short supply and export will confer a benefit in other areas.

7.7. Was Chagos already adequately protected?

Chagos in 2010 already had extensive and comprehensive protection within both a legislative and enforcement regime. Ignoring the "conservation areas" on Diego Garcia, the areas under Strict Nature Reserve protection extended to 1374 km^2 (ESM Table 3-1). To this day, the EPPZ which was created to protect the Great Chagos Bank remains no more than a paper proclamation with no publicly stated purpose and no additional enforcement. The 640,000 km^2 FCMZ was widely acknowledged as the best managed and enforced fishery zone in the Indian Ocean. All of this is evidence of conservation working, albeit not perfectly and with the continuing failure to prevent IUU fishing by small vessels mainly from Sri Lanka. The newly proclaimed MPA faces exactly the same challenges and problems but with the cessation of licenced fishing the capacity of fishermen who have paid for the right to be in the Chagos area and to report infringements by those who have not, has been lost and the cost of enforcement has been substantially increased through the loss of revenue from fishing licences.

7.8. How can achievements be measured?

Prior to the cessation of commercial fishing in BIOT, it was possible to monitor pelagic and inshore fish stocks from fishing vessel records, inspections and the observer programme. The 20-year database allowed trends to be examined and the fisheries management to be adjusted proactively. The irony of the no-take MPA is that further data are no longer collected and yet there is no fishery-independent baseline and no research programme in place to monitor pelagic stocks and species. For inshore reef fish, there is available a sporadic and spatially incomplete database but again no future monitoring programme. The BIOT Administration and UK government commit no direct funds of their own to scientific research in BIOT (other than permitting the use of the patrol vessel by some groups), which due to the remoteness and inaccessibility is both expensive and difficult. The political will to create a "green legacy" with the MPA does not seem to have translated into a commitment to scientific research. The sole fishery monitoring that still exists is of the boat recreational fishery on Diego Garcia, while the larger destructive shore fishery remains unmonitored and unregulated.

In advocating the establishment of pelagic protected areas, Game et al. (2009) emphasised a need for a clear framework for monitoring and understanding the response of the system. In the case of the Chagos MPA, the hurried implementation and lack of additional resource allocation mean that little other than speculation will be available to measure its effectiveness.

8. CONCLUSION

The declaration of the Chagos MPA in April 2010 marked the creation of the eighth such LSMPA and the largest in the world at that time. It was also the first, and remains the only MPA, to claim protectionist rights throughout an entire EEZ. Its creation was the result of aggressive advocacy from NGOs and selected scientists, and a government anxious to claim a green legacy, whose officials also secretly expounded benefits of increased military security and the weakening of a humanitarian lobby for the territory's former inhabitants. The public consultation and subsequent decision were conducted with the minimum of information, and in circumstances which were non-conducive to informed scientific debate. As a consequence much of the scientific evidence in support of the designation has escaped scrutiny, particularly that relating to the fishery stock benefits.

This review has shown that the decision to announce the MPA was a hasty political whim of a UK Foreign Secretary who was about to leave office. It was against the advice of his own officials and of the BIOT's expert fisheries advisers. In particular, what now emerges from an examination of the scientific evidence is that the MPA is likely to confer no meaningful benefit for the protection of Indian Ocean tropical tuna stocks, or for other species which comprise the fishery by-catch. For the coral reefs and their associated well-managed commercial reef fishery, which are widely acknowledged to be in a "near pristine" state, the MPA adds no additional protection over and above that which already existed under domestic and international law since 2003.

If the Chagos MPA is to be considered a successful addition to the areas of the world's oceans coming under increased environmental protection, then it defies the principles which require a proper examination of the available evidence, and instead is an example of one where intense advocacy and political motivation have triumphed (Leenhardt et al., 2013).

Future prospects for the protection of the Chagos are not demonstrably better than before the declaration of the MPA, and because of the impact of displaced effort, they may even be worse. IUU fishing remains the primary threat to reefs and the inshore waters and the cause attributed to significant declines in the reef shark populations. In the offshore areas, the "protection" against IUU fishing afforded by licenced fishing vessels has been lost,

together with the associated source of licence revenue. There has been neither the political will nor the finance to improve the enforcement of the
640,000 km^2 MPA, to finance scientific research in the area, or to recognise
the true administrative costs of such a LSMPA. The exclusion of Diego
Garcia from the protection of the MPA currently means that the most
depleted and degraded reefs and waters in the territory remain unprotected
and subject to continued overfishing and interference by the resident US
military.

Finally, the unilateral declaration with little regard for stakeholders has
resulted in a spate of legal actions in the English courts and before an international tribunal, has further disenfranchised the former inhabitants of the
Chagos, and led to a significant diplomatic dispute between the UK and
the Republic of Mauritius.

ACKNOWLEDGEMENTS

We are most grateful to Chas Anderson and Steve Newman for their contribution to
Section 5.4, and to Chas Anderson and Vassili Papastavrou for their contribution to
Section 5.5 (including in both cases the ESM material at http://dx.doi.org/10.1016/
B978-0-12-800214-8.00003-7). We also thank Julien Million for providing an updated ver-
sion of Figure 3.8 and Jérôme Bourjea for Figure 3.10. The chapter was also improved by the
helpful comments of three reviewers.

REFERENCES

Abraham, G., 2011. Paradise claimed: disputed sovereignty over the Chagos Archipelago.
    S. Afr. Law J. 128, 63–99.
Administration, B.I.O.T., 2009. BIOT Marine Reserve Proposal: Implications for US Activ-
    ities in Diego Garcia and British Indian Ocean Territory (BIOT). FCO, London.
Alliance, Ocean, 2009. The Voyage of the Odyssey. Ocean Alliance, Gloucester, Maine.
Amande, M.J., Bez, N., Konan, N.D., Murua, H., de Molina, A.D., Chavance, P., et al.,
    2011. Areas with High Bycatch Of Silky Sharks (*Carcharhinus falciformis*) in the Western
    Indian Ocean Purse Seine Fishery. IOTC–2011–WPEB07–29.
Anderson, R.C., 1996. Observations of cetaceans in the Chagos Archipelago, February–
    March 1996. Unpublished report.
Anderson, C., Sheppard, C., Spalding, M., Crosby, R., 1998. Shortage of sharks at Chagos.
    Shark News 10, 1–3.
Balmford, A., Gravestock, P., Hockley, N., McClean, C.J., Roberts, C.M., 2004. The
    worldwide costs of marine protected areas. Proc. Natl. Acad. Sci. U.S.A.
    101, 9694–9697.
Bancoult (1), 2000. R v Secretary of State for the Foreign and Commonwealth Office [2000]
    WL 1629583.
Bancoult (3), 2012. R v Secretary of State for Foreign and Commonwealth Affairs [2012]
    EWHC 2115.
Baroness Warsi, 2013. Chagos Islands. Hansard 5 February 2013: Column WA30.
BIOT Constitution Order, 2004. British Indian Ocean Territory (Constitution) Order.
BIOT Ordinance No. 1, 1970. The Protection and Preservation of Wildlife Ordinance.

BIOT Ordinance No. 2, 1971. The Fishery Limits Ordinance.

BIOT Ordinance No. 5, 2007. The Fisheries (Conservation and Management) Ordinance.

BIOT Ordinance No. 6, 1994. The Diego Garcia Conservation (Restricted Area) Ordinance.

BIOT Proclamation, 1969. Proclamation No. 1 of 1969. FCO, London.

BIOT Proclamation, 1991. Proclamation No. 1 of 1991. FCO, London.

BIOT Proclamation, 2010. Proclamation No. 1 of 2010. FCO, London.

BIOT Statutory Instrument No. 1, 2006. The Fishing (Amendment) Regulations.

BIOT Statutory Instrument No. 4, 1998. The Strict Nature Reserve Regulations.

Caveen, A.J., Gray, T.S., Stead, S.M., Polunin, N.V.C., 2013. MPA policy: what lies behind the science? Mar. Policy 37, 3–10.

CEN, 2010. Consultation on Whether to Establish a Marine Protected Area in the British Indian Ocean Territory. Chagos Environment Network, London, UK.

Chagos Conservation Trust, 2009. The Chagos Archipelago: Its Nature and the Future. Chagos Conservation Trust, London.

Clark, S., Crapper, D., Holland, A., Little, B., Warren, J., 2002. Resettlement of the Salomons and Peros Banhos Atolls—a preliminary feasibility study—June 2000, London.

Clayton, S., 2010. Re: Marine Protected Area: Next steps. E-mail to Joanne Yeadon. 18:06 hrs, 30 March 2010, FCO, London.

Clermont, S., Chavance, P., Delgado, A., Murua, H., Ruiz, J., Ciccione, S., et al., 2012. EU Purse Seine Fishery interaction with Marine Turtles in the Atlantic and Indian Oceans: a 15 year analysis: IOTC-2012-WPEB-35.

CNRJ, 2011. Diego Garcia—Final Governing Standards. Commander US Navy Region, Japan.

Dalzell, P., Adams, T.J.H., Polunin, N.V.C., 1996. Coastal fisheries in the Pacific Islands. Oceanogr. Mar. Biol. Annu. Rev. 34, 395–531.

de Boer, M.N., Baldwin, R., Burton, C.L.K., Eyre, E.L., Jenner, K.C.S., Jenner, M.N.M., et al., 2002. Cetaceans in the Indian Ocean Sanctuary: A Review. Whale and Dolphin Conservation Society, Chippenham, Wiltshire, IWC SC/54/05:1–60.

De Santo, E.M., Jones, P.J.S., Miller, A.M.M., 2011. Fortress conservation at sea: A commentary on the Chagos marine protected area. Mar. Policy 35, 258–260.

Deslarzes, K.J.P., Evans, D.J., Smith, S.H., 2005. Marine Biological Survey at United States Navy Support Facility, Diego Garcia, British Indian Ocean Territory, July/August 2004. Navy Support Facility, Diego Garcia.

DiMento, J.M., 2006. Beyond the Water's Edge: United States National Security and the Ocean Environment. PhD thesis, Tufts University, Medford, MA.

Dulvy, N.K., Baum, J.K., Clarke, S., Compagno, L.J.V., Cortés, E., Domingo, A., et al., 2008. You can swim but you can't hide: the global status and conservation of oceanic pelagic sharks and rays. Aquat. Conserv. Mar. Freshwat. Ecosyst. 18, 459–482.

Dumbraveanu, D., Sheppard, C.R.C., 1999. Areas of substrate at different depths in the Chagos Archipelago. In: Sheppard, C.R.C., Seaward, M.R.D. (Eds.), Ecology of the Chagos Archipelago. Linnean Society, London, pp. 35–44.

Dunne, R.P., Johnson, M., 2011. The Chagos Archipelago—conservation and humanity can go hand in hand. Ocean Challenge 18, 32.

Edgar, G.J., Stuart-Smith, R.D., Willis, T.J., Kininmonth, S., Baker, S.C., Banks, S., et al., 2014. Global conservation outcomes depend on marine protected areas with five key features. Nature 506, 216–220.

EG&G Environmental Consultants, 1980. Environmental Survey of Construction and Dredging Related Activities on Diego Garcia, Indian Ocean. Commander, Pacific Division, Naval Facilities Engineering Command, Pearl Harbor, Hawaii.

Erickson, A.S., Walter III, L.C., Mikolay, J.D., 2010. Diego Garcia and the United States' Emerging Indian Ocean Strategy. Asian Security 6, 214–237.

Exchange of Notes, 1966. Exchange of notes between the Government of the United Kingdom of Great Britain and Northern Ireland and the Government of the United States of America concerning the availability for defence purposes of the British Indian Ocean Territory. Treaty Series No. 15 (1967).

Exchange of Notes, 1976. Exchange of notes between the Government of the United Kingdom of Great Britain and Northern Ireland and the Government of the United States of America concerning a United States Naval Support Facility on Diego Gracia, British Indian Ocean Territory. Treaty Series No. 19 (1976).

Falklands Islands Government, 2013. Fisheries. http://www.falklands.gov.fk/self-sufficiency/commercial-sectors/fisheries/ (accessed 9 December 2013).

FCO, 2002. Chagos Islands. Hansard 15 Oct 2002: Column 529W.

FCO, 2009. Consultation on Whether to Establish a Marine Protected Area in the British Indian Ocean Territory. Foreign and Commonwealth Office, London.

FCO Press Release, 2011. Foreign Secretary announces additional funding for Overseas Territories FCO, London.

Filmalter, J.D., Dagorn, L., Soria, M., 2010. Double Tagging Juvenile Silky Sharks to Improve our Understanding of the Behavioural Ecology: Preliminary Results, IOTC-2010-WPEB-10.

First Tier Tribunal (Information Rights), 2012. The Chagos Refugees Group in Mauritius and Chagos Social Committee (Seychelles) v Information Commissioner & Foreign and Commonwealth Office, Case No: EA/2011/0300.

First Tier Tribunal (Information Rights), 2013. Dunne v Information Commissioner & Foreign and Commonwealth Office, Case No: EA/2012/0257.

First Tier Tribunal (Information Rights), 2014. Sand v Information Commissioner & Foreign and Commonwealth Office, Case No: EA/2012/0196.

Foundation, Bertarelli, 2013. Tagging of Pelagics—British Indian Ocean Territory. Bertarelli Foundation, Switzerland.

Gaines, S.D., White, C., Carr, M.H., Palumbi, S.R., 2010. Designing marine reserve networks for both conservation and fisheries management. Proc. Natl. Acad. Sci. U.S.A 107, 18286–18293.

Game, E.T., Grantham, H.S., Hobday, A.J., Pressey, R.L., Lombard, A.T., Beckley, L.E., et al., 2009. Pelagic protected areas: the missing dimension in ocean conservation. Trends Ecol. Evol. 24, 360–369.

Game, E.T., Grantham, H.S., Hobday, A.J., Pressey, R.L., Lombard, A.T., Beckley, L.E., et al., 2010. Pelagic MPAs: the devil you know. Trends Ecol. Evol. 25, 63–64.

Gifford, R., Dunne, R.P., 2014. A dispossessed people: the depopulation of the Chagos Archipelago 1965–1973. Popul. Space Place 20, 37–49.

Gould, M., 2009. Re: BIOT. E-mail to Colin Roberts. 16:3 hrs, 7 May 2009, FCO, London.

Graham, N.A.J., McClanahan, T.R., 2013. The last call for marine wilderness? Bioscience 63, 397–402.

Graham, N.A.J., Spalding, M.D., Sheppard, C.R.C., 2010. Reef shark declines in remote atolls highlight the need for multi-faceted conservation action. Aquat. Conserv. Mar. Freshwat. Ecosyst. 20, 543–548.

Graham, N.A.J., Ainsworth, T.D., Baird, A.H., Ban, N.C., Bay, L.K., Cinner, J.E., et al., 2011. From microbes to people: tractable benefits of no-take areas for coral reefs. Oceanogr. Mar. Biol. Annu. Rev. 49, 105–136.

Graham, N.J., Pratchett, M., McClanahan, T., Wilson, S., 2013. The status of Coral Reef fish assemblages in the Chagos Archipelago, with implications for protected area management and climate change. In: Sheppard, C.R.C. (Ed.), Coral Reefs of the United Kingdom Overseas Territories. Springer, Netherlands, pp. 253–270.

Hallier, J.-P., Million, J., 2009. The Contribution of the Regional Tuna Tagging Project—Indian Ocean to IOTC Stock, Assessment, IOTC-2009-WPTT-24.

Hallier, J.-P., Million, J., 2012. The Indian Ocean Tuna Tagging Programme. Grand Baie, Mauritius.

Halpern, B.S., 2003. The impact of marine reserves: do reserves work and does reserve size matter? Ecol. Appl. 13, 117–137.

Halpern, B.S., Walbridge, S., Selkoe, K.A., Kappel, C.V., Micheli, F., D'Agrosa, C., et al., 2008. A global map of human impact on marine ecosystems. Science 319, 948–952.

Harrison, H.B., Williamson, D.H., Evans, R.D., Almany, G.R., Thorrold, S.R., Russ, G.R., et al., 2012. Larval export from marine reserves and the recruitment benefit for fish and fisheries. Curr. Biol. 22, 1–6.

Huang, H.-W., Liu, K.-M., 2010. Bycatch and discards by Taiwanese large-scale tuna long-line fleets in the Indian Ocean. Fish. Res. 106, 261–270.

IOTC, 2013. Report of the fifteenth session of the IOTC working party on tropical tunas IOTC-2013-WPTT15-R[E].

Kadagi, N.I., Harris, T., Conway, N., 2011. East Africa Billfish Conservation and Research: Marlin, Sailfish and Swordfish Mark-Recapture Field Studies, IOTC-2011-WPB09-10_Pres.pdf.

Kaplan, D.M., Chassot, E., Gruss, A., Fonteneau, A., 2010. Pelagic MPAs: The devil is in the details. Trends Ecol. Evol. 25, 62–63.

Kaplan, D.M., Chassot, E., Amande, J., Dueri, S., Demarcq, H., Dagorn, L., et al., 2014. Spatial management of Indian Ocean tropical tuna fisheries: potential and perspectives. ICES J. Mar. Sci. http://dx.doi.org/10.1093/icesjms/fst233,22.

Kareiva, P., Watts, S., McDonald, R., Boucher, T., 2007. Domesticated nature: shaping landscapes and ecosystems for human welfare. Science 316, 1866–1869.

Kohler, N.E., Turner, P.A., Hoey, J.J., Natanson, L.J., Briggs, R., 2002. Tag and recapture data for three pelagic shark species: blue shark (Prionace glauca), shortfin mako (Isurus xyrinchus), and porbeagle (Lamna nasus) in the North Atlantic Ocean. Collective Volume of Scientific Papers, ICCAT 54, 1231–1260.

Koldewey, H., Gollock, M., Harding, S., Curnick, D., Harrison, L., Rogers, A., 2010a. Benefits to Fisheries of the Chagos Archipelago/British Indian Ocean Territory as a No-Take Marine Reserve. Zoological Society of London, London.

Koldewey, H.J., Curnick, D., Harding, S., Harrison, L.R., Gollock, M., 2010b. Potential benefits to fisheries and biodiversity of the Chagos Archipelago/British Indian Ocean Territory as a no-take marine reserve. Mar. Pollut. Bull. 60, 1906–1915.

Leenhardt, P., Cazalet, B., Salvat, B., Claudet, J., Feral, F., 2013. The rise of large-scale marine protected areas: Conservation or geopolitics? Ocean Coast. Manag. 85, 112–118.

Lunn, J., Mills, C., 2013. Disputes Over the British Indian Ocean Territory: A Survey House of Commons Library, London, Research Paper 13/31.

Mangi, S., Hooper, T., Rodwell, L., Simon, D., Snoxell, D., Spalding, M., et al., 2010. In: Establishing a Marine Protected Area in the Chagos Archipelago: Socio-economic Considerations: Report of Workshop Held 7 January 2010, Royal Holloway, University of London, UK.

Marsac, F., Le Blanc, J.-L., 1999. Oceanographic Changes During the 1997–1998 El Niño in the Indian Ocean and Their Impact on the Purse Seine Fishery, IOTC WPTT99-0.

Martin, S., Mees, C., Edwards, C., Nelson, L., 2011. A Preliminary Investigation into the Effects of Indian Ocean MPAs on Yellowfin Tuna, Thunnus albacares, with Particular Emphasis on the IOTC Closed Area, IOTC-2011-SC14-40[E].

Martin, S.M., Moir Clark, J., Pearce, J., Mees, C.C., 2013. Catch and Bycatch Composition of Illegal Fishing in the British Indian Ocean Territory (BIOT), IOTC–2013–WPEB09–46 Rev_1.

Mata'irea Blogspot, 2009. 26 July 2009—salomon atoll Chagos. http://matairea.blogspot.co.uk/search/label/Chagos (accessed 4 March 2014).

McClanahan, T., 2006. Challenges and accomplishments towards sustainable reef fisheries. In: Cote, I.M., Reynolds, J.D. (Eds.), Coral Reef Conservation. Cambridge University Press, Cambridge, pp. 147–182.

Mees, C., Polunin, N.V.C., 1994. Reef fish communities and management of local fishing in BIOT, with special reference to Diego Garcia 12–28 July 1994, London.

Mees, C.C., King, A., Pilling, G.M., Barry, C.J., 1999a. British Indian Ocean Territory—Fisheries Conservation and Management Zone—The Inshore Fishery in 1998, with Summary Details of the Recreational Fishery on Diego Garcia. MRAG Ltd., London.

Mees, C.C., Pilling, G.M., Barry, C.J., 1999b. Commercial inshore fishing activity in the British Indian Ocean Territory. In: Sheppard, C.R.C., Seaward, M.R.D. (Eds.), Ecology of the Chagos Archipelago. Linnean Society, London, pp. 327–345.

Mees, C., Fonteneau, A., Nishida, T., Dagorn, L., Robinson, J., Mosquiera, I., et al., 2009. The Potential Role of Pelagic Marine Protected Areas for Tropical Tunas in the Indian Ocean, IOTC-2009-SC-INF18.

Mortimer, J.A., 2000. Diego Garcia Marine Turtle Conservation Assessment: British Indian Ocean Territory. Final Report on the Fieldwork. Fauna & Flora International, Cambridge.

Mortimer, J.A., 2006. Status of Leatherback Turtles in United Kingdom—British Indian Ocean Territory (BIOT). IOSEA Marine Turtle MoU, Bangkok.

Mortimer, J.A., 2007. Sea Turtles of the Chagos Archipelago—Populations in Recovery. IOSEA Marine Turtle MoU, Bangkok.

Mortimer, J.A., Broderick, D., 1999. Population genetic structure and developmental migrations of sea turtles in the Chagos Archipelago and adjacent regions inferred from mtDNA sequence variation. In: Sheppard, C.R.C., Seaward, M.R.D. (Eds.), Ecology of the Chagos Archipelago. Linnean Society, London, pp. 185–194.

Mortimer, J.A., Day, M., 1999. Sea turtle populations and habitats in the Chagos Archipelago, British Indian Ocean Territory. In: Sheppard, C.R.C., Seaward, M.R.D. (Eds.), Ecology of the Chagos Archipelago. Linnean Society, London, pp. 159–172.

Mortimer, J.A., Day, M., Broderick, D., 2002. Sea turtle populations of the Chagos Archipelago, British Indian Ocean Territory. Published, In: Proceedings of the Twentieth Annual Symposium on Sea Turtle Biology and Conservation, 2002 Orlando, Florida.

MRAG Ltd., 1996. Marine Protected Areas and Others. Fax to BIOT Administration. 27 March, MRAG Ltd., London.

MRAG Ltd., 1998a. FCO briefing notes—British/Mauritian Fisheries Commission, Mauritius, 11–12 June 1998, London.

MRAG Ltd., 1998b. Results of a Pilot Study undertaken in 1997 to assess the recreational fishery of Diego Garcia, London.

MRAG Ltd., 1999. BIOT recreational fishery survey—May–October 1998—Final report—Report to the Commissioner for the British Indian Ocean Territory, London.

MRAG Ltd., 2003. Report from the 2002–2003 Tuna Scientific Observer Program—British Indian Ocean Territory—December 2002–February 2003, London.

MRAG Ltd., 2004. BIOT Administration—Fisheries Briefing & Guidance 2004. MRAG Ltd., London.

MRAG Ltd., 2009. MRAG comments on the proposal to designate the British Indian Ocean Territory (BIOT) Fisheries Conservation Management Zone (FCMZ) as a marine reserve, London.

MRAG Ltd., 2010. Consultation on whether to establish a marine protected area in the British Indian Ocean Territory, London.

MRAG Ltd., 2011. British Indian Ocean Territory (Chagos Archipelago)—marine protected area—a summary of activities October 2010–September 2011—background paper UK01, London.

Muths, D., Le Couls, S., Evano, H., Grewe, P., Bourjea, J., 2013. Multi-genetic marker approach and spatio-temporal analysis suggest there is a single panmictic population of swordfish *Xiphias gladius* in the Indian Ocean. PLoS One 8, e63558.

NAVFAC Pacific, 2005. Diego Garcia Integrated Natural Resources Management Plan. Naval Facilities Engineering Command, Pacific, Hawaii.

Nel, R., Wanless, R.M., Angel, A., Mellet, B., Harris, L., 2013. Ecological risk assessment and productivity—susceptibility analysis of sea turtles overlapping with fisheries in the IOTC region. Unpublished report to IOTC and IOSEA marine turtle MoU.

Nelson, J., 2010. Cheering the Chagos protected area. Foreign Service Journal October 2010.

Nelson, J., Bradner, H., 2010. The case for establishing ecosystem-scale marine reserves. Mar. Pollut. Bull. 60, 635–637.

Nugent, P., McDonagh, J., Mees, C.C., 2010. British Indian Ocean Territory (Chagos archipelago)—Fisheries Conservation and Management Zone—The Inshore Fishery in 2009, London, BSFC Background Paper UK03.

Pala, C., 2010. Islands champion tuna ban. Nature 468, 739–740.

Palumbi, S.R., 2004. Marine reserves and ocean neighborhoods: the spatial scale of marine populations and their management. Annu. Rev. Environ. Resour. 29, 31–68.

Papanicolopulu, I., 2011. Current legal developments Mauritius/United Kingdom. Int. J. Mar. Coast. Law. 26, 667–678.

Pearce, J., 1996. A review of the British Indian Ocean Territory Fisheries Conservation and Management Zone Tuna Fishery, 1991–1995, London.

Permanent Court of Arbitration, 2014. The Republic of Mauritius v. The United Kingdom of Great Britain and Northern Ireland [Online]. Permanent Court of Arbitration, The Hague. http://www.pca-cpa.org/showpage.asp?pag_id=1429 (accessed 6 July 2014).

Pianet, R., Chavance, P., Murua, H., Delgado de Molina, A., 2009. Quantitative Estimates of the By-catches of the Main Species of the Purse Seine Fleet in the Indian ocean, 2003–2008, IOTC-2009-WPEB-21.

Posford Haskoning Ltd., 2002. Feasibility study for the resettlement of the Chagos Archipelago: Phase 2B, London.

PREM 13/3320, 1965. Record of a conversation between the Prime Minister and the Premier of Mauritius, Sir Seewoosagur Ramgoolam, at No. 10 Downing Street at 10 am on Thursday, 23 September 1965, National Archives, London.

Price, A.R.G., Harris, A., Mcgowan, A., Venkatachalam, A.J., Sheppard, C.R.C., 2010. Chagos feels the pinch: assessment of holothurian (sea cucumber) abundance, illegal harvesting and conservation prospects in British Indian Ocean Territory. Aquat. Conserv. Mar. Freshwat. Ecosyst. 20, 117–126.

Price, A.R.G., Evans, L.E., Rowlands, N., Hawkins, J.P., 2013. Negligible recovery in Chagos holothurians (sea cucumbers). Aquat. Conserv. Mar. Freshwat. Ecosyst. 23, 811–819.

Ramsar, 1971. Convention on wetlands of International importance especially as waterfowl habitat, Ramsar (Iran), 2 February 1971. United Nations Treaty Series 996, 245.

Republic of Maldives, 2010. Commission on the limits of the continental shelf—submission by the Republic of Maldives. Executive summary: MAL-ES-DOC, July 2010.

Republic of Mauritius, 2009a. Joint Communique, Port Louis, Mauritius.

Republic of Mauritius, 2009b. Ministry of Foreign Affairs, Regional Integration and International Trade No. 1197/28/10 dated 10 November 2009, Port Louis, Mauritius.

Republic of Mauritius, 2009c. United Nations Convention on the Law of the Sea—preliminary information submitted by the republic of Mauritius concerning the extended continental shelf in the Chagos Archipelago region pursuant to the decision contained in SPLOS/183.

Republic of Mauritius, 2013. Mauritius submission for an extended continental shelf in the Chagos Archipelago Region, New York.

Reygondeau, G., Maury, O., Beaugrand, G., Fromentin, J.M., Fonteneau, A., Cury, P., 2012. Biogeography of tuna and billfish communities. J. Biogeogr. 39, 114–129.

Robbins, W.D., Hisano, M., Connolly, S.R., Choat, J.H., 2006. Ongoing collapse of coral-reef shark populations. Curr. Biol. 16, 2314–2319.

Roberts, C., 2009. Subject: British Indian Ocean Territory: The World's largest Marine Reserve? Memorandum to Private Secretary to Foreign Secretary. 5 May, FCO, London.

Roberts, C., 2012. Marine ecology: reserves do have a key role in fisheries. Curr. Biol. 22, R444–R446.

Romanov, E., 2001. Bycatch in the tuna purse-seine fisheries of the western Indian Ocean. Fish. Bull. 100, 90–105.

Sand, P.H., 2007. 'Green' Enclosure of Ocean Space—Déjà Vu? Mar. Pollut. Bull. 54, 374–376.

Sand, P.H., 2009a. Diego Garcia: British-American legal black hole in the Indian Ocean? J. Environ. Law 21, 113–137.

Sand, P.H., 2009b. United States and Britain in Diego Garcia: The Future of a Controversial Base. Palgrave Macmillan, New York.

Sand, P.H., 2012. Fortress conservation trumps human rights? The 'marine protected area' in the Chagos Archipelago. J. Environ. Dev. 21, 36–39.

Sand, P.H., 2013. The Chagos Archipelago cases: nature conservation between human rights and power politics. Global Community Yearbook Int. Law Jurisprudence 26, 125–149.

Sheppard, C.R.C., 1980. Coral fauna of Diego Garcia lagoon, following harbour construction. Mar. Pollut. Bull. 11, 227–230.

Sheppard, C., 2010. Marine protected areas and pelagic fishing: the case of the Chagos Archipelago. Mar. Pollut. Bull. 60, 1899–1901.

Sheppard, C.R.C., 2011. Protecting the Chagos Archipelago—a last chance for Indian Ocean reefs? Ocean Challenge 18, 26–31.

Sheppard, C.R.C., Spalding, M., 2003. Chagos Conservation Management Plan. BIOT Administration. FCO, London.

Sheppard, C., Tamelander, J., Turner, J., 2009. The Chagos Archipelago (British Indian Ocean Territory): legal black hole or environmental bright spot? A reply to sand. J. Environ. Law 21, 291–293.

Sheppard, C.R.C., Ateweberhan, M., Bowen, B.W., Carr, P., Chen, C.A., Clubbe, C., et al., 2012. Reefs and islands of the Chagos Archipelago, Indian Ocean: why it is the world's largest no-take marine protected area. Aquat. Conserv. Mar. Freshwat. Ecosyst. 22, 232–261.

Sheppard, C.C., Bowen, B., Chen, A., Craig, M., Eble, J., Fitzsimmons, N., et al., 2013. British Indian Ocean territory (the Chagos Archipelago): setting, connections and the marine protected area. In: Sheppard, C.R.C. (Ed.), Coral Reefs of the United Kingdom Overseas Territories. Springer, Netherlands, pp. 223–240.

Sibert, J., Senina, I., Lehodey, P., Hampton, J., 2012. Shifting from marine reserves to maritime zoning for conservation of Pacific bigeye tuna (*Thunnus obesus*). Proc. Natl. Acad. Sci. U.S.A 109 (44), 18221–18225.

Simmonds, M., 2013. Environment Protection: British Overseas Territories. Hansard 2 September 2013: Column 291W.

Spalding, M.D., 2000. The Status of Commercially Important Reef Fishes of the Chagos Archipelago. Cambridge Coastal Research Group, Cambridge.

Stafford, K.M., Bohnenstiehl, D.R., Tolstoy, M., Chapp, E., Mellinger, D.K., Moore, S.E., 2004. Antarctic-type blue whale calls recorded at low latitudes in the Indian and eastern Pacific Oceans. Deep-Sea Res. I Oceanogr. Res. Pap. 51, 1337–1346.

Stafford, K.M., Chapp, E., Bohnenstiel, D.R., Tolstoy, M., 2011. Seasonal detection of three types of "pygmy" blue whale calls in the Indian Ocean. Mar. Mamm. Sci. 27, 828–840.

Stevenson, R., 2010. Whether to Establish a Marine Protected Area in the British Indian Ocean Territory. Consultation Report. FCO, London.

The Independent, 2010. Billionaire saves marine reserve plans, 12 September 2010, London.

Townsend, C.H., 1935. The distribution of certain whales as shown by logbook records of American whaleships. Zool. Sci. Contrib. N. Y. Zool. Soc. 19, 1–50.

Tucker, T., Doughty, B.T., 1988. Naval facilities, Diego Garcia, British Indian Ocean Territory: management and administration. ICE Proc. 84, 191–215.

US Navy Support Facility Diego Garcia, 2011a. Operation of recreational boating, fishing and swimming program: Diego Garcia local coordinator instruction 1710.17D, Diego Garcia.

U.S. Navy Support Facility Diego Garcia, 2011b. Two Casualties Reported During Fishing Trip in Diego Garcia, BIOT DIEGO GARCIA, British Indian Ocean Territory (Sept. 10, 2011). US Navy Support Facility Diego Garcia, Diego Garcia. https://www.facebook.com/notes/us-navy-support-facility-diego-garcia/this-is-an-official-news-release/286785031336203.

United Nations, 1965. UN General Assembly, Resolution 2066 (XX) Question of Mauritius, 16 December 1965, Resolution 2066 XX.

United Nations, 2004. M.Z.N. 46. 2004. LOS (Maritime Zone Notification) 12 March 2004—deposit by the United Kingdom of Great Britain and Northern Ireland of the list of geographical coordinates of points pursuant to article 75, paragraph 2, of the Convention, United Nations, New York.

United Nations, 2008a. M.Z.N. 63. 2008. LOS (Maritime Zone Notification) 27 June 2008—deposit by the Republic of Mauritius of charts and lists of geographical coordinates of points, pursuant to article 16, paragraph 2, and article 47, paragraph 9, of the Convention, Law of the Sea Information Circular 28.

United Nations, 2008b. Mauritius—(a) Maritime Zones (Baselines and Delineating Lines) Regulations 2005—The Maritime Zones Act 2005—Regulations made by the Prime Minister under sections 4, 5 and 27 of the Maritime Zones Act 2005. Law Sea Bull. 67, 13–38.

US Department of Commerce, 2008. Papahanaumokuakea Marine National Monument Proclamation Provisions. Fed. Regist. 73, 73592–73605.

US Department of Defense, 2013. Base Structure Report: Fiscal Year 2012 Baseline. Government Printing Office, Washington, DC.

US Embassy (London), 2009. HMG floats proposal for marine reserve covering the Chagos Archipelago (British Indian Ocean Territory)—Confidential. To: Secretary of State Washington, dated 15 May 2009.

Vine, D., 2009. Island of Shame: The Secret History of the U.S. Military Base on Diego Garcia. Princeton University Press, Princeton, New Jersey.

Vine, D., Harvey, P., Sokolowski, S.W., 2012. Compensating a people for the loss of their homeland: Diego Garcia, the Chagossians, and the Human Rights Standards Damages Model. Northwest J. Int. Hum. Rights 11, 152–185.

Webster, P.J., Moore, A.M., Loschnigg, J.P., Leben, R.R., 1999. Coupled ocean-temperature dynamics in the Indian Ocean during 1997–98. Nature 401, 356–360.

Williamson, P., 2009. Marine Conservation in the British Indian Ocean Territory: Science Issues and Opportunities. UK Natural Environment Research Council, Swindon.

Worm, B., Barbier, E.B., Beaumont, N., Duffy, J.E., Folke, C., Halpern, B.S., et al., 2006. Impacts of biodiversity loss on ocean ecosystem services. Science 314, 787–790.

Wray, P., Martin, K.R., 1983. Historical Whaling Records from the Western Indian Ocean. International Whaling Commission, Cambridge, Special Issue 5.

Yeadon, J., 2009. British Indian Ocean Territory (BIOT): Public Consultation on Proposed Marine Protected Area. 29 October, Memorandum to Private Secretary to Foreign Secretary. FCO, London.

Yeadon, J., 2010a. British Indian Ocean Territory (BIOT): Proposed Marine Protected Area (MPA): Next Steps. Memorandum to Private Secretary to the Foreign Secretary. FCO, London.

Yeadon, J., 2010b. British Indian Ocean Territory: MPA: Next Steps: Mauritius. 31 March, Memorandum to Private Secretary to the Foreign Secretary. FCO, London.

CHAPTER FOUR

Marine Managed Areas and Associated Fisheries in the US Caribbean

Michelle T. Schärer-Umpierre*,†, Daniel Mateos-Molina†,‡,1,
Richard Appeldoorn†, Ivonne Bejarano†,
Edwin A. Hernández-Delgado§,¶, Richard S. Nemeth‖,
Michael I. Nemeth†, Manuel Valdés-Pizzini#, Tyler B. Smith‖

*Interdisciplinary Center for Coastal Studies, University of Puerto Rico, Mayagüez, Puerto Rico
†Department of Marine Sciences, University of Puerto Rico, Mayagüez, Puerto Rico
‡Departamento de Ecología e Hidrología, Universidad de Murcia, Campus de Espinardo, Murcia, Spain
§Center for Applied Tropical Ecology and Conservation, Coral Reef Research Group, University of Puerto Rico, San Juan, Puerto Rico
¶Department of Biology, University of Puerto Rico, San Juan, Puerto Rico
‖University of the Virgin Islands, Center for Marine and Environmental Studies, St. Thomas, U. S. Virgin Islands
#Department of Social Sciences, University of Puerto Rico, Mayagüez, Puerto Rico
1Corresponding author: e-mail address: dmateos5@gmail.com

Contents

Abstract

The marine managed areas (MMAs) of the U.S. Caribbean are summarized and specific data-rich cases are examined to determine their impact upon fisheries management in the region. In this region, the productivity and connectivity of benthic habitats such as mangroves, seagrass and coral reefs is essential for many species targeted by fisheries. A minority of the 39 MMAs covering over 4000 km^2 serve any detectable management or conservation function due to deficiencies in the design, objectives, compliance or enforcement. Fifty percent of the area within MMA boundaries had no-take regulations in the U.S. Virgin Islands, while Puerto Rico only had 3%. Six case studies are compared and contrasted to better understand the potential of these MMAs for fisheries management. Signs of success were associated with including sufficient areas of essential fish habitat (nursery, spawning and migration corridors), year-round no-take regulations, enforcement and isolation. These criteria have been identified as important in the conservation of marine resources, but little has been done to modify the way MMAs are designated and implemented in the region. Site-specific monitoring to measure the effects of these MMAs is needed to demonstrate the benefits to fisheries and gain local support for a greater use as a fisheries management tool.

Keywords: U.S. Caribbean, Coral reef fishes, Co-management, No-take zone, Enforcement, Fisheries resources, Compliance, Habitat connectivity

1. INTRODUCTION

1.1. Bio-physical settings

Puerto Rico (PR) and the U.S. Virgin Islands (USVI) are located on the Puerto Rican Bank along with the British Virgin Islands at the eastern extreme of the Greater Antilles (Figure 4.1). These two U.S. territories and surrounding waters out to 200 nautical miles (Nm) compose the U.S. Caribbean exclusive economic zone (EEZ). The Puerto Rican Shelf is delimited by the Mona Passage to the west, the Anegada Passage to the east and the Puerto Rico Trench to the north, which is the deepest point in the Atlantic Ocean. South of the Puerto Rican Shelf, the Virgin Island Trough separates the insular platform of St. Croix with depths up to 4200 m. Shallow marine ecosystems of the U.S. Caribbean islands are characterized by coastal coral reefs, near-shore fringing mangroves and extensive seagrass beds. Mangroves and coastal wetlands provide important ecological services, such as sediment and nutrient retention, storm buffering and, along with seagrasses, include areas of fish and shellfish nursery habitat essential for the juvenile phases of many commercially important species. Coral reefs extend throughout the insular platforms and provide the greatest area of

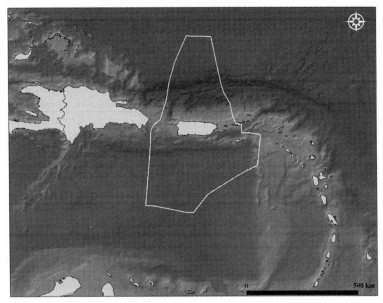

Figure 4.1 The U.S. Caribbean exclusive economic zone.

benthic habitat to depths of 80 m in sites with clear waters. Various marine species considered endangered or threatened (IUCN) are present such as sea turtles, manatees, whales and a few coral and fish species (Waddell and Clarke, 2008; Weil, 2005).

1.2. Marine fisheries

The marine fisheries of PR and the USVI resemble those of other Caribbean Islands with similar geography and history. Aboriginal people of the Caribbean are known to have subsisted on near-shore marine resources prior to the 1500s, since archeological studies of the *Taíno* Indian middens have uncovered evidence of a diversity of molluscs, fish and sea turtles. Although limited in technology, there is some description of fishing weirs and traps used by the *Taíno* and adopted by early settlers of the islands. The impacts of artisanal fisheries upon Caribbean marine ecosystems were significant prior to the 1800s (Hawkins and Roberts, 2004; Jackson, 1997; Wing and Wing, 2001). Since the nineteenth century, the impacts of human development upon coastal and marine ecosystems have been extensive including pollution and sedimentation from agriculture, military activities, industrialization and shifts in land–use patterns (Valdes–Pizzini et al., 2012).

These are important points to bear in mind when studying the region's fisheries due to the shifting baseline syndrome (Pauly, 1995).

More recently, the fisheries of Caribbean islands suffered changes in technology that allowed expansion in terms of fishing grounds, effort and target species. The most notable of these advances include refrigeration, motorized vessels, the use of monofilament lines for nets and line fishing and GPS. Despite some incentives by the local governments to encourage the development of industrialized commercial fleets, much of the modern artisanal fisheries are limited to the insular platform (depths less than 100 fathoms), although some forays to neighboring islands were common in the recent past. This fleet is characterized by vessels less than 12 m in length and harvests are relatively small-scale, geared towards local markets as fresh product and generally not for exportation. Multiple species are targeted simultaneously and forays are multi-geared with hook and line or traps as the main gears followed by diving and nets. Fishers today target highly valued species inhabiting coral reefs and associated habitats of the insular platform or seek pelagic species in deep waters beyond the shelf break.

Snappers (Lutjanidae) and groupers (Epinephelidae) of shallow habitats produced the bulk of landings, however larger bodied species have been substituted by smaller sized snappers, groupers, as well as grunts (Haemulidae), wrasses (Labridae) and smaller parrotfishes (Scaridae). Spiny lobster (*Panulirus argus*) and conch (*Strombus gigas*) are highly important fishery targets, and some fishers have specialized gear to target deep-water snappers (*Etelis oculatus, Lutjanus buccanella* and *Lutjanus vivanus*) and pelagic species (*Coryphaena hippurus* and *Acanthocybium solandri*) due to their high value. Prior to 1990, many coral reef species that form spawning aggregations were targeted directly during that time (Beets and Friedlander, 1999) which led to the extirpation of some known aggregations (Olsen and La Place, 1978) and the commercial extinction of others such as the Nassau grouper (*Epinephelus striatus*).

The number of fishers (let alone the diverse categories it encompasses) in the U.S. Caribbean (PR/USVI) is difficult to estimate. In the 1970s and 1980s, the number of fishers was estimated at 2000. Recent censuses and surveys show a decline of 45% since the 1970s. The current estimate of fishers is of 1100 for both territories combined, with approximately 350 for the USVI (Kojis and Quinn, 2011; Matos-Caraballo and Agar, 2011). This is quite similar to the first census of Puerto Rican fishers conducted in 1803, when 1500 fishers were counted (Valdés-Pizzini and Schärer-Umpierre, 2011). Although the socio-economic importance of fisheries in the U.S. Caribbean has fluctuated since then, it has remained a relatively minor part of the economy of these islands in comparison with agriculture (mostly sugar cane), industrial development, housing construction and tourism. In 2008, it was estimated that PR commercial fishers landed six million US$ for two million pounds of finfish and shellfish (Tonioli and Agar, 2011); however, this is probably an underestimate due to underreporting and unlicensed fishers (García-Quijano, 2009).

1.3. Fisheries management

Fisheries management within the U.S. Caribbean is quite complex due to the multiple levels of governance in the geo-political arena. At one level, the U.S. federal government has direct jurisdiction of the EEZ and indirect participation elsewhere in agreements with the local territorial governments. Federal jurisdiction starts at the seaward limit of the territorial jurisdiction, which in the USVI is 3 Nm and in PR to 9 Nm from the shore. The difference in the extent of territorial limits is due to the amendment of the Jones Act (1917) in 1980 that extended the PR limit to three leagues (~9 Nm). In PR, the first fishery law was enacted in 1936, but until the early 1970s no fisheries data were systematically collected. During the mid-1970s, the first comprehensive fisheries studies were conducted under the auspices of the Department of Agriculture of PR, and in 1976, the Caribbean Fishery Management Council (CFMC) was established to oversee fisheries in the U.S. EEZ. In 1998, the PR Fisheries Act was established to manage fisheries by the Department of Natural and Environmental Resources (DNER); however, the fisheries regulations which included MMAs for fisheries purposes were only implemented in 2004. Today, the territorial governments of PR and the USVI manage their fisheries resources in coordination with the CFMC, and in MMAs with shared jurisdiction both agencies agree on the regulations to be implemented on a site-by-site basis.

2. MARINE MANAGED AREAS

Some of the first evidence of efforts to protect the marine resources including gear limitations, seasonal closures and the establishment of protected areas in PR occurred during Spanish rule (Valdés-Pizzini and Schärer-Umpierre, 2011). In 1918, the U.S.-appointed governor to PR designated the first MMA to protect coastal mangrove forests due to demand for charcoal (Aguilar-Perera et al., 2006). Today, the designations of MMAs within the U.S. Caribbean have a variety of legal bases and different levels of jurisdiction. At the federal level, MMAs are designated based on the Magnuson

Stevens Fishery Conservation and Management Act and reauthorizations, which are managed by the CFMC. In the USVI, there are additional federal-level designations such as by the Department of the Interior, which has designated terrestrial and marine protected areas managed by the National Park Service (NPS) and extensions to these that were based on Presidential Proclamation (National Monuments).

Within territorial jurisdictions, MMAs are designated through two main mechanisms, administrative and legislative. In the case of PR, DNER recommends sites of ecological and cultural value to the Planning Board, which then finalizes the administrative process with a formal legal designation, but no regulations. The bulk of the MMAs in PR were designated in 1978 as part of the Coastal Zone Management Program (CZMP). Alternatively, the legislatures of PR and the USVI can designate marine reserves (equivalent to no-take zones) or protected areas (marine parks). This has been the case for at least five MMAs designated recently in PR, and at least two in the USVI, many of which are no-take zones (NTZs) where fishing is prohibited year-round. In PR, the regulations for MMAs can be based on the Forestry Law or the Fisheries Law. Currently, there is no specific overriding legislation or any coordinated plan to designate networks of MMAs. In the USVI, the CZMP is also the basis for some designations through the Department of Planning and Natural Resources. However, there is no established law to unify the selection, designation and management of MMAs in the region (Gardner, 2002).

For the purpose of this volume, we define MMAs as those areas of marine waters designated by legal mechanisms (local or federal) including submerged marine areas within the boundaries of the designation. We excluded coastal lagoons or wetlands located inland, unless they have a seaward extension with submerged marine areas within the MMA's boundaries. Data were extracted from published documents, maps and data obtained directly from management agencies as well as the National Marine Protected Area Inventory (NOAA, 2014). The extent of the marine submerged area within the limits of each designation was calculated in a geographical information system and the results are summarized in Table 4.1 and displayed in Figures 4.2 and 4.3.

Thirty-nine MMAs met the criteria defined above, 26 located in the territorial waters of PR, 7 in USVI and 6 in the EEZ (or combination). Approximately 4035 km^2 of submerged habitats are incorporated in these MMAs, 10% of this total area has some type of no-take regulation that protects fisheries resources during some part of the year. The USVI had a greater proportion (50%) of area of MMAs with no-take regulation compared to PR

Table 4.1 Summary data for all marine managed areas in the U.S. Caribbean including the site where it occurs: exclusive economic zone (EEZ), Puerto Rico (PR) or U.S. Virgin Islands (USVI)

Name	Site	Composition	Year	Type	Timing	Marine area	No-take area
Abrir la Sierra	EEZ	S	1996	S–NTZ	Dec. to Feb.	29.5	29.5
Bajo de Sico	EEZ	S	1996	S–NTZ	Oct. to Mar.	31.4	31.4
Grammanik Bank	EEZ	S	2005	S–NTZ	Feb. to Apr.	1.5	1.5
Hind Bank MCD	EEZ	S	1989	NTZ	Year-round	44.6	44.6
Lang Bank	EEZ	S	1993	S–NTZ	Dec. to Feb.	11.7	11.7
Tourmaline Bank	EEZ	S	1993	S–NTZ	Dec. to Feb.	31.4	31.4
Arrecife de Isla Verde	PR	S	2013	NTZ	Year-round	0.94	0.94
Arrecifes de Guayama	PR	S	1980	NR		8.1	
Arrecifes de la Cordillera	PR	M	1980	NR		99.9	
Arrecifes de Tourmaline	PR	S	1998	NR		74.6	
Bahías Bioluminiscentes de Vieques	PR	M	1998	NR		79.6	
Boquerón State Forest	PR	M	1998	NR		172.7	
Cabezas de San Juan	PR	M	1998	NR		266.9	
Caja de Muertos	PR	M	1980	NR/NTZ	Year-round	55.1	0.39
Canal Luis Peña	PR	S	1999	NTZ	Year-round	6.3	6.3

Continued

Table 4.1 Summary data for all marine managed areas in the U.S. Caribbean including the site where it occurs: exclusive economic zone (EEZ), Puerto Rico (PR) or U.S. Virgin Islands (USVI)—cont'd

Name	Site	Composition	Year	Type	Timing	Marine area	No-take area
Caño La Boquilla	PR	M	2002	NR		105.8	
Corredor Ecologico del Noreste	PR	M	2013	NR		263.1	
Cueva del Indio	PR	M	1998	NR		15.5	
Finca Belvedere	PR	M	2003	NR		40	
Guánica State Forest	PR	M	1985	NR		14.2	
Hacienda La Esperanza	PR	M	1998	NR		50.6	
Isla de Desecheo	PR	S	2000	NTZ	Year-round	7.4	7.4
Isla de Mona & Monito	PR	M	1997	NR/ NTZ	Year-round	1512.7	81
Jobos Bay	PR	M	1981	NERR		9.8	
La Parguera	PR	M	1998	NR		324	
Pantano Cibuco	PR	M	1998	NR		19.9	
Punta Cucharas	PR	M	2007	NR		13.8	
Punta Guaniquilla	PR	M	2002	NR		8.6	
Punta Petrona	PR	M	1979	NR		30.9	
Punta Yeguas	PR	M	2001	NR		262.4	
Río Espíritu Santo	PR	M	1998	NR		117.7	
Tres Palmas de Rincón	PR	S	2004	NTZ	Year-round	0.89	0.89
Buck Island Reef	USVI	M	2001	NM	Year-round	76.9	76.9
Mutton Snapper Spawning Aggregation	USVI	S	1993	S-NTZ	Mar. to Jun.	8.9	8.9

Table 4.1 Summary data for all marine managed areas in the U.S. Caribbean including the site where it occurs: exclusive economic zone (EEZ), Puerto Rico (PR) or U.S. Virgin Islands (USVI)—cont'd

Name	Site	Composition	Year	Type	Timing	Marine area	No-take area
Salt River	USVI	M	1992	NP	Year-round	3.3	3.3
St. Croix East End	USVI	M	2003	MP/NTZ	Year-round	150.3	12.9
St. Thomas East End Reserves	USVI	S	2011	MP/NTZ	Year-round	9.3	9.3
Virgin Islands Coral Reef	USVI	S	2001	NM	Year-round	51.4	51.4
Virgin Islands	USVI	M	1962	NP		23.5	

The composition of the area may be submerged (S) or mixed (M) if it includes a terrestrial portion, the year of establishment (Year) and the type; seasonal no-take zone (S-NTZ), no-take zone (NTZ), marine conservation district (MCD), natural reserve (NR), marine reserve (MR), national estuarine research reserve (NERR), national park (NP), national monument (NM) and marine park (MP). Timing indicates what part of the year the no-take zone is in effect. All areas calculated in km^2.

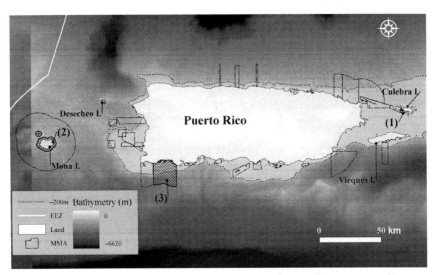

Figure 4.2 Marine managed areas in Puerto Rico, numbers indicate case studies.

(3%). In EEZ waters, 70% of the ecosystems within MMAs are designated with seasonal no-take regulations and only one site (30% of the area) has year-round protection: the Hind Bank Marine Conservation District (MCD) south of St. Thomas.

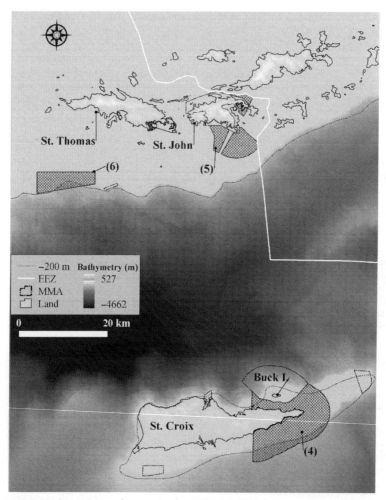

Figure 4.3 Marine managed areas in the U.S. Virgin Islands, numbers indicate case studies.

Overall, 12 MMAs have year-round, NTZs within boundaries delineating 295 km^2 of submerged habitats, whereas six sites have seasonal no-take regulations covering 114 km^2. Twenty-one sites covering approximately half of the area (2001 km^2) lacked no-take regulations. Most of the seasonal no-take regulations of the MMAs encompass the months of December through June with the aim of protecting species of shallow water groupers or snappers that aggregate to spawn at this time.

3. CASE STUDIES

3.1. Canal Luis Peña Natural Reserve

The PR Planning Board designated Canal Luis Peña Natural Reserve (CLPNR) in Culebra Island (east of PR) as the first no-take in PR in 1999. It was designated to encourage the recovery of fish communities by protecting habitats within former U.S. Navy training grounds. The original concept proposed by local fishers, base communities and the local government of Culebra Island was to develop a co-management, participatory process. The site has historically encompassed multiple uses ranging from military training activities, fishing and coral extraction for construction (nineteenth and twentieth century) to recreational uses. Although a management plan was published in 2008, very few of the proposed management actions have been implemented and no co-management agreement has been established.

Before the no-take designation, fish communities were depauperate, with low densities and low biomass of most commercially important species (Hernández-Delgado, 2000). Three years after no-take designation abundance and biomass increased, including a 515% increase in total fish biomass (Figure 4.4), 438% in herbivore biomass and 1249% in piscivore biomass

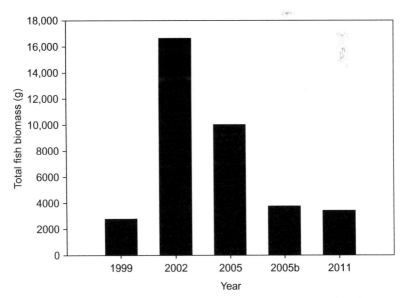

Figure 4.4 Trend in fish biomass (g) within the core area of the CLPNR after designation. Data from year 2005 was collected before and after (2005b) the coral bleaching event.

(Hernández-Delgado et al., 2006). *Epinephelus guttatus*, an important fishery target, were found to be significantly larger inside the reserve, but this could be linked to a greater proportion of appropriate habitat within the boundaries (López-Rivera and Sabat, 2009). However, lack of governance and sustained enforcement resulted in a subsequent decline and no net recovery of fish nor coral reef benthic communities (Hernández-Delgado and Suleimán-Ramos, 2014). Massive coral bleaching in 2005 caused further fish biomass decline with no net recovery. The CLPNR represents an excellent case study of the significance of designating a NTZ supported by local fishers and base communities that resulted in the rapid recovery of fish communities. However, this recovery could not be sustained due to the lack of compliance coupled with damaging land-use practices in Culebra Island and climate change-related impacts, which have resulted in a declining trend of coral reef resources.

3.2. Mona and Monito Islands Natural Reserve

Mona and Monito Islands are located in the middle of the Mona Passage west of PR, a known partial biogeographic barrier (Baums et al., 2006; Dennis et al., 2005). These islands out to 9 Nm from shore comprise the largest MMA in the U.S. Caribbean with a significant, year-round, NTZ (extending to 1 Nm from shore). Although the Nature Reserve was designated in 1986, the NTZ was not established until 2004 (Aguilar-Perera et al., 2006). In 2007, the boundaries of the NTZ were amended to incorporate key grouper spawning aggregation sites. The DNER has jurisdiction over this area, and only government staff inhabit the island permanently. Visitors arrive by charter or private vessel for camping, hiking, recreational hunting and fishing with permits issued by the DNER. Although this site has been inhabited only sporadically (*Taínos* up to 1400s; guano miners in the 1700s), it was the focus of intense fishing for sea turtles and grouper spawning aggregations prior to its designation. More recently, commercial fishing for deep-water snappers and recreational fishing for pelagic species have been the most common uses of the MMA. A recent study evaluated the performance of some coral reef fishes to the NTZ five years after designation (Mateos-Molina et al., 2014). A 50% increase in abundance and biomass was detected for small life stages and species. Large bodied groupers and snappers, considered top predators, were still rare and did not show any trends in abundance or biomass. One of the smaller groupers, the coney (*Cephalopholis fulva*), doubled its abundance and biomass in no-take areas as well as outside. This response may be due to increased availability of ecological niches due to the lack of larger sized predators or the reproductive strategy of this species. At this offshore location, the lack of larval connectivity from areas with resident populations may be limiting the recovery of larger coral reef fishes.

3.3. La Parguera Natural Reserve

The area around La Parguera, in southwestern PR represents a complex marine ecosystem consisting of extensive mangrove stands and sea grass beds inshore, protected by the most well-developed series of coral reefs around the island (Morelock et al., 1977). This is augmented by a broad extent of submerged patch reefs, extensive shelf-edge reef, and well-developed mesophotic coral ecosystems that extend to depths of 80 m or more (Sherman et al., 2010). The abundance and spatial proximity of these habitats render La Parguera as one of the most diverse and productive areas of the coastal environments around PR, with a high degree of ecological connectivity, especially exhibited through the movement of fishes (Appeldoorn et al., 2009). Concern for the protection of La Parguera dates from the 1960s with the recognition of the importance of Bioluminescent Bay. However, management plans were not enacted until 1978 when the CZMP was approved and the PR Planning Board designated La Parguera Natural Reserve (LPNR) with an inland extent of 1 km and seaward extent of 3 Nm, later expanded to 9 Nm in 1998. During the early 1980s, La Parguera was the focus of a contentious debate over the establishment of a U.S. National Marine Sanctuary that was not designated (Fiske, 1992). In 1995, PR adopted the La Parguera Special Planning Area, which extended from the shoreline to the top of nearby hills defining the majority of the proximal watersheds. Yet, no regulations or zones were established within the NR, despite a concerted, multiyear, multi-institutional effort to establish a NTZ (Valdés-Pizzini and Schärer-Umpierre, 2014). Neither was any special effort given to increase enforcement of island-wide fisheries regulations in light of the recognized importance of La Parguera's ecosystem.

There have been substantial changes documented in LPNR in the last 30 years, most likely due directly or indirectly, to anthropogenic stressors including urban development, fishing and tourism. Fishing pressure has resulted in the loss of most large-bodied fishes and spawning aggregations, reducing the frequency of occurrence of 10 out of 13 species sampled in the early 1980s and again in 2000s (Figure 4.5). The common large parrotfishes and groupers are no longer present, while medium-sized red hind

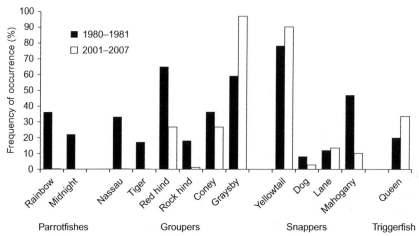

Figure 4.5 Frequency of occurrence (%) in visual surveys of reef fishes from the early 1980s (Kimmel, 1985) and the mid-2000s (Pittman et al., 2010). Values for the latter are multiplied by a factor of 3.33 to adjust for differences in survey times: 50 min for the initial study; 15 min for the latter.

(*E. guttatus*), rock hind (*E. adscensionis*) and mahogany snapper (*Lutjanus mahogoni*) were observed considerably less frequently. There has been little change in the more pelagic yellowtail snapper (*Ocyurus chrysurus*), while the graysby (*Cephalopholis cruentata*), the smallest grouper, has become more frequent in occurrence. Many formally abundant species on the shelf are now only found off the steep insular slope (Bejarano et al., 2014) in mesophotic depths.

3.4. St. Croix East End Marine Park

The East End Marine Park is located over the 10 km of the easternmost shelf of St. Croix (USVI) and extends 3 Nm from shore. It was established in 2003 as the first multi-use marine park managed by the USVI government for natural and cultural resources by protecting, replenishing and sustaining healthy populations of key species, habitats and biodiversity (The Nature Conservancy, TNC, 2002). This park is used for multiple recreational and commercial activities, including ecotourism, camping, swimming, snorkeling, diving, boating and fishing. Historically, this site was used for fishing of conch and lobster, as well as netfishers catching parrotfish, among other species (Valdés-Pizzini et al., 2010). In the past, large-bodied fish used the northeastern habitats of the island and there was a Nassau grouper spawning aggregation site in East End. However, intensive and unregulated fishing led to the extirpation of this aggregation by 1971 (Olsen and La Place, 1978). There are four different types of managed areas within the park although zoning was designed based on limited data (Island Resources Foundation, 2002), which did not allow the establishment of an ecological baseline to measure their subsequent performance. Recent research assessing the distribution, diversity and status of the marine environment within the park and land-based stressors has helped fill this gap (Pittman et al., 2013). Although determining the ecological performance of areas closed to fishing requires future monitoring, outcomes so far suggest that the current zoning contributes little to the protection and replenishment of fished populations. Assemblages within NTZs are mostly small-bodied and juvenile fishes that are not primary target species of the fishery, while fish biomass and adult density of fished species is high in areas open to fishing (Pittman et al., 2013). Historical data (Pittman et al., 2008) show that large-bodied fish (e.g. tiger grouper) used the habitats in the northeastern part of the island, but only a few adult Nassau groupers (0.3% of survey sites) were sighted within the park. Thus, the current zoning needs to be reviewed and modified so that it can be based on robust data, as well as include a more complete range of species and habitats. Alternative strategies for the conservation of priority species need to be developed, or current objectives need to be redefined to achieve more realistic replenishment and protection goals.

3.5. Virgin Islands Coral Reef National Monument

The Virgin Islands Coral Reef National Monument (VICRNM) consists of five areas offshore of the island of St. John, which is adjacent to the Virgin Islands National Park, and both are administered by the U.S. NPS. The VICRNM was established in 2001 by Presidential Order to increase efforts by the U.S. federal government mandated by the Coral Reef Conservation Act. All extractive uses are prohibited within the VICRNM with the exception of fishing for baitfish in Hurricane Hole Bay and fishing with hook and line in a slot south of St. John, activities that require special permits. The unprotected region, which splits the larger area in two (Figure 4.3), was specifically requested by fisherman to allow continued access to an important fishing area. Additional management efforts include the prohibition of vessel anchoring within the reserve, and mooring buoys are maintained by the NPS for this reason. The NPS has been intensively investigating marine resources with an emphasis on coral reef ecosystems for the last two decades leading to a focused effort, since 2001, to monitor the composition and abundance of the reef fish assemblage (Friedlander et al., 2013a,b). The results to date indicate a lack of difference in the abundance of fishery species within or outside of the VICRNM. Results suggest that the absence of ecological criteria in the designation process resulted in more complex habitat outside of the reserve (Monaco et al., 2009), conditions that have prevented the expected recovery of depleted fish populations.

3.6. Hind Bank Marine Conservation District

The Hind Bank MCD, south of St. Thomas (Figure 4.3), was first established in 1990 as a seasonal NTZ to protect a commercially important spawning aggregation of red hind (Beets and Friedlander, 1999). The MCD became a much-expanded permanent closed area in 1999, with fishing and anchoring prohibited throughout, thus protecting critical coral reef habitats. The MCD extends 12 km westward along the southern edge of the Puerto Rican Bank into the southern extent of the Virgin Passage. The majority of the seafloor is less than 50 m in depth and is composed of topographically complex coral reefs that are the center of the largest known mesophotic coral reef complex in the U.S. Caribbean (Smith et al., 2010). These coral reefs are dominated by dense *Orbicella* spp. (Armstrong et al., 2006) and were classified by Smith et al. (2010) into three distinct habitat types: high coral banks, flat basin and hillock basin. The hillock basin is unusual in that it is a highly heterogeneous area containing thousands of 2–10 m high coral knolls. Historically, the MCD contained an important Nassau grouper (*E. striatus*) spawning aggregation site that was extirpated in the late 1970s due to overfishing (Olsen and La Place, 1978). Other species of grouper, snappers and jacks also form spawning aggregations within the MCD. This MMA has resulted in an increase in the density of larger size classes of commercially important groupers, snappers and parrotfishes (Nemeth and Quandt, 2005), and also facilitated the recovery of the red hind, whose biomass increased 550% between 1997 and 2003 (Nemeth, 2005). Today, red hind represents one of the most common reef fish in the St. Thomas commercial fishery. The increases in the spawning stock at this site and the movement patterns of tagged red hind up to 33 km away suggest that this MMA has benefitted the fisheries of downstream areas, namely, the platform west of St. Thomas, eastern coast of PR and the offshore islands of Vieques and Culebra (Nemeth, 2005). Continued protection of this important area will enhance other fishery resources of the region and continue to protect the critical coral reef habitat that they rely upon.

4. DISCUSSION

Edgar et al. (2014) classified marine protected areas according to five criteria (NEOLI): no-take, enforced, old (>10 years), large (>100 km^2) and isolated (based on habitat discontinuities). They suggested that conservation benefits increased as more criteria were met, but that significant improvements required at least four of these criteria. In the U.S. Caribbean, most MMAs do not meet more than one of these criteria.

4.1. Area of habitat

The size of an MMA is important due to the extent and diversity of habitats within it as well as the proportion of fish and invertebrate populations that are protected from anthropogenic impacts. In the U.S. Caribbean, insular shelves are narrow, fish distributions are closely coupled to habitat and fishing generally operates at small spatial scales; hence, it has been proposed that MMAs that encompass areas from the shore to edge of the insular shelf are best for protecting the diversity of habitats necessary to support all species across all life stages, especially for fishes that undergo ontogenetic migrations (Appeldoorn et al., 2011; Mumby, 2006; Mumby et al., 2004; Pittman et al., 2004). The study of fish movements and habitat use at the landscape level has evidenced the spatial scales of protection necessary to include nursery, migration and spawning habitats. However, in the case of Red Hind Bank MCD, which does not include shallow nursery habitats, the large area of prime coral reef habitat within the MMA has led to increases in the biomass and spawning stock of red hind. Therefore, the inclusion of spawning aggregations in NTZs is critical for those species with long-distance migrations to spawning sites. Only the Mona and Monito Island Natural Reserve includes a diversity of habitats ranging from shore to shelf, including multi-species spawning aggregations within the NTZ, although populations of many of these larger bodies fishes have not shown signs of recovery. LPNR also includes a large area of the habitat matrix from shore to shelf break and had spawning aggregation sites at the deeper sites, but lack of regulations controlling fishing activities have precluded the recovery of over-fished populations.

4.2. Seasonal versus year-round

The main rationale for the seasonal no-take regulations in the U.S. Caribbean is to protect a particular species during spawning aggregations.

During this time, which may last a few months, individuals remain in high density and are highly vulnerable to fishing. Also, due to hyper-stability (Erisman et al., 2011) much of the spawning stock can be removed without overt signs of overfishing, such as decreased rates of capture. Most of the seasonal no-take MMAs (five of seven initially) were designed to protect the red hind during its reported spawning season from December to February. Research studies evaluating the effect of the seasonal NTZ regulations have shown mixed results. Significant increasing trends in population characteristics over time were found at the Red Hind Bank MCD, which was seasonally protected (1990) before becoming year-round no-take (1999), and at Lang Bank, St. Croix, only seasonally protected since 1993 (Nemeth et al., 2006). Data suggest the recovery of red hind spawning populations at both sites with seasonal no-take, however, the rate of change detected at the Red Hind Bank MCD were much higher after the site became a year-round no-take, including a doubling of maximum density during the spawning aggregation. The inclusion of the time periods prior to and after spawning may have benefitted those individuals that migrated through the MCD as well as protected the spawning stock during times that spawning may have been after the closed season, given the natural variability in temporal dynamics of spawning aggregations. The protection of a large area of continuous prime habitat year-round for over 15 years has led to the recovery of the red hind populations and benefitted local fisheries. The duration of protection may explain differences in the degree of recovery between here and Lang Bank in St. Croix (Nemeth et al., 2006, 2007). In contrast, the three seasonally closed MMAs around red hind aggregation sites off the west coast of PR did not show any increases in abundance, with directed fishing effort for red hind increasing substantially over the subsequent decade before, during (outside closed areas) and after the period of closures when red hind may still have been on the spawning grounds. PR has since adopted an island-wide closed season for red hind during these months (December to February).

4.3. Level of enforcement

Management consists of both planned regulations and enforcement as necessary components; without one, the other is incomplete and relatively meaningless. Governance structures are necessary for achieving this balance. We have already shown that many MMAs in the U.S. Caribbean were not adequately designed and regulated to achieve their respective goals.

Unfortunately, no-take areas in the U.S. Caribbean also present nonexistent or ineffective enforcement and therefore fail to achieve compliance with local regulations and other federal and territorial statutes. The economic cost of this lack of compliance with regulations is a significant constraint to evidencing the effects, if any, of the NTZs. In addition to the measures of the biological, fishery and ecological indicators, the studies of NTZ effectiveness require some measure of compliance and law enforcement interventions, which may confound results and conclusions. Areas with high levels of compliance require little enforcement, but traditional measures of governance that limit the uses of marine or natural environments have not been evidenced in the Caribbean. Most of the regulations are top-down, directed from external bodies governing elsewhere or lack meaningful public participation in fisheries management.

The political history of the Caribbean primed the local surrogate governments and the metropolitan authorities with the task of managing, unilaterally, the resources and making the appropriate decisions, without the participation of the diverse communities of users. Increased political autonomy in recent years also meant an increase in power over those decisions and actions that framed the current status of fisheries management throughout the region. Arguably, the incorporation of the human dimension—a critical parameter in Ecosystem-Based Management—has not been a priority, or a key element in the process. And yet, the main problems are social and need to be addressed before any success is attainable in management (Appeldoorn et al., 2005; Valdes-Pizzini et al., 2012). However, the region (and PR/USVI) is changing, with a number of experiments in community-based management, highly participatory schemes and co-management experiences with MMAs, which also include—with the expected tensions—local communities and organizations, as well as powerful Island-based and international NGOs.

Co-management approaches among different enforcement entities (government, NGOs and local communities) appear as a viable alternative to the low enforcement capacity. Good examples of these are two of the smallest NTZs in PR, Tres Palmas Marine Reserve in Rincón and Arrecifes Isla Verde Marine Reserves where neighbors initiated the designation and have taken up much of the education and enforcement responsabilities, involving local municipal police to capture poachers when the DNER agency law enforcement agents are unavailable. Similarly, community involvement played an important role in the establishment and early success at CLPNR, but failure to follow-up with the planned co-management structure ultimately led to its collapse. St. Croix East End Marine Park, on the other hand, has an enforcement program, but its success is limited by inefficient coordination among the different parts involved in the management of the MMA.

5. CONCLUSION

Multiple MMAs in the U.S. Caribbean demonstrate a diverse range of sizes, ages and regimes yet provide limited evidence to measure their impact upon fisheries resources. Year-round NTZ designation including prime habitat and spawning aggregations can provide the greatest benefit to fisheries as long as local communities favor compliance and enforcement is coordinated. The limited data available suggest that fisheries management in the region with MMAs as a tool will require evidence that can be used to support their application to benefit local economies. Most areas existing today represent mere extensions of coastal managed areas without specific management goals or actions. Assessments exist for only a handful of MMAs, and these primarily show little impact. In most cases, it is clear that the MMAs are under-performing in large part from not meeting the multiple criteria identified as key by Edgar et al. (2014). Thus, assessing the overall potential of MMAs to improve resource condition should not be based on the overall record but rather directed toward those areas where proper design, management and monitoring exist. Here, there are some bright spots. The Red Hind Bank MCD to some extent meets all five criteria and has been an unqualified success relative to its original goal of protecting spawning red hind. Similarly, the Mona and Monito Island Natural Reserve has shown some improvement after only 5 years. While enforcement is poor, its distance from PR serves to reduce fishing effort relative to the main platform. Recovery of this MMA may be slowed by limited connectivity to other areas. Historically, enforcement of MMAs, even for NTZs has been poor to nonexistent, and this has been identified as the primary problem determining their effectiveness. Recent efforts to establish governance programs and co-management may improve this situation over time. In island settings with narrow shelves, total area within an MMA may not be as important as capturing critical habitat, such as spawning sites and migration corridors or ensuring protection from the shore to the shelf-edge (Appeldoorn et al., 2011). From this perspective, we view essential fish habitat (EFH) not on a species basis (since all habitats are used across all species) but rather on a multi-species basis where key areas of diverse habitat promote connectivity among habitats and productivity across species (Cerveny et al., 2011). Issues of total area can be ameliorated if MMAs are planned not in isolation but within a context of a network design, with areas characterized by high multi-species EFH serving as core areas. We conclude that the success of MMAs in achieving management goals can only really be assessed when both proper biological and socio-economic criteria were used to design, implement and manage MMAs. However, establishing the necessary criteria has come from practical trial and error as much as from scientific theory, and the lessons learned here should be used to guide future efforts. Lastly, assessments of MMA efficiency must consider the effects of water quality and climate change, especially when considering corals and other benthic resources. Therefore, the criteria of Edgar et al. (2014) may benefit from the addition of habitat quality criteria for effective MMA design.

REFERENCES

Aguilar-Perera, A., Schärer, M.T., Valdés-Pizzini, M., 2006. Marine protected areas in Puerto Rico: historical and current management approaches. Ocean Coast. Manage. 49, 961–975.

Appeldoorn, R.S., Kimmel, J.J., Meyers, S., Sadovy, Y., Valdés-Pizzini, M., 2005. Fisheries management policy in Puerto Rico: a progress report. Proc. Gulf Caribb. Fish. Inst. 47, 111–122.

Appeldoorn, R.S., Aguilar-Perera, A., Bouwmeester, B.L.K., Dennis, G.D., Hill, R.L., Merten, W., Recksiek, C.W., Williams, S.J., 2009. Movement of fishes (Grunts: Haemulidae) across the coral reef seascape: a review of scales, patterns and processes. Caribb. J. Sci. 45, 304–316.

Appeldoorn, R.S., Ruíz, I., Pagan, F.E., 2011. From habitat mapping to ecological function: incorporating habitat into coral reef fisheries management. Proc. Gulf Caribb. Fish. Inst. 63, 10–17.

Armstrong, R.A., Singh, H., Torres, J., Nemeth, R.S., Can, A., Roman, C., Eustice, R., Riggs, L., Garcia-Moliner, G., 2006. Characterizing the deep insular shelf coral reef habitat of the Hind Bank Marine Conservation district (US Virgin Islands) using the Seabed autonomous underwater vehicle. Coast. Shelf Res. 26, 194–205.

Baums, I., Paris, C., Chérubin, L., 2006. A bio-oceanographic filter to larval dispersal in a reef-building coral. Limnol. Oceanogr. 51, 1969–1981.

Beets, J., Friedlander, A., 1999. Evaluation of a conservation strategy: a spawning aggregation closure for red hind, *Epinephelus guttatus*, in the U.S. Virgin Islands. Environ. Biol. Fishes 55, 91–98.

Bejarano, I., Nemeth, M.I., Appeldoorn, R.S., 2014. Fishes associated with mesophotic coral ecosystems at La Parguera, Puerto Rico. Coral Reefs 33, 313–328.

Cerveny, K., Appeldoorn, R.S., Recksiek, C.W., 2011. Managing habitat in coral reef ecosystems for fisheries: just what is essential? Proc. Gulf Caribb. Fish. Inst. 63, 23–36.

Dennis, G.D., Smith-Vaniz, W.F., Colin, P.L., Hensley, D.A., McGehee, A., 2005. Shore fishes from islands of the Mona Passage, Greater Antilles with comments on their zoogeography. Caribb. J. Sci. 41, 716–743.

Edgar, G.J., Stuart-Smith, R.D., Willis, T.J., Kininmonth, S., Baker, S.C., Banks, S., Barrett, N.S., Becerro, M., Bernard, A.T.F., Berkhout, J., Buxton, C.D., Campbell, S.J., Cooper, A.T., Davey, M., Edgar, S.C., Försterra, G., Galván, D.E., Irigoyen, A.J.,

Kushner, D.J., Moura, R., Parnell, P.E., Shears, N.T., Soler, G., Strain, E.M., Thomson, R.J., 2014. Global conservation outcomes depend on marine protected areas with five key features. Nature 506, 216–220.

Erisman, B.E., Allen, L.G., Claisse, J.T., Pondella, D.J., Miller, E.F., Murray, J.H., 2011. The illusion of plenty: hyperstability masks collapses in two recreational fisheries that target fish spawning aggregations. Can. J. Fish. Aquat. Sci. 68, 1705–1716.

Fiske, S., 1992. Sociocultural aspects of establishing marine protected areas. Ocean Coast. Manag. 17, 25–46.

Friedlander, A.M., Jeffrey, C.F.G., Hile, S.D., Pittman, S.J., Monaco, M.E., Caldow, C., 2013a. Coral Reef Ecosystems of St. John, U.S. Virgin Islands: Spatial and Temporal Patterns in Fish and Benthic Communities (2001-2009). NOAA Technical Memorandum 152, Silver Spring, MD.

Friedlander, A.M., Monaco, M.E., Clark, R., Pittman, S.J., Beets, J., Boulon, R., Callender, R., Christensen, J., Hile, S.D., Kendall, M.S., Miller, J., Rogers, C., Stamoulis, K., Wedding, L., Roberson, K., 2013b. Fish Movement Patterns in Virgin Islands National Park, Virgin Islands Coral Reef National Monument and Adjacent Waters. NOAA Technical Memorandum NOS NCCOS 172, Silver Spring, MD.

García-Quijano, C., 2009. Managing complexity: ecological knowledge and success in Puerto Rican small-scale fisheries. Hum. Org. 68, 1–17.

Gardner, L., 2002. Management framework for a system of marine protected areas for the U.S. Virgin Islands. University of the Virgin Islands and Department of Planning and Natural Resources. USVI.

Hawkins, J.P., Roberts, C.M., 2004. Effects of artisanal fishing on Caribbean Coral Reefs. Conserv. Biol. 18, 215–226.

Hernández-Delgado, E.A., 2000. Effects of anthropogenic stress gradients in the structure of coral reef epibenthic and fish communities. Ph.D. Dissertation, Department of Biology, University of Puerto Rico, San Juan, PR.

Hernández-Delgado, E.A., Suleimán-Ramos, S.E., 2014. E.S.A. coral species listing: a roadblock to community-based engagement in coral reef conservation and rehabilitation across the U.S Caribbean? Reef Encounter 29 (1), 11–15.

Hernández-Delgado, E.A., Rosado-Matías, B.J., Sabat, A.M., 2006. Management failures and coral decline threatens fish functional groups recovery patterns in the Luis Peña Channel No-Take Natural Reserve, Culebra Island PR. Proc. Gulf Caribb. Fish. Inst. 57, 577–605.

Jackson, J.B., 1997. Reefs since Columbus. Coral Reefs 16, 23–32.

Kimmel, J., 1985. A characterization of Puerto Rican fish assemblages. Department of Marine Science, University of Puerto Rico, Mayagüez, PR. PhD. Dissertation.

Kojis, B., Quinn N., 2011. Census of the marine commercial fishers of the United States Virgin Islands. Department of Planning and Natural Resources Division of Fish and Wildlife. USVI.

López-Rivera, M.M., Sabat, A.M., 2009. Effects of a marine fishery reserve and habitat characteristics in the abundance and demography of the red hind grouper, *Epinephelus guttatus*. Caribb. J. Sci. 45, 348–362.

Mateos-Molina, D., Schärer-Umpierre, M.T., Appeldoorn, R.S., García-Charton, J.A., 2014. Measuring the effectiveness of a Caribbean oceanic island no-take zone with an asymmetrical BACI approach. Fish. Res. 150, 1–10.

Matos-Caraballo, D., Agar, J.J., 2011. Census of active commercial fishermen in Puerto Rico: 2008. Mar. Fish. Rev. 73, 13–27.

Monaco, M.E., Friedlander, A.M., Caldow, C., Hile, S.D., Menza, C., Boulon, R.H., 2009. Long-term monitoring of habitats and reef fish found inside and outside the US Virgin Islands Coral Reef National Monument: a comparative assessment. Caribb. J. Sci. 45, 338–347.

Morelock, J., Schneidermann, N., Bryant, W.R., 1977. Shelf reefs, southwestern Puerto Rico. In: Frost, Stanley H., Weiss, Malcolm P., Saunders, John B. (Eds.), Reefs and Related Carbonates—Ecology and Sedimentology. In: Studies in Geology 4, American Association Petroleum Geologists, Tulsa, Okla, pp. 17–25.

Mumby, P.J., 2006. Connectivity of reef fish between mangroves and coral reefs: algorithms for the design of marine reserves at seascape scales. Biol. Conserv. 128, 215–222.

Mumby, P.J., Alasdair, J., Edwards, A.J., Arias-González, J.E., Lindeman, K.C., Blackwell, P.G., Gall, A., Gorczynska, M.I., Harborne1, A.R., Pescod, C.L., Renken, H., Wabnitz, C.C.C., Llewellyn, G., 2004. Mangroves enhance the biomass of coral reef fish communities in the Caribbean. Nature 427, 533–536.

Nemeth, R.S., 2005. Population characteristics of a recovering US Virgin Islands red hind spawning aggregation following protection. Mar. Ecol. Prog. Ser. 286, 81–97.

Nemeth, R.S., Quandt, A., 2005. Differences in fish assemblage structure following the establishment of the Marine Conservation District, St. Thomas, U. S. Virgin Islands. Proc. Gulf Caribb. Fish. Inst. 56, 367–381.

Nemeth, R.S., Herzlieb, S., Blondeau, J., 2006. Comparison of two seasonal closures for protecting red hind spawning aggregations in the US Virgin Islands. In: Proceedings of the 10th International Coral Reef Conference. 4, 1306–1313.

Nemeth, R.S., Blondeau, J., Herzlieb, S., Kadison, E., 2007. Spatial and temporal patterns of movement and migration at spawning aggregations of red hind, Epinephelus guttatus, in the U.S.Virgin Islands. Environ. Biol. Fish. 78 (4), 365–381.

NOAA, 2014. Marine Protected Areas Inventory. Accessed March of 2014, http://marineprotectedareas.noaa.gov/dataanalysis/mpainventory/.

Olsen, D.A., La Place, J.A., 1978. A study of a Virgin Islands grouper fishery based on a breeding aggregation. Proc. Gulf Caribb. Fish. Inst. 31, 130–144.

Pauly, D., 1995. Anecdotes and the shifting baseline syndrome of fisheries. Trends Ecol. Evol. 10, 430.

Pittman, S.J., Christensen, J.D., Caldow, C., Menza, C., Monaco, M.E., 2004. Predictive mapping of fish species richness across shallow-water seascapes in the Caribbean. Ecol. Model. 204, 9–21.

Pittman, S.J., Hile, S.D., Jeffrey, C.F.G., Caldow, C., Kendall, M.S., Monaco, M.E., Hillis-Starr, Z., 2008. Fish Assemblages and Benthic Habitats of Buck Island Reef National Monument (St. Croix, U.S. Virgin Islands) and the Surrounding Seascape: A Characterization of Spatial and Temporal Patterns. NOAA Technical Memorandum NOS NCCOS 71, Silver Spring, MD.

Pittman, S.J., Hile, S.D., Jeffrey, C.F.G., Clark, R., Woody, K., Herlach, B.D., Caldow, C., Monaco, M.E., Appeldoorn, R., 2010. Coral Reef Ecosystems of Reserva Natural La Parguera (Puerto Rico): Spatial and Temporal Patterns in Fish and Benthic Communities (2001-2007). NOAA Technical Memorandum NOS NCCOS 107, Silver Spring, MD.

Pittman, S.J., Dorfman, D.S., Hile, S.D., Jeffrey, C.F.G., Edwards, M.A., Caldow, C., 2013. Land-Sea Characterization of the St. Croix East End Marine Park, U.S. Virgin Islands. NOAA Technical Memorandum NOS NCCOS 170, Silver Spring, MD.

Island Resources Foundation. 2002. Resource Description Report. University of the Virgin Islands and Department of Planning and Natural Resources. USVI.

Sherman, C., Nemeth, M.I., Ruíz, H., Bejarano, I., Appeldoorn, R., Pagán, F., Schärer, M., Weil, E., 2010. Geomorphology and benthic cover of mesophotic coral ecosystems of the upper insular slope of southwest Puerto Rico. Coral Reefs 29, 347–360.

Smith, T.B., Blondeau, J., Nemeth, R.S., Calnan, J.M., Kadison, E., Gass, J., 2010. Benthic structure and cryptic mortality in a Caribbean mesophotic coral reef bank system, the Hind Bank Marine Conservation District, U.S. Virgin Islands. Coral Reefs 29, 289–308.

The Nature Conservancy (TNC). 2002. St. Croix East End Marine Park management plan. University of the Virgin Islands and Department of Planning and Natural Resources. USVI.

Tonioli, F.C., Agar, J.J., 2011. Synopsis of Puerto Rican Commercial Fisheries. NOAA Technical Memorandum NMFS-SEFSC-622.

Valdés-Pizzini, M., Schärer-Umpierre, M.T., 2011. Una mirada al mundo de los Pescadores: una perspectiva global. Una publicación de asuntos de política pública del Programa Sea Grant de la Universidad de Puerto Rico. UPRSG-G-209.

Valdés-Pizzini, M., Schärer-Umpierre, M.T., 2014. People, Habitats, Species, and Governance: An Assessment of the Social-Ecological System of La Parguera, Puerto Rico. Interdisciplinary Center for Coastal Studies, University of Puerto Rico, Mayagüez, PR.

Valdés-Pizzini, M., Agar, J.J., Kitner, K., García-Quijano, C., Trust, M., Forrestal, F., 2010. Cruzan fisheries: a rapid assessment of the historical, social, cultural and economic processes that shaped coastal communities' dependence and engagement in fishing in the island of St. Croix, U.S. Virgin Islands. NOAA Series on U.S. Caribbean Fishing Communities, NOAA Technical Memorandum NMFS-SEFSC-597.

Valdes-Pizzini, M., Schärer-Umpierre, M.T., García-Quijano, C.G., 2012. Connecting humans and ecosystems in Tropical Fisheries: social sciences and the ecosystem-based fisheries management in Puerto Rico and the Caribbean. Caribb. Stud. 40, 95–128.

Waddell, J.E., Clarke, A.M. (Eds.), 2008. The State of Coral Reef Ecosystems of the United States and Pacific Freely Associated States: 2008. Silver Spring, MD, NOAA Technical Memorandum NOS NCCOS 73. NOAA/NCCOS Center for Coastal Monitoring and Assessment's Biogeography Team.

Weil, E., 2005. Puerto Rico. In: Miloslavich, P., Klein, E. (Eds.), Caribbean Marine Biodiversity: The Known and the Unknown. DEStech Publications Inc., Pennsylvania, USA.

Wing, S.R., Wing, E.S., 2001. Prehistoric fisheries in the Caribbean. Coral Reefs 20, 1–8.

CHAPTER FIVE

Understanding the Scale of Marine Protection in Hawai'i: From Community-Based Management to the Remote Northwestern Hawaiian Islands

Alan M. Friedlander*,†, Kostantinos A. Stamoulis*, John N. Kittinger‡, Jeffrey C. Drazen§, Brian N. Tissot¶

*Fisheries Ecology Research Laboratory, Department of Biology, University of Hawaii, Honolulu, HI, USA
†National Geographic Society, Washington, DC, USA
‡Conservation International, Betty and Gordon Moore Center for Science and Oceans, Honolulu, HI, USA
§Department of Oceanography, University of Hawaii, Honolulu, HI, USA
¶Marine Laboratory, Humboldt State University, Trinidad, CA, USA

Contents

Advances in Marine Biology, Volume 69
ISSN 0065-2881
http://dx.doi.org/10.1016/B978-0-12-800214-8.00005-0

153

Abstract

Ancient Hawaiians developed a sophisticated natural resource management system that included various forms of spatial management. Today there exists in Hawai'i a variety of spatial marine management strategies along a range of scales, with varying degrees of effectiveness. State-managed no-take areas make up less than 0.4% of nearshore waters, resulting in limited ecological and social benefits. There is increasing interest among communities and coastal stakeholders in integrating aspects of customary Hawaiian knowledge into contemporary co-management. A network of no-take reserves for aquarium fish on Hawai'i Island is a stakeholder-driven, adaptive management strategy that has been successful in achieving ecological objectives and economic benefits. A network of large-scale no-take areas for deepwater (100–400 m) bottomfishes suffered from a lack of adequate data during their initiation; however, better technology, more ecological data, and stakeholder input have resulted in improvements and the ecological benefits are becoming clear. Finally, the Papahānaumokuākea Marine National Monument (PMNM) is currently the single largest conservation area in the United States, and one of the largest in the world. It is considered an unqualified success and is managed under a new model of collaborative governance. These case studies allow an examination of the effects of scale on spatial marine management in Hawai'i and beyond that illustrate the advantages and shortcomings of different management strategies. Ultimately a marine spatial planning framework should be applied that incorporates existing marine managed areas to create a holistic, regional, multi-use zoning plan engaging stakeholders at all levels in order to maximize resilience of ecosystems and communities.

Keywords: Hawai'i, MPAs, Scale, Community-based management, Aquarium fishery, Marine spatial planning, Overfishing, Governance

ABBREVIATIONS

BRFA bottomfish restricted fishing area
CBSFA community–based subsistence fishing area
DAR Division of Aquatic Resources
DLNR Department of Land and Natural Resources
FMA Fisheries Management Area
FRA fish replenishment area
KIR Kaho'olawe Island Reserve
MHI Main Hawaiian Islands
MLCD marine life conservation district
MMA marine managed area
MPA marine protected area
MSP marine spatial planning
NOAA National Oceanic and Atmospheric Administration
NWHI Northwestern Hawaiian Islands
PMNM Papahānaumokuākea Marine National Monument
USFWS United States Fish and Wildlife Service
WHFC West Hawai'i Fishery Council
WPRFMC West Pacific Regional Fisheries Management Council

1. INTRODUCTION

1.1. Bio-physical description

The Hawaiian Archipelago consists of two regions: the populated main Hawaiian Islands (MHI), and the mostly uninhabited atolls, islands, and banks of the Northwestern Hawaiian Islands (NWHI). The archipelago extends from the island of Hawai'i (19°N) northwest to Kure Atoll (28°N), a distance of over 2500 km (Figure 5.1). This vast expanse is connected by geological origin and geographic isolation, and is subject to large spatial gradients in oceanography, erosion, and geomorphology (Grigg, 1997; Juvik et al., 1998). The MHI consist of eight high volcanic islands that range in age from active lava flows on the east side of Hawai'i Island to 7 million-year-old Kaua'i (Juvik et al., 1998). Beginning at Nihoa and Mokumanamana (Necker Island) (about 7 and 10 million years old, respectively) and extending to Midway and Kure atolls (both about 28 million years old), the NWHI represents the older portion of the emergent Hawaiian Archipelago (Grigg, 1997; Grigg et al., 2008).

The Hawaiian Archipelago resides in the middle of the North Pacific Subtropical Gyre and is exposed to large open ocean swells and strong trade winds that have a major impact on the structure of the nearshore marine ecosystems, with distinctive communities being sculpted by these dynamic natural processes (Dollar, 1982; Gove et al., 2013; Grigg, 1983). At the northern end of the chain, Kure is the world's highest latitude atoll and is located at the "Darwin Point" where coral accretion is balanced by losses due to bioerosion, mechanical erosion, and subsidence (Grigg, 1982, 1997). Circulation is primarily from east to west and intensifies to the south, however, in the lee of the islands, surface currents driven by wind combine with large-scale ocean currents to yield more complicated flow patterns such as eddies (Flament et al., 1996; Lobel and Robinson, 1986).

The Hawaiian Archipelago occupies its own province in the tropical Indo-West Pacific region (Briggs and Bowen, 2012). The geographic isolation of Hawai'i has resulted in some of the highest endemism of any tropical marine ecosystem on the Earth (Jokiel, 1987; Kay and Palumbi, 1987; Randall, 1998). Some of these endemics are dominant components of the nearshore marine community, resulting in a unique ecosystem that has extremely high biodiversity and conservation value (DeMartini and Friedlander, 2004; Maragos et al., 2004).

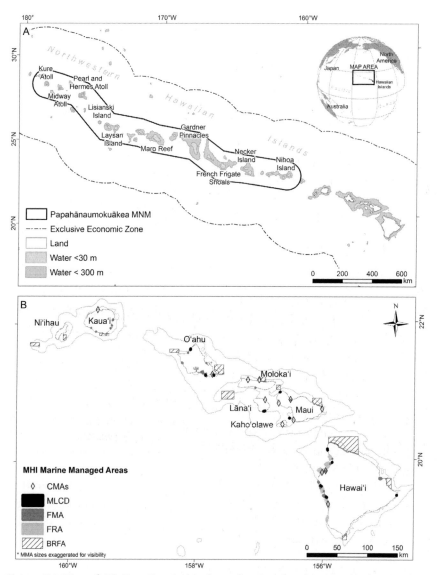

Figure 5.1 Map of (A) Hawaiian Archipelago showing Papahānaumokuākea Marine National Monument (PMNM), (B) Main Hawaiian Islands (MHI) showing locations of marine managed areas. These include community-based co-management areas (CMA), state marine life conservation districts (MLCDs), state Fisheries Management Areas (FMA), Fish Replenishment Area network (FRA), and the bottomfish restricted fishing areas (BRFA).

Hawai'i provides habitat for a wide variety of species protected by federal acts and state statutes. Seabird colonies in the NWHI constitute one of the largest and most important assemblages of seabirds in the world, with approximately 14 million birds representing 21 species (Friedlander et al., 2008; Harrison, 1990). The Hawaiian monk seal (*Monachus schauinslandi*) is the only endangered pinniped occurring entirely within US waters, with a current population estimated at only 1200 seals—a decrease of about 60% since the 1950s (Antonelis et al., 2006). The Hawaiian green turtle stock comprises a single closed genetic stock that is endemic to the Hawaiian Archipelago (Bowen et al., 1992). Stock size in the 1970 were estimated to be at 20% of pre-exploitation biomass but protection since that time has resulted in population levels >80% of pre-exploitation levels with a 5% intrinsic growth rate (Chaloupka and Balazs, 2007).

1.2. Historical use and management of marine resources

Hawaiians of old (pre-western contact, <AD 1778) developed sophisticated and complex management systems for marine resource use (Friedlander et al., 2013; 'Ī'ī, 1993; Kahā'ulelio, 2006; Kamakau, 1976; Malo, 1951). These societies depended on fishing and gathering for survival, which motivated them to acquire a sophisticated understanding of the factors that caused limitations and fluctuations in their marine resources. In traditional Hawaiian society, the basic unit of land division and socioeconomic organization was the ahupua'a, which generally encompassed a watershed catchment unit that included interior uplands through valleys into the sea and was managed adaptively according to resource availability, life cycles, and fluctuations (Kaneshiro et al., 2005; Kirch, 1989). Ahupua'a units were nested within districts (moku) that were hierarchical and roughly corresponded to bio-physical attributes of island ecosystems (e.g. windward/leeward and wet/dry districts of islands; Malo, 1951).

At the local (ahupua'a) and district (moku) levels, fishing activities were strictly regulated by a system of rules that were embedded in socio-political structures and religious systems (the kapu system) (Malo, 1951; Poepoe et al., 2007). While the basic unit of land management was the ahupua'a, the basic unit of marine resource management and harvesting was the moku, or district (McGregor, 2007). Under this management regime, Hawaiian communities were able to maintain a high level of productivity and fisheries yield over several centuries prior to Western contact (McClenachan and Kittinger, 2013).

Following Western contact, a variety of socio-political factors led to the demise of the traditional system of resource management in the late eighteenth to early nineteenth centuries (Friedlander et al., 2013; Ralston, 1984; Seaton, 1974). The annexation by the United States and the Organic Act of 1900 that followed resulted in the erosion of traditional fishing rights, which ultimately created open-access to coastal fisheries for residents and non-residents alike (Kosaki, 1954; Tanaka, 2008). The early 1900s also saw the centralization of economic activities and fisheries markets in Honolulu and large increases in the commercial landing of marine resources (Bell and Higgens, 1939; Cobb, 1901).

Just prior to World War II, commercial fishing in Hawai'i was a multi-million dollar industry that employed hundreds directly and thousands indirectly. Subsistence and commercial fishing pressure increased due to the post-war growth in population, increases in boat ownership, introduction of export-driven fisheries (e.g. aquarium trade, tuna), and other technological advances, such as refrigeration, which still continue today (Kittinger, 2010; Schug, 2001). Following statehood, Hawai'i saw a rapid growth in tourism, an increasingly urban resident population, and the continued development of shoreline areas for tourism and recreation, which resulted in changes in the character of the coastal fisheries as they became dominated by recreational anglers and a greater number of part-time commercial fishers who curtailed their fishing to take advantage of more lucrative economic activities (Friedlander, 2004; Shomura, 2004).

1.3. Contemporary use and management of marine resources

In 2012, Hawai'i's fishing industry generated US $91.5 million from 13.3 million kg of fish, ranking it twelfth in value among US states (National Marine Fisheries Service, 2014). Residents of Hawai'i have the highest per capita seafood consumption in the United States with an annual total of >17.6 million kg (Geslani et al., 2012). The longline fishery for pelagic species, primarily bigeye tuna (*Thunnus obesus*) and swordfish (*Xiphias gladius*), accounts for the vast majority of the catch by value (National Marine Fisheries Service, 2014). Longline fishing is prohibited within 80–120 km from shore, depending on the location and time of year. Although pelagic fisheries are by far the most important economically, Hawai'i's non-pelagic fisheries have substantial cultural, subsistence, commercial, and recreational value (Lowe, 2004; Pooley, 1993).

Much of Hawai'i's marine habitat is deep (>100 m) in contrast to continental regions elsewhere that have broad shelves. These deeper waters show high consistency in hydrographic conditions since they are below the permanent thermocline. Dramatic changes in biological communities are observed with depth but relatively few changes occur horizontally (Chave and Mundy, 1995; Yeh and Drazen, 2009). In the 1960s, aggregations of pelagic armorhead (*Pentaceros richardsoni*) were found at the Hancock Seamounts to the NW of Kure Atoll and heavily exploited at depths up to 500 m (Uchida and Tagami, 1984). This fishery collapsed by the early 1980s and briefly switched to alfonsinos, *Beryx decadactylus* (to about 1000 m), and pink coral for the jewellery trade (Clark and Koslow, 2008). Other seamounts nearby were also exploited until bottom trawling within the Hawaiian Archipelago was banned in 2004 (Hawai'i Administrative Rules, 2004). Throughout the archipelago, a hook-and-line fishery for deep water snappers and groupers has existed for decades from depths of 100–400 m (Haight et al., 1993), and is the second-most valuable commercial fishery in Hawai'i. In addition to the commercial catch, the non-commercial catch for this fishery from 1950 to 2005 was estimated to be over two times higher than reported commercial landings (Hospital and Beavers, 2012; Zeller et al., 2008).

Nearshore fisheries constitute a mix of commercial, recreational, and subsistence sectors that land a diverse catch. The Hawai'i marine aquarium fishery is one of the state's most lucrative nearshore fisheries with an annual reported value of over $2 million (Walsh et al., 2013). Although the true economic value of this fishery was estimated to be two to five times higher than reported values in the past (Cesar and van Beukering, 2004; Walsh et al., 2003), recent analysis indicate under-reporting by collectors is not significant (Walsh et al., 2013). The major coastal commercial fishery in Hawai'i by weight is the net fishery for bigeye scad (akule, *Selar crumenopthalmus*), along with mackerel scad (opelu, *Decapterus* spp.). This fishery accounts for nearly 80% of the entire coastal catch, with commercial fishers reporting nearly 388,000 kg of akule and opelu landed in 2010.

It is difficult to separate nearshore fisheries into sectors, as fishers can engage in multiple activities—both commercial and non-commercial—in a single trip (Glazier, 2007). Non-commercial fishing includes subsistence/consumptive, recreational, and cultural fishing and gathering activities that occur in open ocean and nearshore coastal zones. Non-commercial fishing is the most prevalent type of extractive activity on most coral reefs in Hawai'i (Geslani et al., 2012; Kittinger, 2013). However, the catch is largely unreported or undocumented and can substantially exceed reported commercial landings (Hospital et al., 2011; Zeller et al., 2008).

Furthermore, recreational and subsistence fishers take more species using a wider range of fishing gear (Friedlander and Parrish, 1997).

Hawai'i's nearshore marine environment provides numerous ecosystem services and is vital to the state's approximately $800 million per year marine tourism industry (Friedlander et al., 2008). The economic value of Hawai'i's coral reefs was estimated at $10 billion with direct economic benefits of $360 million per year in 2002 (Cesar and van Beukering, 2004). Hawai'i's near-shore resources also have cultural importance for the Native Hawaiian community. The continuance of subsistence fishing activities and associated socio-cultural practices are critical to the transfer of Native Hawaiian culture to subsequent generations (Kikiloi and Graves, 2010; McGregor et al., 1998, 2003).

Despite their economic and cultural significance, reefs near urbanized areas have declined due to a variety of human-mediated pressures. Reef fish populations and their associated fisheries have declined dramatically around Hawai'i due to intensive fishing pressure, land-based pollution, destruction of habitat, invasive species, and other threats. These are driven by a growing human population, export-driven markets for resources, access to technological innovations (e.g. motorized boats and freezers for storing catch), and introduction of new and overly efficient fishing techniques (e.g. inexpensive monofilament gill nets, SCUBA, GPS) (Friedlander, 2004; Friedlander et al., 2003, 2013; Shomura, 1987; Smith, 1993; Williams et al., 2008). Furthermore, there is poor compliance with state fishing laws and regulations and insufficient enforcement, which is partially attributed to lack of resources, capacity, and political will (Tanaka et al., 2012; Tissot et al., 2009).

1.4. Marine protected areas in Hawai'i

Today, myriad state and federal authorities provide for the management of Hawai'i's coastal resources (Lowry et al., 1990) that primarily rely on top-down governance approaches implemented by government resource agencies and managers (Kittinger, 2013). The State of Hawai'i has numerous marine protected areas (MPAs) and other marine managed areas (MMAs)—natural area reserves, fisheries management areas, marine life conservation districts (MLCDs), various protective subzones, military defensive areas, and National Park coastlines (Figure 5.1B). Hawai'i established its first MPAs over 45 years ago. Since that time, many MPAs and MMAs have been created with varying levels of protection ranging from complete 'no-take' areas to areas that allow a wide variety of activities to

Table 5.1 Percent of total area by island restricted to fishing in the main Hawaiian Islands nearshore marine (0–18 m) waters by gear type

Location	No/negligible fishing/access	Some fishing permitted	Laynet	Spear	Pole and line	Throw-net	AQ fishing
MHI total	**4.8**	**95.2**	**72.5**	**94.9**	**94.7**	**94.4**	**92.0**
Hawai'i	0.2	99.8	85.5	98.9	96.7	96.5	81.9
Kahoʻolawe	100.0	–	–	–	–	–	–
Kauaʻi	5.9	94.1	93.9	94.1	94.1	94.0	93.9
Lānaʻi	–	100.0	96.5	96.5	100.0	96.5	96.5
Maui	1.7	99.3	–	98.3	98.3	98.3	98.3
Molokaʻi	–	100.0	99.9	100.0	100.0	100.0	99.9
Molokini	100.0	–	–	–	–	–	–
Niʻihau	–	100.0	100.0	100.0	100.0	100.0	100.0
Oʻahu	6.3	93.7	67.0	93.3	93.7	93.3	93.0

AQ, aquarium fish fishery.

occur within their boundaries (Table 5.1). Designation of many of these areas was not based on comprehensive biological selection criteria or a systematic ecological assessment. Rather, the existing system was built piecemeal and is reflective of various needs to manage user conflicts, safeguard protected species, or on the wishes of local communities (Friedlander et al., 2007a,b). Below, we present five case studies detailing different spatial scales of marine management in Hawai'i, which are organized starting from local scale (state-managed MMAs: <1 km^2) to archipelagic scale (Marine National Monument: >100,000 km^2). This comparison enables a detailed examination of the effect of scale on various aspects of marine protection in Hawai'i with potential applications across the globe.

2. MPA CASE STUDIES

2.1. Marine managed areas

2.1.1 Establishment

Within the MHI, there are at least 33 state-managed areas that limit fishing activities in nearshore marine waters, with an average area of 1.0 (0.01–6.2) km^2 and a total area of 33.8 km^2 (Figure 5.2A). Hanauma Bay Nature Reserve was established in 1967 and is likely the most visited MPA

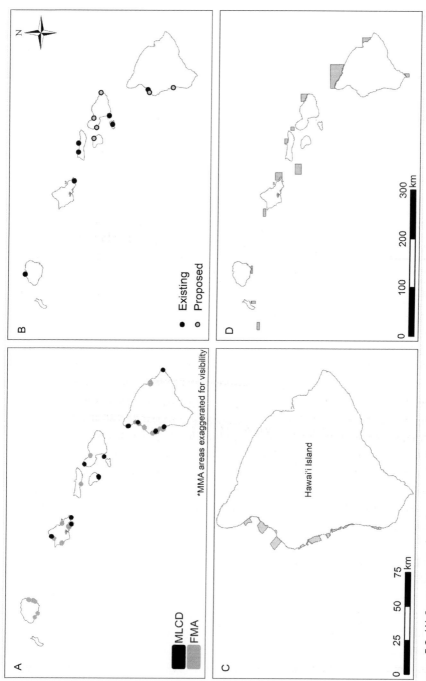

Figure 5.2 (A) State managed marine areas in the main Hawaiian Islands including marine life conservation districts (MLCDs) and Fisheries Management Areas (FMAs), (B) locations of community managed/co-management areas, (C) West Hawai'i Aquarium Fish MPA network (FRAs), and (D) bottomfish restricted fishing areas (BFRAs).

in the world with more than 1 million annual visitors in a 0.4-km^2 area. In addition to 11 MLCDs (areas designed to conserve and replenish marine life) and 22 Fisheries Management Areas (FMAs—areas designed to resolve conflicts among users, including fishers), members of the public have limited or no-access to the shoreline and nearshore waters within and around military or security areas on O'ahu and Kaua'i (Pearl Harbor, Kaneohe Bay Marine Corps Base Hawai'i, Barking Sands Pacific Missile Range Facility, and Honolulu Reef Runway) or in Hawai'i Volcanoes National Park on Hawai'i Island.

MLCDs are established by the state's Department of Land and Natural Resources (DLNR), as authorized by statute. Suggestions for areas to be included in the MLCD system may come from the State Legislature or general public. In addition, the DLNR 's Division of Aquatic Resources (DAR) regularly conducts surveys of marine ecosystems throughout the state, and may recommend MLCD status for areas that appear particularly promising. Criteria for designating MLCDs include: (1) the marine life and its potential for increase, (2) its "pristine state", (3) compatibility with existing uses within and adjoining the MLCD, (4) geological features that provide well-defined boundaries for enforcement, and (5) the site's ability to support public safety and accessibility from the shoreline (DAR, 1992).

The large number of restricted-access or restricted-fishing areas in the MHI gives the impression of a substantial network of actively managed and protected marine areas, but in reality the majority of these areas are small, and nearly all allow some or several forms of fishing within their boundaries. MLCDs are the most restrictive of protected area designations in the State of Hawaii, but some types of fishing are permitted within 6 of the 11 existing MLCDs. The proportion of nearshore MHI waters in no-take and negligible-take areas including fully protected MLCDs, extremely limited access reserves, and no-access zones is only 4.8% (Table 5.1). The large majority of this is in military and security no-access zones around O'ahu and Kaua'i, or in the Kaho'olawe Island Reserve (KIR). Therefore, the extent of complete no-take areas on other islands is extremely limited, with only 0.4% of nearshore MHI waters less than 18 m depth (an approximation of inshore habitats that are the primary targets for fishing of reef and reef-associated species) are within no-take MPAs. Nearly, 70% of nearshore waters are not spatially managed for fishing or specially restricted in any way (Table 5.2).

2.1.2 Ecological performance

A comprehensive examination of existing MLCDs showed that areas fully protected from fishing had higher fish biomass, larger overall fish size,

Table 5.2 Marine managed and restricted-access areas containing nearshore (0–18 m) marine waters in the main Hawaiian Islands

Location	Area (km²)	No-take MMA	State regulated areas-partial closure MMA	Lay gill-net prohibited area	Little/ no-access	Restricted access	No spatial management
MHI total	**1074.8**	**0.4**	**5.3**	**22.9**	**2.7**	**2.1**	**68.7**
Hawai'i	174.7	0.2	19.0	8.1	–	3.1	72.9
Kahoʻolawe	18.1	–	100.0	–	–	–	0.0
Kaua'i	166.6	–	0.4	–	5.9	3.8	90.0
Lāna'i	32.2	–	3.5	–	–	–	96.5
Maui	133.6	1.7	–	100.0	–	–	0.0
Moloka'i	141.7	–	0.1	–	–	–	99.9
Molokini	0.3	100.0	–	–	–	–	100.0
Ni'ihau	83.3	–	–	–	–	–	100.0
O'ahu	324.7	0.5	1.2	30.4	5.9	3.2	63.3

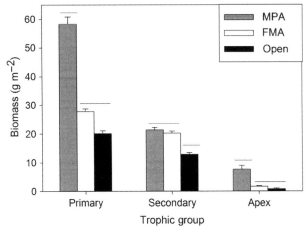

Figure 5.3 Comparisons of fish biomass by trophic group among marine managed areas and areas open to fishing in Hawai'i. MLCDs, marine life conservation districts; FMA, Fisheries Management Areas. (Overall ANOVA—$F_{2,2255} = 10.42$, $P < 0.05$). Primary, primary consumer; Secondary, secondary consumer; Apex, apex predator. Horizontal lines above bars show management types that are not significantly different for each trophic group at $\alpha = 0.05$ (Tukey's HSD tests). *Adapted from Friedlander et al. (2007a).*

and higher biodiversity than adjacent areas of similar habitat quality (Friedlander et al., 2007a,b). Overall fish biomass was 2.6 times greater in the MLCDs compared to open areas. In addition, apex predators and other trophic groups were more abundant and larger in the MLCDs (Figure 5.3), illustrating the effectiveness of these closures in conserving fish populations within their boundaries. The differences in biomass among management types for all three trophic groups reflects the fact that in Hawai'i, overfishing occurs at all trophic levels with targeted species occurring across all trophic group. Habitat type, protected area size, and level of protection from fishing were all important determinates of MLCD effectiveness with respect to their associated fish assemblages (Friedlander et al., 2007a,b). Overall, MLCDs protected from fishing that had high habitat complexity and good habitat quality (e.g. high coral cover and low macroalgae cover) had higher values for most fish assemblage characteristics. Areas that only provided partial protection from fishing due to rotating closures or other means were no more effective than areas completely open to fishing (Williams et al., 2006).

MPAs can supplement adjacent fisheries through increased production and export of pelagic eggs and larvae, and net emigration of adults and juveniles, otherwise known as spillover (Gaines et al., 2010; McClanahan and Mangi, 2000; Russ, 2002). Stamoulis and Friedlander (2013) measured adult spillover of fish species from Pūpūkea-Waimea MLCD on the north shore of O'ahu and found a significant negative gradient of resource fish biomass across the protected area boundary extending nearly 1 km into the fished area.

MLCDs in Hawai'i were established to support the state's conservation and education objectives, not to enhance fish stocks. As a consequence, most of the MLCDs are currently too small to provide noticeable fisheries benefits (Friedlander et al., 2007a). Their small size and limited habitat types do not allow for the entire fish assemblage to function in a natural manner com-pared to large and relatively pristine areas such as the NWHI. Closing areas to fishing is far from a new idea in the management of marine resources. Pacific Islanders traditionally used a variety of closures that were often imposed to ensure large catches for special events or as a cache for when resources on the usual fishing grounds ran low (Johannes, 1978, 1981; Jupiter et al., 2012; Ruddle, 1996). Rotational closures have been less suc-cessful in contemporary Hawai'i where there are few or no controls on effort once the area is open to fishing. The Waikīkī -Diamond Head FMA rota-tional closure has been an ecological failure with fish biomass tending to increase slightly during the 1- to 2-year closure periods, but the scale of these increases is insufficient to compensate for declines during open periods (Williams et al., 2006). The net effect was that, between 1978 and 2002, total biomass declined by around two-thirds. Coincident with this decline was the virtual disappearance of larger fishes (>40 cm) of fishery-target groups. This management action has created a 'derby' mentality where fishing effort is greatly intensified in a rush to fish once these areas are re-opened.

2.1.3 Socio-economic performance

Marine ecosystems generate a wide range of goods and services that benefit Hawaiian society, including supporting important livelihood and food pro-visioning functions, as well as cultural practices, customs, and traditions. Declining reef health threatens the societal benefits that these ecosystems provide (Bell et al., 2011; Sadovy, 2005). Currently, less than 1% of the state's budget is directed towards natural resource management, despite a high reliance on ecosystem health to support the state economy's depen-dency on tourism. Hanauma Bay MLCD, for example, receives more than 1 million visitors and generates more than $35 million annually. The net benefits (including direct and indirect expenditures and future willingness to pay) greatly exceeds the net-costs to society (Figure 5.4; Cesar and van Beukering, 2004).

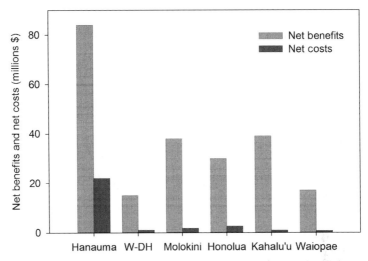

Figure 5.4 Net benefits and costs for six marine managed areas (MMAs) in Hawai'i. W-DH, Waikīkī Diamond Head Fisheries Management Area (FMA). Cost estimates combine the investment costs and recurrent costs associated with expenditures in services, education/awareness, monitoring/assessment, enforcement/compliance, and other costs such as infrastructure. Overall benefits were estimated by aggregating recreational, fishery, and educational benefit values. *Adapted from van Beukering and Cesar (2004).*

The economic value of six MMAs in Hawai'i ranges from $6 million for the Waikīkī -Diamond Head Fisheries Management Area to $650 million for Hanauma Bay MLCD (van Beukering and Cesar, 2004). A cost-benefit analysis of management options among MMAs found that the economic benefit–cost ratio was very high ($\overline{X} = 18.8 \pm 13.0$, range: 3.8–40.5; van Beukering and Cesar, 2004). This means that assuming people pay their stated willingness-to-pay value on entry to the MMA, over time the benefits outweigh the costs by a factor of 40, although the absolute values should be treated with caution.

Data from nearly 1000 users at Pūpūkea MLCD showed that most users had protectionist (i.e. biocentric, nature-centered) value orientations towards reefs (Needham et al., 2008). Overall satisfaction was extremely high and despite moderate-to-high crowding at some sites, most users encountered fewer people than their maximum tolerance. A similar study at Molokini MLCD with over 1000 users surveyed found that almost all had biocentric values towards the environment in general and protectionist-specific values towards coral reefs (Szuster and Needham, 2010). Results showed that overall satisfaction of visitors to Molokini MLCD was extremely high, although a large proportion were dissatisfied with the inability to escape crowds and that they did not learn about the history of the area or Hawaiian culture.

2.1.4 Overview and future prospects

There is strong opposition to the creation of additional MPAs in Hawai'i by the large and vocal fishing community. Fishing is a large part of the local culture and fishers often view MPAs as having a direct negative impact on their activities—this is often exacerbated by prevailing opinions that land-based pollution comprises a larger threat to nearshore marine ecosystems than overexploitation (e.g. for Maunalua Bay, Kittinger, 2013). While most marine reserves in Hawai'i are either too small or poorly placed to generate significant fishery benefits, spillover of juvenile and adult fishes from protected areas can be a very tangible benefit of MPAs that may serve to improve perception of this type of management among fishers (Russ and Alcala, 1996). Williams et al. (2009) and Stamoulis and Friedlander (2013) provide examples of adult fish spillover from MPAs in Hawai'i. This information should be made accessible to the fishing public along with education about the less tangible, though greater fisheries benefits, provided by larval export from MPAs (Palumbi, 2004; Sladek Nowlis and Friedlander, 2005). The negative perception of MPAs in Hawai'i is perhaps the greatest obstacle to the use of this valuable management tool. This issue can be addressed through public relations efforts and a re-branding of MPAs to something more palatable to fishers. In addition, managers need to engage the fishing community in an equitable stakeholder participatory approach, which includes alternatives such as technical measures, evaluated on a case-by-case basis.

To realize the full fishery benefits of MPAs in Hawai'i, substantial increases in size and number of protected areas will need to occur. To support higher fish biomass and greater numbers and diversity of species, future protected area in the MHI should include a mosaic of habitats with a range of complexities and depths to accommodate the wide range of species found in Hawaiian waters (Friedlander et al., 2007a). In addition, consideration should be given to the habitat requirements and life histories of the species being protected, the level of fishing and other pressures on the resources in adjacent areas, and the degree of enforcement (Foley et al., 2013).

The State of Hawai'i should systematically create a statewide network of MMAs encompassing existing MMAs that utilizes an ecosystem-based approach and direct community stewardship. Public participation from the beginning of the process and long-term community co-management with DLNR is essential for success. The State Legislature must exhibit the vision to move forward expeditiously and provide a welcoming venue for all stakeholders committed to healthy ocean ecosystems (Antolini et al., 2003).

2.2. Community-based management

2.2.1 Establishment

In Hawai'i, there is increased interest among communities and coastal stakeholders in integrating aspects of Native Hawaiian knowledge systems and customary practices into contemporary management (Kittinger et al., 2012). Communities have increasingly explored the development of co-management partnerships between state resource management agencies and community groups to incorporate aspects of traditional ecological knowledge and customary marine tenure and to devolve some management authority to local scales where it was traditionally based (Friedlander et al., 2013). Communities can enter into a co-management relationship with the State of Hawai'i either through the legislative process (e.g. as a stand-alone legislative act) or by working directly with DLNR through its administrative rule-making process to establish a community-based subsistence fishing area (CBSFA), for the purposes of reaffirming and protecting fishing practices customarily and traditionally exercised for purposes of Native Hawaiian subsistence, culture, and religion (Kittinger et al., 2012). The state of Hawai'i passed legislation for the designation of CBSFAs in 1994 with the intent of revitalizing local fisheries through customary Hawaiian practices and tenure. The CBSFA legislation was specifically directed towards Native Hawaiian communities "for the purpose of reaffirming and protecting fishing practices customarily and traditionally exercised for purposes of Native Hawaiian subsistence" (Hawaii Revised Statutes, HRS, 2005: Chapter 188–22.6).

Co-management can take many forms but generally involves shared management authority and responsibility between resource users or community groups at the local level and governmental agencies (Berkes, 2010). For the purpose of this analysis, we considered two categories of co-management areas: existing co-management areas which have been designated as CBSFAs though are awaiting approval of their management plans (Hā'ena, Kaua'i and Mo'omomi, Moloka'i) or areas where state or federal management co-exists with community stewardship (e.g. Kalaupapa, Moloka'i; Kaho'olawe; 'Ahihi-Kina'u, Maui) and co-management areas which are proposed through the CBSFA legislature or other MMA mechanisms (e.g. Wailuku & Hana, Maui; Maunalei, Lāna'i; Ka'ūpūlehu, Hawai'i) (Figure 5.2B). We used a standard depth range of 0–18 m bounded by watershed boundaries or those specified in management plans to map co-management areas. The existing co-management areas have an average area of 10.7 km^2 and a total area of 74.5 km^2, and the proposed areas make up another 29.2 km^2.

The island of Kaho'olawe is a special case; it was a *de facto* marine reserve during the US Military bombing era, and since 1990, it has been under the administration of the state's Kaho'olawe Island Reserve Commission (KIRC), with only limited take of marine life permitted for cultural, spiritual, and subsistence purposes in an 18-km^2 area, making it the largest area protected from most fishing in the MHI (Friedlander et al., 2013). For the purpose of this comparison, we considered Kaho'olawe as an existing co-management area. Ni'ihau is the smallest inhabited island in Hawai'i and is privately owned with a resident population of about 130 Native Hawaiians. Ni'ihau has no stores, and inhabitants fish and farm for subsistence (Tava and Keale, 1990). Although no formal rules have yet to be established on Ni'ihau, the community has developed general guidelines for permitted fishing activities through local peer pressure, and those visiting from outside are encouraged to follow these guidelines (Friedlander et al., 2013). Because of the lack of a formal management plan, we did not include Ni'ihau in our estimate of total existing co-management areas though it would double the estimate of community managed areas with a nearshore area of 83.3 km^2.

2.2.2 Ecological performance

Scientific surveys of various locations around Hawai'i show that locations under community-based management with customary stewardship harbour fish biomass equal to or greater than that found in many MPAs in Hawai'i and substantially greater than areas open to fishing (Friedlander et al., 2002, 2003, 2013). These results are consistent with findings by McClanahan et al. (2006) when comparing MPAs and collaborative management areas in Indonesia and Papua New Guinea. Owing to the lack of formal rules associated with many of these community managed areas, enforcement is typically through informal means including self-regulation and via local peer pressure and site-based monitoring of activities and resource condition. A number of these locations are in remote areas with limited access, thus allowing the community greater control over these resources and also potentially reducing overall fishing pressure.

2.2.3 Socio-economic performance

Despite interest from more than 19 communities, in the nearly 20 years since the act allowing designation of CBFSAs was passed, only two communities have successfully designated CBSFAs, and none currently have an approved management plan (Higuchi, 2008; Kittinger et al., 2012; Levine and Richmond, 2014). Nonetheless, community interest in co-management remains quite high (Ayers and Kittinger, 2014). This interest derives from several sources—first, by transferring some authority to the local level, co-management is more aligned with traditional forms of government, which endowed local resource managers with the authority to develop and implement place-based management efforts (Higuchi, 2008; Kittinger et al., 2012; Poepoe et al., 2007). In this way, co-management is viewed by many community members as more legitimate than top-down forms of governance. Second, local management can be more responsive to community needs. For example, one of the basic functions of nearshore fisheries in Hawai'i and elsewhere in the Pacific is to provide a source of seafood (Vaughan and Vitousek, 2013). As with MPAs, co-management areas can be highly managed, but unlike MPAs, they provide opportunities for harvest, providing food provisioning and cultural services to communities. A burgeoning literature documents these important functions (e.g. Cinner and Aswani, 2007; Kittinger, 2013; Vaughan and Vitousek, 2013), and community-based management can be tailored to meet community goals for fisheries. In addition, co-management areas lend themselves well to adaptive management because their rate of change is limited only by the capacity of the co-managers to accept it.

Co-management planning can also carry significant social costs. As the co-management planning process is arduous, it requires significant resources from communities, the state, and bridging organizations such as non-governmental organizations. Further, the process can be stymied by a variety of factors, including lack of human, financial, and organizational capacity to successfully engage in the planning and implementation process. In Hawai'i, there are two instances—Mo'omomi and Hā'ena—where communities self-organized, built consensus around a management plan, and collectively acted to achieve a modicum of decision-making over resource rules in their area (Friedlander et al., 2013; Poepoe et al., 2007). Despite the presence of enabling legislation, and in some cases extraordinary community effort and collective action, co-management in Hawai'i has been hindered by a lack of capacity in communities and at the state management agency, institutional culture and rigidity at the partner resource management agency, and an ambiguous, complicated administrative rule-making process (Ayers and Kittinger, 2014).

2.2.4 Overview and future prospects

Unfortunately, implementation of the CBSFA legislation has not lived up to expectations due to many challenges and has so far failed to be fully implemented in any community (Levine and Richmond, 2014; Ayers and Kittinger, 2014), although interest in developing local-state partnerships and devolving authority to community levels still remains very high among coastal stakeholders. Despite numerous obstacles to formal governmental authorization, a number of communities are currently strengthening local influence and accountability for local marine resources through revitalization of local traditions and resource knowledge (Friedlander et al., 2013).

The return to the local scale of management represents a form of contemporary adaptation of traditional management practices to modern governance contexts (Poepoe et al., 2007). There are several important challenges that hinder effective implementation of co-management legislation and policy. These include developing a standard operating procedure for the State of Hawai'i to engage fruitfully with communities, developing a viable model of practise to build community capacity to plan for and engage in co-management, and resourcing these efforts through a diverse set of partnerships and funding mechanisms (Ayers and Kittinger, 2014; Gutiérrez et al., 2011; Levine and Richmond, 2014; Ostrom et al., 2007).

Despite these challenges, a variety of community-based initiatives have emerged to ensure multigenerational knowledge-sharing and to build capacity across the state to protect and perpetuate traditional knowledge. Non-profit organizations, state and federal agencies, and communities are working in concert towards these ends, and communities are taking advantage of a great number and variety of legal and policy mechanisms to partner with the State of Hawai'i in collaborative management initiatives. These recent actions provide promise for future co-management of fisheries in Hawai'i.

2.3. West Hawai'i aquarium fish MPA network

2.3.1 Establishment

In 1999, an MPA network was implemented to protect against declines of reef fish harvested for the live aquarium trade around the island of Hawai'i (Tissot and Hallacher, 2003; Figure 5.2C). This network was implemented on the west coast of Hawai'i Island (hereafter West Hawai'i) to reduce conflict between aquarium fishers and other marine resource users (e.g. dive operators, recreational divers) as well as encourage sustainable marine resource management (Capitini et al., 2004; Tissot, 2005). The network, contained within the West Hawai'i Regional Fishery Management Area, comprised nine fish replenishment areas (FRAs), where take of any reef fishes for the aquarium trade was illegal, and when combined with existing MPAs, these FRAs closed 35.2% of the total coastline to aquarium fishing (Tissot et al., 2009). The FRAs have an average area of 17.1 (1.8–40.1) km^2 and combined area of 153.9 km^2.

The MPA network was established by recommendations from a community-based team of stakeholders, the West Hawai'i Fishery Council (WHFC), and the Hawai'i DAR. The WHFC through a collaborative dispute resolution process proposed the location of the FRAs in West Hawai'i (Capitini et al., 2004). Because one goal of West Hawai'i's MPA network was to reduce user conflict between aquarium fishers and other groups, primarily dive charter operators and the tourism industry (Stevenson and Tissot, 2013), the placement of the MPAs was based on both conflict "hotspots" and expert testimony (Capitini et al., 2004). This approach resulted in establishing many of the MPAs within West Hawai'i's west catch zone, where high human population densities, tourist infrastructure, and major ports exist.

2.3.2 Ecological performance

The creation of a network of no-take areas for aquarium fishes in West Hawai'i in 1999 has increased the abundance of targeted aquarium species, while at the same time increasing the value of the fishery (Tissot et al., 2004a, 2009; Figure 5.5). Overall, the top 20 species of aquarium fishes increased 24% between pre-(1999–2000) and post-MPA implementation (2010–2012). The two top targeted species, yellow tang (*Zebrasoma flavescens*) and goldring surgeonfish (*Ctenochaetus strigosus*), which together account for 92% of the total aquarium catch in the state showed significant increases over time (88% and 37%, respectively), as did three other important aquarium fish species (Walsh et al, 2013). Tissot et al. (2004b) found that habitat quality, FRA size (especially reef width), and density of adult fishes were associated with significant recovery of fish stocks. Of particular importance are areas of high finger coral (*Porites compressa*) cover, which is critical habitat for juvenile yellow tang and young-of-year of other important fishery species (Walsh, 1987). In addition, variation in the abundance and distribution of both

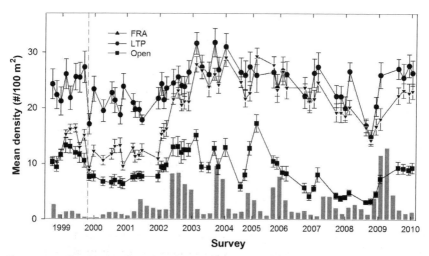

Figure 5.5 Changes in the abundance of yellow tang over time in West Hawai'i, 1999–2010 in three study area categories. FRA, fish replenishment areas established in 1999, LTP, long-term protected sites (closed ≥10 year prior to 1999), and Open areas, areas open to aquarium collecting. Histograms are the abundance of YOY (Young-of-the year) (William Walsh and Brian Tissot, unpublished data).

juvenile and adult habitats, and geomorphology of the reef, may influence effectiveness of these FRAs (Ortiz and Tissot, 2008, 2012).

Research has demonstrated both adult spillover and larval seeding of yellow tangs in the FRA network. Surveys of adult fishes within–relative to outside–MPAs show that densities at MPA boundaries in areas open to fishing are significantly higher than fished areas at distance from boundary areas (Williams et al., 2009). Moreover, genetic analyses (parentage based on microsatellite DNA) has demonstrated both general northward within–island larval dispersal and seeding via larval connectivity among local populations of yellow tang (Christie et al., 2010). Such studies of population connectivity are crucial for understanding how MPA networks function at the metapopulation level and how to design effective MPA networks at both island– and archipelago–level scale (Grorud–Colvert et al., 2014).

2.3.3 Socioeconomic performance
Analysis of catch reports and fisher interviews indicate that the West Hawaii FRA network significantly displaced fishing effort from the central to the northern and southern coastal regions of the island farther from easy access ports (Stevenson et al., 2013). Estimated catch revenues and experimental catch per unit effort were statistically greater as distance from port increased.

Both perceived fishing cost and travel time increased significantly post–MPA network implementation. Although the MPA network displaced fishing effort, fisher socioeconomic well-being was not compromised; likely by expanding their operating range, as well as favourable market factors that helped offset potential economic losses. Although there is evidence of adult yellow tang spillover and larval dispersal from within West Hawai'i's MPA network (Christie et al., 2010; Williams et al., 2009), the gradual decline in their abundance in waters remaining open to fishing (Figure 5.5) suggests that fishing mortality is likely greater than the rates of replenishment provided by the network. Fishers indicated their economic status was unchanged or marginally improved since the MPA network was implemented (Stevenson et al., 2013). Therefore, although the MPA network had a negative impact on distance travelled and cost, these attributes were perhaps offset by exogenous factors (e.g. price increase for fishes), such that the net change for economic status was constant or marginally positive, and thus may have stabilized other socio-economic well-being attributes.

Therefore, in addition to changes in fishing tactics that occurred post–MPA network (Stevenson et al., 2011), fishers were able to either maintain or potentially increase their fishing yield because the waters where they re-allocated effort to were either underexploited or more biologically productive than the pre-MPA fishing sites. It is possible that the redistribution of fishing effort synergistically acted with favourable market forces to influence fisher socio-economic well-being post–MPA network; however, the long-term viability of the fishery and the management strategy are yet to be determined.

The effectiveness of the FRA network has also been associated with an increase in the productivity of the aquarium fishery. Since 2000, the total catch and value of this fishery have increased by 39% and 59%, respectively. Approximately 79% of the fish caught in the state and 68% of the total aquarium catch value presently comes from Hawai'i Island (Walsh et al., 2013). There has also been an increase in permit holders, in the number of active fishers, and improvements in fishing effectiveness that could also account for some or all of these changes (Stevenson et al., 2011).

2.3.4 Overview and future prospects

The FRAs are considered a successful case of the MPA network implementation and a marine conservation success due in part to the unique nature of the aquarium fish fishery in West Hawai'i and the fact that the FRAs prohibit only one type of fishing, rather than attempting to prohibit all take, and the therefore excluding the broader fishing community (Rossiter and Levine, 2014). Moreover, the key targeted fish, yellow tang, reproduce quickly and have relatively small home ranges (Claisse et al., 2009), allowing for rapid recovery after collection. In addition, aquarium fish fishers are a small and somewhat marginalized group, the fishery is not considered a cultural right that needs to be protected, and revenues and livelihoods are restricted to a small number of fishers. These factors help simplify enforcement and compliance with FRA regulations. Clear scientific guidelines, careful planning and design, and extensive long-term involvement of local stakeholders in co-management with the state government have all contributed to the success of the FRAs (Rossiter and Levine, 2014). Social conflicts, however, have continued, necessitating the state's adoption of additional technical measures for West Hawai'i, including prohibited species lists and restrictions on scuba-spearfishing (Dawson, 2014). Ongoing adaptive management is an additional hallmark of effective and sustainable management and one that bodes well for the future of the West Hawai'i's FRAs.

A decade after the FRAs were established, surveys indicate that these MPAs were moderately effective in reducing conflict; however, encounters between stakeholders continued to occur and dive operators perceived aquarium fish fishing as a serious threat to the coral reef ecosystem (Stevenson and Tissot, 2013). Moreover, polarized value orientations towards the aquarium fish trade confirmed pervasive social values conflict indicating that MPAs were inadequate for resolving long-term conflict between groups who hold highly dissimilar value orientations towards the use of marine resources. Future marine spatial planning (MSP) and MPA siting processes should include stakeholder value and conflict assessments to avoid and manage tensions between competing user groups (Stevenson and Tissot, 2013).

2.4. Bottomfish restricted fishing areas
2.4.1 Establishment
In 1998, following a steady decline in catch rates and evidence that the two most commercially valuable species in the bottomfish fishery (ehu—*Etelis carbunculus* and onaga—*Etelis coruscans*) may be overfished, the State of Hawai'i DLNR implemented 19 bottomfish restricted fishing areas (BRFA) throughout the MHI. The Magnuson-Stevens Act imposed a mandate on the Western Pacific Regional Fisheries Management Council (WPRFMC) to restore the stocks of species listed as overfished to healthy levels within a 10-year time period. Since most of the MHI bottomfishing

grounds are within state rather than federal waters, WPRFMC turned to DAR to address this problem. The BRFAs were spread throughout the MHI and were designed to protect 20% of the designated 0–400 m essential fish habitat for onaga and ehu (Parke, 2007). The closure of these areas took effect on June 1, 1998 and their effectiveness, in terms of the quantity and type of habitat protected and their effect on commercial landings, was subsequently reviewed in 2005 (Moffitt et al., 2006). Only 5% of preferred habitat (e.g. hard bottom high relief, structurally complex substrates) was believed to occur within the boundaries of the BRFAs, and DAR's commercial catch data analysis furthermore indicated that modifications to the BRFA system were warranted. In 2007, as a result of ongoing overfishing, additional restrictions were imposed, including a 6–month seasonal closure, reduced non–commercial bag limits, mandatory permits, vessel markings, and a revision of the BRFAs that reduced the number of restricted areas to 12 but increased the area protected to include more essential bottomfish habitat based on comprehensive multi-beam sonar habitat mapping since the original 1998 closures (Figure 5.2D; Parke, 2007; Moore et al., 2013; Sackett et al., 2014). The average area of the BRFAs is 172.6 (40.8–907.3) km^2, with a total area of 2071.9 km^2. Because the BRFAs were designed as boxes for ease of navigation, they extend into deeper waters and only 710 km^2 occur at depths between 100 and 400 m where bottomfish are found (Parke, 2007).

2.4.2 Ecological performance

A monitoring programme has been in place for a subset of the BRFAs since 2007. Due to the great depths of this fishery, monitoring is accomplished using autonomous stereo baited camera systems (Merritt et al., 2011). Two of the BRFA's boundaries remained unchanged since 1998. While no differences were detected in species relative abundance between these two zones and neighbouring fished areas, evaluation of size-frequency distributions found that two commercially valuable species (onaga and opakapaka—E. coruscans and Pristipomoides filamentosus) were significantly larger inside the BRFA at Ni'ihau, the most remote of the MHI (Moore et al., 2013). No positive effects of protection were observed for the second monitored BRFA located off Hawai'i Island, which when established in 1998 did not include sufficient area of preferred habitat and is also close to the second largest port in the state.

The deep bottomfish populations inside KIR (18 km^2) were compared to neighbouring fished areas, and results suggested positive local effects of protection, with diversity of commercially harvested species higher inside the reserve (Drazen et al., 2010). Protection at KIR may have eliminated or reduced selective harvest and therefore increased diversity. Furthermore, the average sizes of many commercially harvested species were larger in KIR and possessed greater proportions of sexually mature fishes compared to fished areas, although onaga, a highly sought after species, were smaller inside KIR.

The most compelling evidence for the ecological effects of the BRFAs comes from comparing time series of the BRFA populations to neighbouring fished areas. A 4-year time series in three BRFAs showed that size increased significantly inside the BRFA for several species but declined or remained unchanged where fishing occurred (Figure 5.6; Sackett et al., 2014). The species showing these trends were also the most economically important of the deep bottomfish (e.g. onaga, opakapaka, ehu). One species, *P. sieboldii* (kalekale), showed a reverse pattern (Figure 5.6A and E). Kalekale are generally not targeted by commercial fishers because of their small body size (Kelley et al., 2006) and declines in length may result from competition with, or predation by, larger target species (Lizaso et al., 2000). Relative abundance showed fewer significant patterns over time with increases for onaga and opakapaka inside two BRFAs while there were little or no changes outside these BRFAs over time.

Differences among the BRFAs were also evident and likely influenced by the duration of protection from fishing. For example, the oldest BRFA (Ni'ihau, protected approximately 14 years) showed more mature fishes inside compared to outside the reserve for each species examined, and species richness in adjacent fished habitats increased while remaining unchanged inside the reserve, possibly due to spillover (Sackett et al., 2014). BRFAs with an intermediate duration of protection (Penguin Bank and Makapu'u) had positive protection effects (i.e. increases in mean fish lengths and relative abundance, Figure 5.6B and F), and the youngest BRFA (Pailolo Channel, protected approximately 4 years, Figure 5.6G) showed little change over the duration of protection. These results are consistent with other studies that suggest that at least 15 years of protection are necessary to see reliable benefits of protection (Molloy et al., 2009; Russ and Alcala, 2010). Nonetheless, the predominant finding of more abundant, larger, and more mature fishes inside the BRFAs compared to outside these zones could suggest that the BRFAs have benefited Hawai'i's deepwater fish populations.

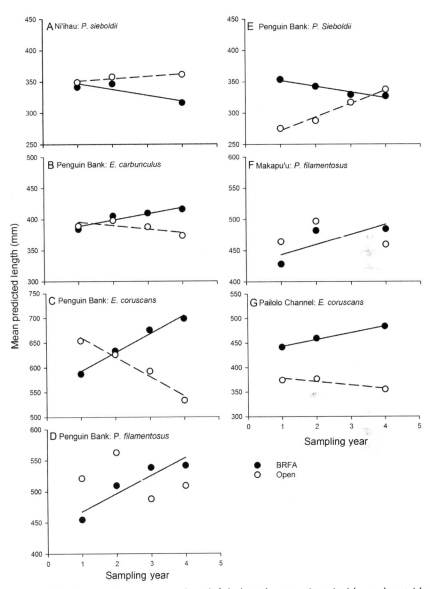

Figure 5.6 Changes in mean predicted fish length over time inside and outside bottomfish restricted fishing areas (BRFAs). Solid lines show significant trends inside BRFAs, and dashed lines show significant trends outside BRFAs ($P < 0.05$). The genera displayed are *Etelis* and *Pristipomoides*. *Adapted from Sackett et al. (2014).*

2.4.3 Socio-economic performance

The bottomfish fishery in the MHI is composed of a complex mix of commercial, recreational, cultural, and subsistence fishermen. Fifty-one per cent of fisherman surveyed in 2010 were 55 years old or more and had targeted bottomfish for an average of 19 years (Hospital and Beavers, 2012). Participants were more likely to identify themselves as Asian or Native Hawaiian/ Pacific Islander relative to the general population of the state of Hawai'i (Hospital and Beavers, 2012). While fishery highliners appear to be able to regularly cover trip expenditures and turn a profit, many supplement their income with other fishing activities. Based on average catch disposition of MHI bottomfish, it is clear that for a large majority of fishery participants, the social and cultural motivations far outweigh the economic benefits (Hospital and Beavers, 2012).

Mail surveys on attitudes and perceptions from 519 bottomfish fishermen found that some fishermen were concerned over a lack of scientific evidence that BRFAs enhance bottomfish populations and how large the BRFAs should be (Hospital and Beavers, 2012). Respondents also questioned why there are both total allowable catch (TAC) management and BFRAs and expressed frustration about a lack of enforcement of the BFRAs. The study highlighted the need for a more thorough evaluation of the protected areas, as well as the need for reliable estimates of recreational catch. Zeller et al. (2008) estimated non-commercial catches were twice as high as reported commercial catches by using adjustment ratios applied to commercial time-series data. Independent estimates of recreational bottomfish catch and effort are needed to better address uncertainty in these key management parameters.

2.4.4 Overview and future prospects

Though the local effects of the BRFAs on commercially harvested species are clear, because of the longevity of these species (e.g. >40 years for opakapaka, Andrews et al., 2012), it is likely that benefits to the fishery in terms of enhanced larval export and adult spillover will take even more time to accrue. Longevity increases, growth rates declines, and other productivity parameters changes with depth, likely increase the time required to observe obvious benefits of MPAs to regional deep water fisheries (Drazen and Haedrich, 2012).

The majority of the commercial fishing industry dislikes or even actively opposes the BRFA system (Hospital and Beavers, 2014). However, despite a lack of active enforcement, positive local effects on commercially harvested species are still observed. Opposition to the BRFAs has been bolstered by NMFS adopting a TAC management scheme in 2006, just prior to the revision of the BRFA boundaries (Hospital and Beavers, 2014). The TAC is based on a stock assessment that principally uses fishery-dependant catch data. In 2010, the quota was increased substantially and it was determined that the stock was no longer in a state of overfishing. The WPRFMC recommended that the BRFAs be eliminated but because they are under the purview of the state no action was taken at that time. This year (2014) under continuing pressure from fishers and WPRFMC, the state may open 6 of the 12 BRFAs to fishing.

2.5. Papahānaumokuākea Marine National Monument

2.5.1 Establishment

Protection of the NWHI began in 1909 with the creation of the Hawaiian Islands National Wildlife Refuge for the purpose of safeguarding nesting seabird colonies from overexploitation (Executive Order 1019). In 2000, President Clinton created the NWHI Coral Reef Ecosystem Reserve and in 2001 initiated the process to designate a National Marine Sanctuary (Executive Orders 13178 and 13196). The state of Hawai'i also recognized the significance of the NWHI in establishing the NWHI State Marine Refuge (Kittinger et al., 2010). In 2006, President Bush established the NWHI Marine National Monument under the authority of the Antiquities Act of 1906 (16 U.S.C. 431). Subsequently renamed the Papahānaumokuākea Marine National Monument (PMNM), it is the single largest conservation area under the US flag, and one of the largest marine conservation areas in the world, encompassing 362,073 km^2 (Figure 5.1B; Toonen et al., 2013).

PMNM includes a number of pre-existing federal conservation areas: the NWHI Coral Reef Ecosystem Reserve, managed by the Department of Commerce through the National Oceanic and Atmospheric Administration (NOAA) Office of National Marine Sanctuaries; Midway Atoll National Wildlife Refuge, Hawaiian Islands National Wildlife Refuge, and Battle of Midway National Memorial, managed by the Department of the Interior through the United States Fish and Wildlife Service (USFWS). These areas remain in place within the Monument, subject to their applicable laws and regulations in addition to the provisions of the Proclamation (Kittinger et al., 2011). The NWHI also includes state of Hawai'i lands and waters, managed by the DLNR as the NWHI Marine Refuge and the State Seabird Sanctuary at Kure Atoll. These areas also remain in place and are subject to their applicable laws and regulations. The governance arrangement for the monument represents a new model in US MPA management, requiring two federal agencies and the State of Hawai'i to collaboratively manage the NWHI (Kittinger et al., 2011).

In 2010, the Monument was inscribed as a UNESCO World Heritage Site for both natural and cultural value. Pursuant to the proclamation, full protections took effect in 2011 with the closure of the last remaining fishery (bottomfish fishery). In January 2010, however, the National Marine Fisheries Service signed an agreement with the remaining bottomfish fishers to surrender their federal fishing permits in exchange for compensation; as a result, all commercial fishing ended in January 2010. Although some fishing effort was re-directed towards the MHI, a number of vessels dropped out of the fishery all together. Extraction is now limited to subsistence take by visiting scientists, residents of Midway Atoll and Native Hawaiian cultural practitioners, as well as minimal extraction for research purposes. Due to the limited number of permitted entries and negligible extraction for research, the monument is primarily considered a no-take reserve.

2.5.2 Ecological performance

The remoteness and protective status of the NWHI have resulted in a relatively undisturbed state compared with the MHI and many other marine-based ecosystems in the world (Friedlander and DeMartini, 2002; Friedlander et al., 2008; Pandolfi et al., 2005; Williams et al., 2008). Because of its remoteness and limited fishing, the NWHI is one of the few places left in the world that is sufficiently pristine to study how unaltered ecosystems are structured, how such ecosystems function, and how they can be most effectively preserved. One of the most striking and unique components of the NWHI ecosystem is the abundance and dominance of large apex predators such as sharks and jacks (Friedlander and DeMartini, 2002), which exert a strong top-down control on the ecosystem (DeMartini and Friedlander, 2006; DeMartini et al., 2005) and have been depleted in most other locations around the world (Myers and Worm, 2003, 2005).

A comparison between NWHI and MHI revealed dramatic differences in the shallow reef fish assemblages with standing stock in the NWHI nearly threefold greater than in the MHI with over 54% of the total fish biomass in the NWHI consisting of apex predators, whereas this trophic group accounted for less than 3% of fish biomass in the MHI (Friedlander and DeMartini, 2002). Recent archaeological evidence suggests that the NWHI serves as a good proxy for past lightly exploited baselines in the MHI, thus supporting the validity of the space-for-time approach used above (Longenecker et al., 2014).

Endemism is remarkably high for shallow reef fishes throughout the archipelago, particularly in the NWHI where endemic species account for 30% of the species present and more than 52% of the numerical standing stock (DeMartini and Friedlander, 2004). Surveys of mesophotic coral reef depths (30–90 m) across the NWHI reveal average endemism of 46% with the relative abundances of endemic reef fishes on mesophotic reefs ranging from 16% at the southernmost end of the NWHI to upwards of 92% at the northernmost end of the NWHI (Kane et al., 2014). This unprecedented level of endemism indicates that mesophotic reefs in the NWHI are reservoirs of biodiversity, and of high conservation value.

PMNM has value as a reference area to assess individual fish stocks and provides guidance for fisheries management in the MHI. Using the NWHI as a reference, an assessment of fish stocks found that over one-quarter (27%) of fished species in the MHI were critically depleted (<10% of unfished abundance) and 42% were below 25% of unfished abundance, which is often considered a threshold for overfishing (Friedlander et al., 2008, 2014).

2.5.3 Socio-economic performance

Interest in commercialization of nearshore fisheries in the NWHI increased in the 1970s with the discovery of the potential for a profitable lobster fishery. Subsequently, the lobster fishery and a bottomfish fishery focusing on demersal species became active in the nearshore ecosystem (Kittinger et al., 2010). Pelagic fisheries also operate in and around the NWHI, but outside of the 50 nautical mile protected area. The lobster fishery became the most lucrative single fishery in the late 1980s but then underwent a steep decline beginning in the early 1990s, eventually leading to a 1–year closure in 1993 (Townsend and Pooley, 1995), then a permanent closure in 2000 with the establishment of the NWHI Coral Reef Ecosystem Reserve. Relative to the overall economy in Hawai'i and even in terms of commercial fisheries, the NWHI bottomfish fishery was rather small; nevertheless, the proposal to close it was controversial. One of the arguments against closure was the importance of the fishery to Hawai'i's economy, providing jobs in commercial fishing and supplying bottomfish to seafood retailers and restaurants. Because demand for Hawai'i-caught bottomfish was found to be highly elastic and widespread substitution with imports, the overall economic loss was quite small (Coffman and Kim, 2009).

Using information gathered from a representative subset of MPAs worldwide, McCrea-Strub et al. (2011) showed that variation in MPA startup costs was significantly related to both MPA size and the duration of the establishment phase. The largest MPA in the sample, PMNM, was also the most expensive to establish ($34.8 million; 2005) (McCrea-Strub et al., 2011). Over 99% of funding was provided by national NGOs and governmental agencies and approximately 20% of the total cost of establishment was allocated towards a compensation programme for NWHI commercial bottomfish and lobster fishermen who were displaced by the creation of PMNM (Kittinger et al., 2011).

2.5.4 Overview and future prospects

Marine ecosystems of the NWHI are being altered by direct effects of climate change including ocean warming, ocean acidification, rising sea level, changing circulation patterns, and increasing severity of storms (Keller et al., 2009). Direct anthropogenic threats include marine debris, ship-based pollution and strike risks, and alien species. Selkoe et al. (2009) mapped impacts in the NWHI and found that ocean temperature variation associated with disease outbreaks had the highest predicted impact overall, followed closely by other climate-related threats.

To address these and other management concerns, PMNM in cooperation with NOAA, USFWS, the State of Hawai'i, the NWHI Coral Reef Ecosystem Reserve Advisory Council, and others worked to design a plan to protect the living, cultural, and historical resources of the region as a public trust (PMNM, 2008). The public played a vital role in shaping the management plan for the proposed national marine sanctuary in the NWHI. This process formally began with public scoping meetings in 2002 and formed the basis for comprehensive management planning for the monument.

The management framework for the monument includes key elements to move towards ecosystem-based management, requiring implementation of multiple steps in a comprehensive and coordinated way. These elements include the legal and policy basis for establishment; the vision, mission, and guiding principles that provide the overarching policy direction; institutional arrangements between co-trustees and other stakeholders; regulations and zoning to manage human activities and threats; goals to guide implementation of action plans and priority management needs; and concepts and direction for moving towards a co-ordinated ecosystem approach to management (PMNM, 2008).

3. DISCUSSION

The case studies of different types of marine spatial management in Hawai'i presented in this chapter encompass a range of scales (Figure 5.7). The concept of scale is critical and influences many aspects of marine spatial management from ecology to human dimensions to governance. Different management objectives are best addressed by different scales of management. An understanding of the effects of scale for MPAs and MPA network design is vital to achieving success in implementation and effectiveness of these managed areas.

Successful MPA implementation and management require a balance between human uses and conservation objectives. Often MPA planning does not sufficiently address human activities in the marine space by failing to fully engaging all stakeholders early and throughout the process (Charles and Wilson, 2009; Mascia, 2003; Stewart et al., 2011). This has been a deficiency of marine spatial management in Hawai'i as it has elsewhere in the

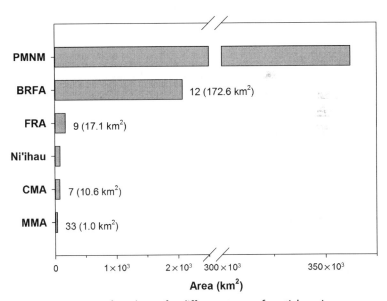

Figure 5.7 Comparison of total area for different types of spatial marine management in Hawai'i. For categories where $N > 1$, the number of MPAs is shown to the right of the bar with average area in parentheses. PMNM, Papahānaumokuākea Marine National Monument; BRFA, bottomfish restricted fishing areas; FRA, West Hawaii Fish Replenishment Areas; Ni'ihau, the island of Ni'ihau; CMA, co-management areas; and MMA, marine managed areas.

world (Tissot et al., 2009). Other shortcomings in MPA planning in Hawai'i and beyond include mismatches of MPA scale to issues and context, inadequate attention to compliance, failure due to degradation of the surrounding ecosystem, and damaging displacement of fishing effort (Agardy et al., 2011; Bergseth et al., 2013; Crowder et al., 2006). The State of Hawai'i has several MMA designations, with various goals surrounding conservation, fisheries management, and multi-use objectives (Figure 5.2A). However, while many of these areas are showing success, most are too small to achieve meaningful benefits outside their boundaries. Furthermore, because there has not been enough focus on compliance and enforcement, the modest successes of these areas are also being eroded by illegal, unreported, and unregulated fishing.

In contrast, community managed areas in Hawai'i are currently undergoing a renaissance. Despite CBSFA legislation, a functional approach to co-management has yet to be achieved; therefore, implementation of these areas has been through an entirely bottom–up approach that has been slow and arduous (Higuchi, 2008; Kittinger et al., 2012). In some cases, the context and objectives for these areas are dependent on the active members of the community, sometimes to the exclusion of other stakeholders or the broader context within which these community managed areas operate (Levine and Richmond, 2014). The West Hawai'i FRA network was based on a collaborative process involving a number of stakeholders. Though it has only been moderately successful in terms of reducing conflict between user groups, it has been very effective in achieving its ecological and economic goals (Stevenson and Tissot, 2013; Tissot et al., 2009). One advantage in the implementation process was that it was limited in scope, only addressing one small fishery (aquarium fish). Nevertheless, it is the best example in Hawai'i of an inclusive, collaborative, stakeholder-driven, participatory MPA planning process.

The BRFAs are similar to the FRA network in that the objectives were focused on only one fishery (bottomfish), however, that is where the similarities end. BRFA design and implementation was a top–down approach wherein DAR-designated areas based on pressure from the WPRFMC with little or no stakeholder involvement or adequate ecological data. While the system has since been evaluated and improved based on better understanding and mapping of bottomfish habitats, there has been little to no stakeholder involvement and fishermen are concerned about the perceived lack of science informing BRFA placement and evaluation of efficacy, as well as expressing frustration over the dearth of enforcement (Hospital and

Beavers, 2014). The establishment of the PMNM was a long process sum-marized previously in this chapter. This is another example of a top-down approach, though via the federal government rather than the state. A simplifying factor was the lack of human habitation in the area, and the few commercial fishing interests were well compensated for ceasing their activities. Governance of this area is complex but well addressed by Kittinger et al. (2010a). By virtue of its size and remoteness, PMNM by most measures is an unqualified success, while the situation in the MHI is a great deal more complicated.

3.1. Effects of scale
3.1.1 Ecological
One of the most important concepts related to scale of MPAs is the biology and ecology of the marine organisms which are to be protected. Neighbourhood sizes of both adult and larval life phases will have direct bearing on the ability of an MPA to protect a given species or suite of species (Palumbi, 2004). Pelagic species can move thousands of kilometers annually, while many reef fishes have home ranges <1 km (Alerstam et al., 2003; Block et al., 2001; Palumbi, 2004). Larval neighbourhood sizes can be even larger than these adult neighbourhoods, but recent studies using genetic and micro-chemical analyses of larval spread show cases where local retention of larvae is surprisingly high, suggesting that marine populations are not uni-versally open over large geographic scales (Palumbi, 2004).

In a review of dispersal distance of propagules of benthic marine organisms, Shanks et al. (2003) showed a bimodal distribution suggesting two evolutionarily stable dispersal strategies: short-distance (<1 km) and long-distance (>20 km). Based on this, they recommend that reserves be designed large enough to contain the short-distance dispersing propagules and be spaced far enough apart that long-distance dispersing propagules released from one reserve can settle in adjacent reserves. The mean area of fully protected MLCDs in Hawai'i ($N=8$) is only 0.26 km^2, likely inhibiting self-recruitment and leaving these areas dependent on larval import. The FRA network however is much more in-line with these guidelines, with an average area of 12.8 km^2 and spaced 1–15 km apart over 150 km of coastline (Figures 5.7 and 5.2C). This design almost certainly promotes propagule sharing among FRAs, which likely contributes to the success of this MPA network in replenishing target species. Christie et al. (2010) found dispersal distances ranging from 15 to 184 km from a genetic parentage analysis of yellow tang confirming the export of larvae from West Hawai'i FRAs at this scale.

The larger the protected area, the smaller its border-to-area ratio, reducing the amount of "edge" habitat that is exposed to outside pressures (Keller et al., 2009; Woodroffe and Ginsberg, 1998). Although Stamoulis and Friedlander (2013) showed a fisheries spillover benefit from a small marine reserve in Hawai'i, small MPAs are unlikely to provide significant export of larvae and therefore the potential benefits to fisheries are limited. Furthermore, spillover of adult fishes from small reserves can reduce reproductive output with negative implications for stock enhancement (Sladek Nowlis and Roberts, 1999). Another drawback of small reserves is that they encompass a limited amount and variety of habitats and thus may not protect a full complement of marine species and/or all life stages of resident species (Palumbi, 2004; Sladek Nowlis and Friedlander, 2005). Thus, large-scale MPAs can be seen as maximizing the potential for achieving ecological objectives (Toonen et al., 2013).

While there are many factors at work, in general small-scale management units in Hawai'i such as community-based fishery managed areas, MLCDs, and FMAs will be able to protect sessile species, small benthic fishes, and some larger benthic fishes. Larger management units such as the West Hawai'i FRA network (if it were to eliminate all types of fishing) would be able to protect and enhance populations of most benthic fishes and some small pelagic species. The PMNM, however, is at a scale sufficient to protect the entire ecosystem including large pelagic fishes and migratory species. While ecologically this may be the most effective scale for spatial management, human dimensions including socio-economic and political considerations necessitate the use of smaller-scale spatial management units in populated locations such as the MHI.

3.1.2 Social

Spatial marine protection also has social goals that are linked with ecological objectives (Cinner et al., 2009b; Halpern et al., 2013; McClanahan et al., 2006; Rossiter and Levine, 2014). In order to promote success, the scale of protection should match the scale of the social, as well as ecological, outcomes desired (Charles and Wilson, 2009). To meet this challenge, researchers are developing innovative methods to assess the social attributes of ocean environments (Koehn et al., 2013), and ocean planning practitioners are increasingly engaging social data in planning practise to help spur inclusive planning processes that can engender better social and ecological outcomes (Kittinger et al., 2014; Le Cornu et al., 2014).

Such approaches will also have to consider the existing institutions, enabling environment, and MMA designations in a given geography. In Hawaii, MLCDs were designed to preserve and replenish marine life, providing opportunities for the public to interact with the marine environment and are popular sites for snorkelling, diving, and underwater photography. FMAs were designed to resolve conflicts among users including fishermen. While the scale of MLCDs ($\overline{X}=0.35\,\text{km}^2$) and FMAs ($\overline{X}=0.72\,\text{km}^2$) are quite small, they are generally sufficient to address their ecological and social objectives. If they were designed to benefit fisheries, a larger scale would be necessary. These MMAs and other areas can be scaled to meet both social and ecological objectives through a systematic approach to assess the cumulative impacts of current activities, the historical condition and current trajectory of nearshore ecosystems, and the current ecological and socioeconomic performance of existing management approaches.

Another critical aspect of MPA scale is the extent of the costs and benefits and the number of people affected. Naturally, the larger the MPA or MPA network, the more people will be impacted (Pollnac and Seara, 2011). The loss of fishing areas is often the primary public concern when implementing MPAs and many fishers in Hawai'i are vehemently opposed to them for this reason. Displacing too many fishers can introduce significant social and economic costs and make MPA establishment politically untenable (Jones, 2009). MPAs tend to have concentrated costs and disbursed benefits producing inequitable social impacts (Halpern et al., 2013). While large MPAs are known to produce ecological benefits across broad scales which benefit a range of ocean users, the costs will be concentrated among a relatively small group of fishermen. This holds true in Hawai'i and one of the guidelines the State of Hawai'i uses when evaluating areas for MLCD designation is that they are '... small enough so that fishermen are not denied the use of unreasonably vast fishing areas' (Division of Aquatic Resources, 2014). For the BFRAs on the other hand, continued pressure from fishermen and WPRFMC may result in the decision to open 6 of the 12 BRFAs to fishing. Due to the remoteness of the NWHI, the economic and social costs of establishing the PMNM were quite low compared to the benefits of designating the world's largest (at the time) MPA (Coffman and Kim, 2009; McCrea-Strub et al., 2011).

For land-related resource systems, very large territories are unlikely to be self-organized given the high costs of management, while small territories do not generate significant flows of valuable products. Thus, moderate-sized areas are most conducive to self-organization (Chhatre and Agrawal, 2008; Ostrom, 2009). Fishers who consistently utilize moderately sized coastal zones are also more likely to organize (Wilson et al., 2007) than fishers who target pelagics in the open ocean (Berkes et al., 2006). This seems to hold true for spatial marine protection strategies in Hawai'i, where government was responsible for implementing MMAs at very large and very small scales (the federal PMNM and state MMAs). However, at intermediate scales, community-based managed areas and the FRA network are both examples of self-organization, or bottom–up approaches, which are gaining momentum across the state.

To ensure success, MPA scale should not exceed institutional capacity (Christie et al., 2009). BRFAs are a case in Hawai'i where the scale exceeds the capacity of DLNR to manage, and enforcement is not occurring. For very large MPAs such as PMNM, the costs of management are proportionate to the spatial scale and likely beyond the capacity of any single institution or government agency. This MPA encompasses a range of institutional jurisdictions, thus a new model of institutional collaborative governance was created (Kittinger et al., 2011). Since the 2010 Convention on Biological Diversity and the establishment of the Aichi Biodiversity targets of protecting at least 10% of coastal and marine areas by 2020, large-scale MPAs have begun to proliferate and ocean governance is moving towards increased collaboration among countries as well as institutions (Toonen et al., 2013). In economic terms, large MPAs are more efficient in terms of establishment and maintenance costs. Though the overall cost to establish the PMNM was higher than any other MPA at the time, the cost per km^2 was among the lowest (McCrea-Strub et al., 2011). Furthermore, the long-term cost of MPA maintenance per km^2 drops significantly as size increases, providing broad economic, conservation, and scientific benefits (Toonen et al., 2013).

3.2. Hawaiian MPAs in the context of large-scale marine spatial planning

There exists in the MHI a patchwork of spatial marine management across a range of scales (Figures 5.1 and 5.7), with varying degrees of effectiveness. With the exception of co-management areas and the FRAs, stakeholder engagement is largely lacking for these management schemes and their efficacy is in question due to a combination of factors including lack of local support and non-compliance inside the boundaries, and ongoing impacts outside. Furthermore, this collection of MMAs does not ensure connectivity among sites, which is crucial to maintaining populations of mobile species and vital connections between local ecosystems, and does not recognize important processes originating offshore that provide linkages between coastal areas (Gaines et al., 2010). Finally, this patchwork of management areas cannot address the full suite of stressors that impact the marine populations and ecosystems in Hawai'i. In order to properly address the many impacts and competing objectives of a myriad of stakeholders, a larger vision is required (Tissot et al., 2009). One which could leverage the attention and resources currently being spent trying to protect this collection of discrete and rather small areas. One solution is a strategic, coordinated, and comprehensive planning effort that could be supported by robust and targeted management within discreet MPAs for which the sum total, within the context of the wider strategic marine plan, could drive effective ecosystem-based management. MSP provides a framework to achieve this goal (Agardy et al., 2011; Ehler and Douvere, 2009; Foley et al., 2010; Gopnick, 2008).

A coordinated, regional plan is not only necessary because of the large scale over which ecosystem dynamics, resource markets, and governance systems operate, but it is also likely more efficient and cost effective (e.g. Kark et al., 2009). MSP does not stand alone; rather it emerges from and builds on existing management frameworks such as integrated coastal management and ecosystem-based management. While regional planning is vital, effective implementation will always occur at the local level. Therefore, balancing the dynamics of regional and local level planning is essential for success (Agardy et al., 2011).

The Hawaiian Islands are an ideal location to apply the MSP framework. There exists an abundance of quality ecological data and the collection of human use data for marine spaces has been prioritized. The Bureau of Ocean Energy Management is currently compiling existing marine ecological data and collecting marine human use data to inform coastal zone planning and management strategies and for analysis of future offshore renewable energy programmes.

4. CONCLUSIONS

There is much resistance to the establishment of MPAs from the fishing sector for a variety of reasons including: loss of fishing areas, displacement or marginalization of subsistence fishers, perceived loss of income and cultural access, and the long lag time before benefits are realized (Cinner et al., 2009a; McClanahan et al., 2005; Pauly, 2009). Although not a panacea for marine fisheries management, MPAs in conjunction with other input and output controls are critical to sustaining fisheries and maintaining ecosystem health.

Spatial management of fisheries in Hawai'i exists along a range of scales (Figure 5.7), and as exemplified by the BRFAs, bigger is not necessarily better. A critical theme underlying MPA success is the participatory nature of the planning process. To put it simply, MPAs which engage stakeholders early and often in the planning process tend to be more effective at achieving ecological and social goals (Agardy et al., 2011; Cinner, 2007; Mascia, 2003; McClanahan et al., 2006). This is exemplified by the community-based co-management areas in Hawai'i and by the FRA network which was created through a stakeholder-driven, participatory process (Stevenson and Tissot, 2013).

Because social costs of MPAs tend to be concentrated while the benefits are disbursed, it is difficult to maintain social equity, especially for larger MPAs (Halpern et al., 2013; Lowry et al., 2009). Larger scales correspond to large numbers of stakeholders at increasing levels of organization (Agrawal, 2001; Baland and Platteau, 1996; Lowry et al., 2009). Thus governance structures need to make better use of the human capacity for complex normative frameworks to ensure that management requirements do not exceed institutional capacity (Christie et al., 2009; Kittinger et al., 2011). The BRFA system in Hawai'i is a case where the scale of management created an imbalance in social equity, and because planning and management requirements exceeded institutional capacity, stakeholder confidence eroded to the point where the state was pressured to open 6 of the 12 areas to fishing.

The establishment of PMNM is a great achievement and an unqualified success by most measures. While the total cost of implementation was extreme, the cost per unit area was among the lowest in Hawai'i, as are the long-term maintenance costs. While large-scale marine reserves provide our best hope for arresting the global decline in biodiversity and addressing global fisheries collapse (Toonen et al., 2013), they are simply not applicable in populated areas such as the MHI. Properly implemented MSP is necessary to balance conservation and social objectives in these complex, socio-ecological systems and achieve effective, equitable, and sustainable outcomes (Agardy et al., 2011; Douvere, 2008; Halpern et al., 2008).

REFERENCES

Agardy, T., di Sciara, G.N., Christie, P., 2011. Mind the gap: addressing the shortcomings of marine protected areas through large scale marine spatial planning. Mar. Policy 35, 226–232. http://dx.doi.org/10.1016/j.marpol.2010.10.006.

Agrawal, A., 2001. Common property institutions and sustainable governance of resources. World Dev. 29, 1649–1672. http://dx.doi.org/10.1016/S0305-750X(01)00063-8.

Alerstam, T., Hedenström, A., Åkesson, S., 2003. Long-distance migration: evolution and determinants. Oikos 103, 247–260. http://dx.doi.org/10.1034/j.1600-0706.2003.12559.x.

Andrews, A.H., DeMartini, E.E., Brodziak, J., Nichols, R.S., Humphreys, R.L., 2012. A long-lived life history for a tropical, deepwater snapper (*Pristipomoides filamentosus*): bomb radiocarbon and lead–radium dating as extensions of daily increment analyses in otoliths. Can. J. Fish. Aquat. Sci. 69, 1850–1869. http://dx.doi.org/10.1139/f2012-109.

Antolini, D., Moffie, K., Paulson, D., 2003. Hawaii marine protected areas governance study. (Commissioned report). State Department of Land and Natural Resources, Division of Aquatic Resources, Honolulu, HI.

Antonelis, G.A., Baker, J.D., Johanos, T.C., Braun, R.C., Harting, A.L., 2006. Hawaiian monk seal (*Monachus schauinslandi*): status and conservation issues. Atoll Res. Bull. 543, 75–101.

Ayers, A.L., Kittinger, J.N., 2014. Emergence of co-management governance for Hawai'i coral reef fisheries. Global Environ. Change 28, 251–262. http://dx.doi.org/10.1016/j.gloenvcha.2014.07.006.

Baland, J.-M., Platteau, J.-P., 1996. Halting Degradation of Natural Resources: Is There a Role for Rural Communities? Food & Agriculture Org, Rome, Italy.

Bell, F.T., Higgens, E., 1939. A plan for the development of the Hawaiian fisheries. (Investigational report no. 4). US Bureau of Fisheries, Washington, DC.

Bell, J.D., Johnson, J.E., Hobday, A.J., 2011. Vulnerability of Tropical Pacific Fisheries and Aquaculture to Climate Change. Secretariat of the Pacific Community, Noumea, New Caledonia.

Bergseth, B.J., Russ, G.R., Cinner, J.E., 2013. Measuring and monitoring compliance in no-take marine reserves. Fish Fish. n/a–n/a. http://dx.doi.org/10.1111/faf.12051.

Berkes, F., 2010. Devolution of environment and resources governance: trends and future. Environ. Conserv. 37, 489–500. http://dx.doi.org/10.1017/S037689291000072X.

Berkes, F., Hughes, T.P., Steneck, R.S., Wilson, J.A., Bellwood, D.R., Crona, B., Folke, C., Gunderson, L.H., Leslie, H.M., Norberg, J., Nyström, M., Olsson, P., Österblom, H., Scheffer, M., Worm, B., 2006. Globalization, roving bandits, and marine resources. Science 311, 1557–1558. http://dx.doi.org/10.1126/science.1122804.

Block, B.A., Dewar, H., Blackwell, S.B., Williams, T.D., Prince, E.D., Farwell, C.J., Boustany, A., Teo, S.L.H., Seitz, A., Walli, A., Fudge, D., 2001. Migratory movements, depth preferences, and thermal biology of Atlantic bluefin tuna. Science 293, 1310–1314. http://dx.doi.org/10.1126/science.1061197.

Bowen, B.W., Meylan, A.B., Ross, T.P., Limpus, C.T., 1992. Global population structure and natural history of the green turtle (*Chelonia mydas*) in terms of matriarchal phylogeny. Evolution 46, 865–881.

Briggs, J.C., Bowen, B.W., 2012. A realignment of marine biogeographic provinces with particular reference to fish distributions. J. Biogeogr. 39, 12–30. http://dx.doi.org/10.1111/j.1365-2699.2011.02613.x.

Capitini, C.A., Tissot, B.N., Carroll, M.S., Walsh, W.J., Peck, S., 2004. Competing perspectives in resource protection: the case of marine protected areas in West Hawai'i. Soc. Nat. Resour. 17, 763–778. http://dx.doi.org/10.1080/08941920490493747.

Cesar, H.S.J., van Beukering, P., 2004. Economic valuation of the coral reefs of Hawai'i. Pac. Sci. 58, 231–242. http://dx.doi.org/10.1353/psc.2004.0014.

Chaloupka, M., Balazs, G., 2007. Using Bayesian state-space modelling to assess the recovery and harvest potential of the Hawaiian green sea turtle stock. Ecol. Model. 205, 93–109. http://dx.doi.org/10.1016/j.ecolmodel.2007.02.010.

Charles, A., Wilson, L., 2009. Human dimensions of Marine Protected Areas. ICES J. Mar. Sci. 66, 6–15. http://dx.doi.org/10.1093/icesjms/fsn182.

Chave, E.H., Mundy, B.C., 1994. Deep-sea benthic fish of the hawaiian Archipelago, Cross Seamount, and Johnston Atoll. Pac. Sci. 48, 367–409.

Chhatre, A., Agrawal, A., 2008. Forest commons and local enforcement. Proc. Natl. Acad. Sci. U.S.A. 105, 13286–13291. http://dx.doi.org/10.1073/pnas.0803399105.

Christie, P., Pollnac, R.B., Oracion, E.G., Sabonsolin, A., Diaz, R., Pietri, D., 2009. Back to basics: an empirical study demonstrating the importance of local-level dynamics for the success of tropical marine ecosystem-based management. Coast. Manag. 37, 349–373. http://dx.doi.org/10.1080/08920750902851740.

Christie, M.R., Tissot, B.N., Albins, M.A., Beets, J.P., Jia, Y., Ortiz, D.M., Thompson, S.E., Hixon, M.A., 2010. Larval connectivity in an effective network of marine protected areas. PLoS ONE 5, e15715. http://dx.doi.org/10.1371/journal.pone.0015715.

Cinner, J.E., 2007. Designing marine reserves to reflect local socioeconomic conditions: lessons from long-enduring customary management systems. Coral Reefs 26, 1035–1045. http://dx.doi.org/10.1007/s00338-007-0213-2.

Cinner, J.E., Aswani, S., 2007. Integrating customary management into marine conservation. Biol. Conserv. 140, 201–216. http://dx.doi.org/10.1016/j.biocon.2007.08.008.

Cinner, J.E., Daw, T., McClanahan, T.R., 2009a. Socioeconomic factors that affect artisanal fishers' readiness to exit a declining fishery. Conserv. Biol. 23, 124–130. http://dx.doi.org/10.1111/j.1523-1739.2008.01041.x.

Cinner, J.E., McClanahan, T.R., Daw, T.M., Graham, N.A.J., Maina, J., Wilson, S.K., Hughes, T.P., 2009b. Linking social and ecological systems to sustain coral reef fisheries. Curr. Biol. 19, 206–212. http://dx.doi.org/10.1016/j.cub.2008.11.055.

Claisse, J.T., Kienzle, M., Bushnell, M.E., Shafer, D.J., Parrish, J.D., 2009. Habitat-and sex-specific life history patterns of yellow tang Zebrasoma flavescens in Hawaii, USA. Mar. Ecol. Prog. Ser. 389, 245–255.

Clark, M.R., Koslow, J.A., 2008. Impacts of fisheries on seamounts. In: Pitcher, T.J. (Ed.), Seamounts: Ecology, Fisheries & Conservation. Blackwell Pub, Oxford; Ames, Iowa, p. 413.

Cobb, J.N., 1901. Commercial fisheries of the Hawaiian Islands. (U.S. Fish Commission Report for 1901). Government Printing Office, Washington, DC.

Coffman, M., Kim, K., 2009. The economic impacts of banning commercial bottomfish fishing in the Northwestern Hawaiian Islands. Ocean Coast. Manage. 52, 166–172. http://dx.doi.org/10.1016/j.ocecoaman.2008.12.003.

Crowder, L.B., Osherenko, G., Young, O.R., Airamé, S., Norse, E.A., Baron, N., Day, J.C., Douvere, F., Ehler, C.N., Halpern, B.S., 2006. Resolving mismatches in US ocean governance. Science 313, 617–618.

Dawson, T., 2014. Abercrombie signs West Hawaii fishing rules. Environ. Hawaii 24, 12 (Gov).

DeMartini, E.E., Friedlander, A.M., 2004. Spatial patterns of endemism in shallow-water reef fish populations of the Northwestern Hawaiian Islands. Mar. Ecol. Prog. Ser. 271, 281–296.

DeMartini, E.E., Friedlander, A.M., 2006. Predation, endemism, and related processes structuring shallow-water reef fish assemblages of the NWHI. Atoll Res. Bull. 543, 237–256.

DeMartini, E.E., Friedlander, A.M., Holzwarth, S.R., 2005. Size at sex change in protogynous labroids, prey body size distributions, and apex predator densities at NW Hawaiian atolls. Mar. Ecol. Prog. Ser. 297, 259–271.

Division of Aquatic Resources, 1992. Marine Life Conservation District Plan. Department of Land and Natural Resources, Division of Aquatic Resources, Honolulu, HI.

Division of Aquatic Resources, 2014. Hawaii Marine Life Conservation Districts. http://state.hi.us/dlnr/dar/mlcd.html.

Dollar, S.J., 1982. Wave stress and coral community structure in Hawaii. Coral Reefs 1, 71–81. http://dx.doi.org/10.1007/BF00301688.

Douvere, F., 2008. The importance of marine spatial planning in advancing ecosystem-based sea use management. Mar. Policy 32, 762–771. http://dx.doi.org/10.1016/j.marpol.2008.03.021.

Drazen, J.C., Haedrich, R.L., 2012. A continuum of life histories in deep-sea demersal fishes. Deep-Sea Res. I Oceanogr. Res. Pap. 61, 34–42. http://dx.doi.org/10.1016/j.dsr.2011.11.002.

Drazen, J.C., Moriwake, V., Demarke, C., Alexander, B., Misa, W., Yeh, J., 2010. Assessing Kaho'olawe Island Reserve's bottomfish populations: a potential benchmark for main Hawaiian Island restricted fishing areas. Prepared for the Kahoolawe Island Reserve Commission. University of Hawaii, Honolulu, HI.

Ehler, C., Douvere, F., 2009. Marine Spatial Planning: A Step-by-Step Approach Toward Ecosystem-Based Management. Intergovernmental Oceanographic Commission and Man and the Biosphere Programme. IOC Manual and Guides No. 53, ICAM Dossier No. 6. UNESCO, Paris.

Flament, P.J., Kennan, S.C., Knox, R.A., Niiler, P.P., Bernstein, R.L., 1996. The three-dimensional structure of an upper ocean vortex in the tropical Pacific Ocean. Nature 383, 610–613.

Foley, M.M., Halpern, B.S., Micheli, F., Armsby, M.H., Caldwell, M.R., Crain, C.M., Prahler, E., Rohr, N., Sivas, D., Beck, M.W., Carr, M.H., Crowder, L.B., Emmett Duffy, J., Hacker, S.D., McLeod, K.L., Palumbi, S.R., Peterson, C.H., Regan, H.M., Ruckelshaus, M.H., Sandifer, P.A., Steneck, R.S., 2010. Guiding ecological principles for marine spatial planning. Mar. Policy 34, 955–966. http://dx.doi.org/10.1016/j.marpol.2010.02.001.

Foley, M.M., Armsby, M.H., Prahler, E.E., Caldwell, M.R., Erickson, A.L., Kittinger, J.N., Crowder, L.B., Levin, P.S., 2013. Improving ocean management through the use of ecological principles and integrated ecosystem assessments. Bioscience 63, 619–631. http://dx.doi.org/10.1525/bio.2013.63.8.5.

Friedlander, A.M., 2004. Status of Hawaii's coastal fisheries in the new millennium, Revised 2004 edition, in: 2001 Fisheries Symposium. Presented at the 2001 Fisheries Symposium, American Fisheries Society, Hawaii Chapter.

Friedlander, A.M., DeMartini, E.E., 2002. Contrasts in density, size, and biomass of reef fishes between the northwestern and the main Hawaiian islands: the effects of fishing down apex predators. Mar. Ecol. Prog. Ser. 230, 253–264.

Friedlander, A.M., Parrish, J.D., 1997. Fisheries harvest and standing stock in a Hawaiian Bay. Fish. Res. 32, 33–50.

Friedlander, A., Poepoe, K., Poepoe, K., Helm, K., Bartram, P., Maragos, J., Abbott, I., 2002. Application of Hawaiian traditions to community-based fishery management. In: Proceedings of the Ninth International Coral Reef Symposium, Bali, 23–27 October 2000, pp. 813–815.

Friedlander, A.M., Brown, E.K., Jokiel, P.L., Smith, W.R., Rodgers, K.S., 2003. Effects of habitat, wave exposure, and marine protected area status on coral reef fish assemblages in the Hawaiian archipelago. Coral Reefs 22, 291–305. http://dx.doi.org/10.1007/s00338-003-0317-2.

Friedlander, A.M., Brown, E., Monaco, M.E., 2007a. Defining reef fish habitat utilization patterns in Hawaii: comparisons between marine protected areas and areas open to fishing. Mar. Ecol. Prog. Ser. 351, 221–233.

Friedlander, A.M., Brown, E.K., Monaco, M.E., 2007b. Coupling ecology and GIS to eval-
uate efficacy of marine protected areas in Hawaii. Ecol. Appl. 17, 715–730. http://dx.
doi.org/10.1890/06-0536.
Friedlander, A., Aeby, G., Balwani, S., Bowen, B., Brainard, R., Clark, A., Kenyon, J.,
Maragos, J., Meyer, C., Vroom, P., Zcbeamzow, J., 2008. The state of coral reef
ecosystems of the northwestern Hawaiian Islands. (NOAA Technical Memorandum
No. NOS NCCOS 73). In: Waddell, J.E., Clarke, A.M. (Eds.), The State of Coral Reef
Ecosystems of the United States and Pacific Freely Associated States: 2008.
NOAA/NCCOS Center of Coastal Monitoring and Assessment's Biogeography
Team, Silver Spring, MD.
Friedlander, A.M., Nowlis, J., Koike, H., 2014. Improving fisheries assessments using historical
data: stock status and catch limits. In: Kittinger, J.N., McClenachan, L., Gedan, K.B.,
Blight, L.K. (Eds.), Applying Marine Historical Ecology to Conservation and Manage-
ment: Using the Past to Manage for the Future. University of California Press, pp. 91–118.
Friedlander, A.M., Shackeroff, J.M., Kittinger, J.N., 2013. Customary marine resource
knowledge and use in contemporary Hawai'i. Pac. Sci. 67, 441–460. http://dx.doi.
org/10.2984/67.3.10.
Gaines, S.D., White, C., Carr, M.H., Palumbi, S.R., 2010. Designing marine reserve net-
works for both conservation and fisheries management. Proc. Natl. Acad. Sci. U.S.A.
107, 18286–18293. http://dx.doi.org/10.1073/pnas.0906473107.
Geslani, C., Loke, M., Takenaka, B., Leung, P., 2012. Hawaii's Seafood Consumption and
Its Supply Sources. Department of Natural Resources and Environmental Management,
University of Hawaii, Manoa, Honolulu, HI.
Glazier, E.W., 2007. Hawaiian fishermen. In: Spindler, G., Stockard, J. (Eds.), Case Studies
in Cultural Anthropology. Thomson-Wadsworth, Belmont, CA, p. 145.
Gopnick, M., 2008. Integrated Marine Spatial Planning in US Waters: The Path Forward.
Marine Conservation Initiative of the Gordon and Betty Moore Foundation.
Gove, J.M., Williams, G.J., McManus, M.A., Heron, S.F., Sandin, S.A., Vetter, O.J.,
Foley, D.G., 2013. Quantifying climatological ranges and anomalies for Pacific coral reef
ecosystems. PLoS ONE 8, e61974. http://dx.doi.org/10.1371/journal.pone.0061974.
Grigg, R.W., 1982. Darwin point: a threshold for atoll formation. Coral Reefs 1, 29–34.
http://dx.doi.org/10.1007/BF00286537.
Grigg, R.W., 1983. Community structure, succession and development of coral reefs in
Hawaii. Mar. Ecol. Prog. Ser. 11, 1–14.
Grigg, R.W., 1997. Paleoceanography of coral reefs in the Hawaiian-Emperor Chain—
revisited. Coral Reefs 16, S33–S38. http://dx.doi.org/10.1007/s003380050239.
Grigg, R.W., Polovina, J., Friedlander, A.M., Rohmann, S.O., 2008. Biology of coral reefs
in the Northwestern Hawaiian Islands. In: Riegl, B.M., Dodge, R.E. (Eds.), Coral Reefs
of the USA, Coral Reefs of the World. Springer, The Netherlands, pp. 573–594.
Grorud-Colvert, K., Claudet, J., Tissot, B.N., Carr, M.H., Day, J.C., Friedlander, A.M.,
Lester, S.E., Lison de Loma, S.E., Malone, D., Walsh, W.J., 2014. Marine protected area
networks: assessing whether the whole is greater than the sum of its parts? PLoS ONE
9 (8), e102298.
Gutiérrez, N.L., Hilborn, R., Defeo, O., 2011. Leadership, social capital and incentives promote successful fisheries. Nature 470, 386–389. http://dx.doi.org/10.1038/
nature09689.
Haight, W.R., Parrish, J.D., Hayes, T.A., 1993. Feeding ecology of deepwater Lutjanid
snappers at Penguin Bank, Hawaii. Trans. Am. Fish. Soc. 122, 328–347. http://dx.
doi.org/10.1577/1548-8659(1993)122<0328:FEODLS>2.3.CO;2.
Halpern, B.S., McLeod, K.L., Rosenberg, A.A., Crowder, L.B., 2008. Managing for cumulative impacts in ecosystem-based management through ocean zoning. Ocean Coast.
Manage. 51, 203–211. http://dx.doi.org/10.1016/j.ocecoaman.2007.08.002.

Halpern, B.S., Klein, C.J., Brown, C.J., Beger, M., Grantham, H.S., Mangubhai, S., Ruckelshaus, M., Tulloch, V.J., Watts, M., White, C., Possingham, H.P., 2013. Achieving the triple bottom line in the face of inherent trade-offs among social equity, economic return, and conservation. Proc. Natl. Acad. Sci. U.S.A. 110, 6229–6234. http://dx.doi.org/10.1073/pnas.1217689110.

Harrison, C.S., 1990. Seabirds of Hawaii: Natural History and Conservation. Cornell University Press, Ithaca, NY.

Hawai'i Administrative Rules (HAR), 2004. Hawaii State Legislature, Honolulu, HI (Chapter 13–94).

Hawaii Revised Statutes (HRS), 2005. Hawaii State Legislature, Honolulu, HI (Chapter 188–22.6).

Higuchi, J., 2008. Propagating cultural Kipuka: the obstacles and opportunities of establishing a community-based subsistence finishing area. U. Haw. L. Rev. 31, 193.

Hospital, J., Beavers, C., 2012. Economic and social characteristics of bottomfish fishing in the main Hawaiian islands. (Administrative report no H-12-01), NMFS-PIFSC, Honolulu, HI.

Hospital, J., Beavers, C., 2014. Catch shares and the main Hawaiian Islands bottomfish fishery: linking fishery conditions and fisher perceptions. Mar. Policy 44, 9–17. http://dx.doi.org/10.1016/j.marpol.2013.08.006.

Hospital, J., Bruce, S.S., Pan, M., 2011. Economic and social characteristics of the Hawaii small boat pelagic fishery. (Administrative report no. H-11-01), US Department of Commerce, National Oceanic and Atmospheric Administration, National Marine Fisheries Service, Pacific Islands Fisheries Science Center, Honolulu, HI.

'I'i, J.P., 1993. Fragments of Hawaiian history. Bishop Museum Press, Honolulu, HI.

Johannes, R.E., 1978. Traditional marine conservation methods in oceania and their demise. Annu. Rev. Ecol. Syst. 9, 349–364. http://dx.doi.org/10.1146/annurev.es.09.110178.002025.

Johannes, R.E., 1981. Working with fishermen to improve coastal tropical fisheries and resource management. Bull. Mar. Sci. 31, 673–680.

Jokiel, P.L., 1987. Ecology, biogeography and evolution of corals in Hawaii. Trends Ecol. Evol. 2, 179–182. http://dx.doi.org/10.1016/0169-5347(87)90016-4.

Jones, P.J.S., 2009. Equity, justice and power issues raised by no-take marine protected area proposals. Mar. Policy 33, 759–765. http://dx.doi.org/10.1016/j.marpol.2009.02.009.

Jupiter, S.D., Weeks, R., Jenkins, A.P., Egli, D.P., Cakacaka, A., 2012. Effects of a single intensive harvest event on fish populations inside a customary marine closure. Coral Reefs 31, 321–334. http://dx.doi.org/10.1007/s00338-012-0888-x.

Juvik, S.P., Juvik, J.O., Paradise, T.R., 1998. Atlas of Hawaii. University of Hawaii Press, Honolulu, HI.

Kaha'ulelio, D., 2006. Ka Oihana Lawai'a: Hawaiian Fishing Traditions. Bishop Museum Press and Awaiaulu Press, Honolulu, HI.

Kamakau, S.M., 1976. The Works of the People of Old: Na hana a ka po'e kahiko. Bishop Museum Press, Honolulu, HI.

Kane, C., Kosaki, R.K., Wagner, D., 2014. High levels of mesophotic reef fish endemism in the Northwestern Hawaiian Islands. Bull. Mar. Sci. 90, 693–703. http://dx.doi.org/10.5343/bms.2013.1053.

Kaneshiro, K.Y., Chinn, P., Duin, K.N., Hood, A.P., Maly, K., Wilcox, B.A., 2005. Hawai'i's mountain-to-sea ecosystems: social–ecological microcosms for sustainability science and practice. Ecohealth 2, 349–360. http://dx.doi.org/10.1007/s10393-005-8779-z.

Kark, S., Levin, N., Grantham, H.S., Possingham, H.P., 2009. Between-country collaboration and consideration of costs increase conservation planning efficiency in the Mediterranean Basin. Proc. Natl. Acad. Sci. 106, 15368–15373. http://dx.doi.org/10.1073/pnas.0901001106.

Kay, A.E., Palumbi, S.R., 1987. Endemism and evolution in Hawaiian marine invertebrates. Trends Ecol. Evol. 2, 183–186. http://dx.doi.org/10.1016/0169-5347(87)90017-6.

Keller, B.D., Gleason, D.F., McLeod, E., Woodley, C.M., Airamé, S., Causey, B.D., Friedlander, A.M., Grober-Dunsmore, R., Johnson, J.E., Miller, S.L., Steneck, R.S., 2009. Climate change, coral reef ecosystems, and management options for marine protected areas. Environ. Manag. 44, 1069–1088. http://dx.doi.org/10.1007/s00267-009-9346-0.

Kelley, C., Moffitt, R., Smith, J.R., 2006. Mega-to micro-scale classification and description of bottomfish essential fish habitat on four banks in the Northwestern Hawaiian Islands. Atoll Res. Bull. 543, 319–332.

Kikiloi, K., Graves, M., 2010. Rebirth of an archipelago: sustaining a Hawaiian cultural identity for people and homeland. Hulili 6, 73–114.

Kirch, P.V., 1989. The Evolution of the Polynesian Chiefdoms. Cambridge University Press, Cambridge, England.

Kittinger, J.N., 2010. Historical ecology of coral reefs in the Hawaiian Archipelago (Ph.D. dissertation). University of Hawaii, Honolulu, HI.

Kittinger, J.N., 2013. Participatory fishing community assessments to support coral reef fisheries comanagement. Pac. Sci. 67, 361–381. http://dx.doi.org/10.2984/67.3.5.

Kittinger, J.N., Dowling, A., Purves, A.R., Milne, N.A., Olsson, P., 2011. Marine Protected Areas, Multiple-Agency Management, and Monumental Surprise in the Northwestern Hawaiian Islands. J. Mar. Bio. 2011, 17 pages. Article ID 241374, http://dx.doi.org/10.1155/2011/241374.

Kittinger, J.N., Duin, K.N., Wilcox, B.A., 2010. Commercial fishing, conservation and compatibility in the Northwestern Hawaiian Islands. Mar. Policy 34, 208–217. http://dx.doi.org/10.1016/j.marpol.2009.06.007.

Kittinger, J.N., Ayers, A.L., Prahler, E.E., 2012. Policy Briefing: Co-Management of Coastal Fisheries in Hawai'i: Overview and Prospects for Implementation. Stanford University, Center for Ocean Solutions & Department of Urban and Regional Planning, University of Hawaii, Manoa, Monterey, CA and Honolulu, HI.

Kittinger, J.N., Koehn, J.Z., Le Cornu, E., Ban, N.C., Armsby, M., Brooks, C., Carr, M.H., Cinner, J.E., Cravens, A., D'iorio, M., Erickson, A., Finkbeiner, E.M., Foley, M.M., Fujita, R., Gelcich, S., Gopnick, M., Hazen, L.J., Lopuch, M., Martin, K.S., Prahler, E.E., Reineman, D.R., Shackeroff, J., White, C., Caldwell, M.R., Crowder, L.B., 2014. A practical approach for putting people into ecosystem-based ocean planning. Front. Ecol. Environ. http://dx.doi.org/10.1890/130267.

Koehn, J.Z., Reineman, D.R., Kittinger, J.N., 2013. Progress and promise in spatial human dimensions research for ecosystem-based ocean planning. Mar. Policy 42, 31–38. http://dx.doi.org/10.1016/j.marpol.2013.01.015.

Kosaki, R.H., 1954. Konohiki Fishing Rights. Legislative Reference Bureau, University of Hawaii, Honolulu, HI.

Le Cornu, E., Kittinger, J.N., Koehn, J.Z., Finkbeiner, E.M., Crowder, L.B., 2014. Current practice and future prospects for social data in coastal and ocean planning. Conserv. Biol. http://dx.doi.org/10.1111/cobi.12310.

Levine, A.S., Richmond, L.S., 2014. Examining enabling conditions for community-based fisheries comanagement: comparing efforts in Hawai'i and American Samoa. Ecol. Soc. 19, 24.

Lizaso, J.S., Goñi, R., Reñones, O., Charton, J.G., Galzin, R., Bayle, J.T., Jerez, P.S., Ruzafa, A.P., Ramos, A.A., 2000. Density dependence in marine protected populations: a review. Environ. Conserv. 27, 144–158.

Lobel, P.S., Robinson, A.R., 1986. Transport and entrapment of fish larvae by ocean mesoscale eddies and currents in Hawaiian waters. Deep-Sea Res. I Oceanogr. Res. Pap. 33, 483–500. http://dx.doi.org/10.1016/0198-0149(86)90127-5.

Longenecker, K., Chan, Y.L., Toonen, R.J., Carlon, D.B., Hunt, T.L., Friedlander, A.M., DeMartini, E.E., 2014. Archaeological evidence of validity of fish populations on unexploited reefs as proxy targets for modern populations. Conserv. Biol.

Lowe, M.K., 2004. The status of inshore fisheries ecosystems in the main Hawaiian Islands at the down of the millenium: cultural impacts, fisheries trends, and management challenges. In: Friedlander, A.M. (Ed.), Status of Hawaiis Coastal Fisheries in the New Millenium. Revised 2004 Edition, Proceedings of the 2001 Fisheries Symposium. American Fisheries Society, Hawaii Chapter, Honolulu, HI.

Lowry, K., Jarman, C., Maehara, S., 1990. Ocean management in Hawaii. Coast. Manag. 18, 233–254. http://dx.doi.org/10.1080/08920759009362113.

Lowry, G.K., White, A.T., Christie, P., 2009. Scaling up to networks of marine protected areas in the Philippines: biophysical, legal, institutional, and social considerations. Coast. Manag. 37, 274–290. http://dx.doi.org/10.1080/08920750902851146.

Malo, D., 1951. Hawaiian Antiquities: Moʻolelo Hawaiʻi. Bishop Museum Press, Honolulu, HI.

Maragos, J.E., Potts, D.C., Aeby, G.S., Gulko, D., Kenyon, J., Siciliano, D., VanRavenswaay, D., 2004. 2000–2002 rapid ecological assessment of corals (Anthozoa) on shallow reefs of the northwestern Hawaiian Islands. Part 1: species and distribution. Pac. Sci. 58, 211–230. http://dx.doi.org/10.1353/psc.2004.0020.

Papahānaumokuākea Marine National Monument, NOAA, FWS, DLNR, 2008. Papahānaumokuākea Marine National Monument Management Plan, Honolulu, HI.

Mascia, M.B., 2003. The human dimension of coral reef marine protected areas: recent social science research and its policy implications. Conserv. Biol. 17, 630–632.

McClanahan, T.R., Mangi, S., 2000. Spillover of expoitable fishes from a marine park and its effect on the adjacent fishery. Ecol. Appl. 10, 1792–1805. http://dx.doi.org/10.1890/1051-0761(2000) 010[1792:SOEFFA]2.0.CO;2.

McClanahan, T.R., Maina, J., Davies, J., 2005. Perceptions of resource users and managers towards fisheries management options in Kenyan coral reefs. Fish. Manag. Ecol. 12, 105–112. http://dx.doi.org/10.1111/j.1365-2400.2004.00431.x.

McClanahan, T.R., Marnane, M.J., Cinner, J.E., Kiene, W.E., 2006. A comparison of marine protected areas and alternative approaches to coral-reef management. Curr. Biol. 16, 1408–1413. http://dx.doi.org/10.1016/j.cub.2006.05.062.

McClenachan, L., Kittinger, J.N., 2013. Multicentury trends and the sustainability of coral reef fisheries in Hawai'i and Florida. Fish Fish. 14, 239–255. http://dx.doi.org/10.1111/j.1467-2979.2012.00465.x.

McCrea-Strub, A., Zeller, D., Rashid Sumaila, U., Nelson, J., Balmford, A., Pauly, D., 2011. Understanding the cost of establishing marine protected areas. Mar. Policy 35, 1–9. http://dx.doi.org/10.1016/j.marpol.2010.07.001.

McGregor, D., 2007. Na Kuaʻaina: Living Hawaiian Culture. University of Hawaii Press, Honolulu, HI.

McGregor, D.P., Minerbi, L., Matsuoka, J., 1998. A holistic assessment method of health and well-being for Native Hawaiian communities. Pac. Health Dialog 5, 361–369.

McGregor, D., Morelli, P., Matsuoka, J., Rodenhurst, R., Kong, N., Spencer, M., 2003. An ecological model of Native Hawaiian well-being. Pac. Health Dialog 10, 106–128.

Merritt, D., Donovan, M.K., Kelley, C., Waterhouse, L., Parke, M., Wong, K., Drazen, J.C., 2011. BotCam: a baited camera system for non-extractive monitoring of bottomfish species. Fish. Bull. 109, 56–67.

Moffitt, R.B., Kobayashi, D.R., DiNardo, G.T., 2006. Status of the Hawaiian bottomfish stocks, 2004. (Administrative Report). National Marine Fisheries Service, Honolulu, HI.

Molloy, P.P., McLean, I.B., Côté, I.M., 2009. Effects of marine reserve age on fish populations: a global meta-analysis. J. Appl. Ecol. 46, 743–751. http://dx.doi.org/10.1111/j.1365-2664.2009.01662.x.

Moore, C.H., Drazen, J.C., Kelley, C.D., Misa, W.F.X.E., 2013. Deepwater marine protected areas of the main Hawaiian Islands: establishing baselines for commercially valuable bottomfish populations. Mar. Ecol. Prog. Ser. 476, 167–183. http://dx.doi.org/10.3354/meps10132.

Myers, R.A., Worm, B., 2003. Rapid worldwide depletion of predatory fish communities. Nature 423, 280–283. http://dx.doi.org/10.1038/nature01610.

Myers, R.A., Worm, B., 2005. Extinction, survival or recovery of large predatory fishes. Philos. Trans. R. Soc., B 360, 13–20. http://dx.doi.org/10.1098/rstb.2004.1573.

National Marine Fisheries Service, 2014. Fisheries economics of the United States: 2012. (NOAA Technical Memorandum No. NMFS-F/SPO-137). US Department of Commerce, National Oceanic and Atmospheric Administration, National Marine Fisheries Service, Pacific Islands Fisheries Science Center.

Needham, M.D., Tynon, J.F., Ceurvorst, R.L., Collins, R.L., Connor, W.M., Culnane, M.J.W., 2008. Recreation carrying capacity and management at Pupukea Marine Life Conservation District on Oahu, Hawaii, (Final project report for Hawaii Division of Aquatic Resources, Department of Land and Natural Resources). Oregon State University, Department of Forest Ecosystems and Society, Corvallis, OR.

Ortiz, D.M., Tissot, B.N., 2008. Ontogenetic patterns of habitat use by reef-fish in a Marine Protected Area network: a multi-scaled remote sensing and in situ approach. Mar. Ecol. Prog. Ser. 365, 217–232.

Ortiz, D.M., Tissot, B.N., 2012. Evaluating ontogenetic patterns of habitat use by reef fish in relation to the effectiveness of marine protected areas in West Hawaii. J. Exp. Mar. Biol. Ecol. 432–433, 83–93. http://dx.doi.org/10.1016/j.jembe.2012.06.005.

Ostrom, E., 2009. A general framework for analyzing sustainability of social-ecological systems. Science 325, 419–422. http://dx.doi.org/10.1126/science.1172133.

Ostrom, E., Janssen, M.A., Anderies, J.M., 2007. Going beyond panaceas. Proc. Natl. Acad. Sci. 104 (39), 15176–15178.

Palumbi, S.R., 2004. Marine reserves and ocean neighborhoods: the spatial scale of marine populations and their management. Annu. Rev. Environ. Resour. 29, 31–68. http://dx.doi.org/10.1146/annurev.energy.29.062403.102254.

Pandolfi, J.M., Jackson, J.B., Baron, N., Bradbury, R.H., Guzman, H.M., Hughes, T.P., Kappel, C.V., Micheli, F., Ogden, J.C., Possingham, H.P., 2005. Are U. S. coral reefs on the slippery slope to slime? Science 307, 1725–1726.

Parke, M., 2007. Linking Hawaii fisherman reported commercial bottomfish catch data to potential bottomfish habitat and proposed restricted fishing areas using GIS and spatial analysis. (NOAA Technical Memorandum). National Marine Fisheries Service, Honolulu, HI.

Pauly, D., 2009. Beyond duplicity and ignorance in global fisheries. Sci. Mar. 73, 215–224. http://dx.doi.org/10.3989/scimar.2009.73n2215.

Poepoe, K., Bartram, P., Friedlander, A., 2007. The use of traditional Hawaiian knowledge in the contemporary management of marine resources. In: Haggan, N., Neis, B., Baird, I. (Eds.), Fishers' Knowledge in Fisheries Science and Management. UNESCO, Paris, France, pp. 117–141.

Pollnac, R., Seara, T., 2011. Factors influencing success of marine protected areas in the Visayas, Philippines as related to increasing protected area coverage. Environ. Manag. 47, 584–592. http://dx.doi.org/10.1007/s00267-010-9540-0.

Pooley, S.G., 1993. Hawai'i's marine fisheries: some history, long-term trends, and recent developments. Mar. Fish. Rev. 55, 7–19.

Ralston, C., 1984. Hawaii 1778–1854: some aspects of Maka'ainana response to rapid cultural change. J. Pac. Hist. 19, 21–40. http://dx.doi.org/10.1080/00223348408572478.

Randall, J.E., 1998. Zoogeography of shore fishes of the Indo-Pacific region. Zool. Stud. 37, 227–268.

Rossiter, J.S., Levine, A., 2014. What makes a "successful" marine protected area? The unique context of Hawaii's fish replenishment areas. Mar. Policy 44, 196–203. http://dx.doi.org/10.1016/j.marpol.2013.08.022.

Ruddle, K., 1996. Traditional management of reef fishing. In: Polunin, N.V.C., Roberts, C.M. (Eds.), Reef Fisheries. In: Chapman & Hall Fish and Fisheries Series. Springer, Netherlands, pp. 315–335.

Russ, G.R., 2002. Yet another review of marine reserves as reef fishery management tools. In: Sale, P.F. (Ed.), Coral Reef Fishes: Dynamics and Diversity in a Complex Ecosystem. Academic Press, Inc., San Diego, CA, pp. 421–443.

Russ, G., Alcala, A., 1996. Do marine reserves export adult fish biomass? Evidence from Apo Island, central Philippines. Mar. Ecol. Prog. Ser. 132, 1–9. http://dx.doi.org/10.3354/meps132001.

Russ, G.R., Alcala, A.C., 2010. Enhanced biodiversity beyond marine reserve boundaries: the cup spillith over. Ecol. Appl. 21, 241–250. http://dx.doi.org/10.1890/09-1197.1.

Sackett, D.K., Drazen, J.C., Moriwake, V.N., Kelley, C.D., Schumacher, B.D., Misa, W.F. X.E., 2014. Marine protected areas for deepwater fish populations: an evaluation of their effects in Hawai'i. Mar. Biol. 161, 411–425. http://dx.doi.org/10.1007/s00227-013-2347-9.

Sadovy, Y., 2005. Trouble on the reef: the imperative for managing vulnerable and valuable fisheries. Fish Fish. 6, 167–185. http://dx.doi.org/10.1111/j.1467-2979.2005.00186.x.

Schug, D.M., 2001. Hawai'i's commercial fishing industry: 1820–1945. Hawaii. J. Hist. 35.

Seaton, S.L., 1974. The Hawaiian kapu abolition of 1891. Am. Ethnol. 1, 193–206. http://dx.doi.org/10.1525/ae.1974.1.1.02a00100.

Selkoe, K.A., Halpern, B.S., Ebert, C.M., Franklin, E.C., Selig, E.R., Casey, K.S., Bruno, J., Toonen, R.J., 2009. A map of human impacts to a "pristine" coral reef ecosystem, the Papahānaumokuākea Marine National Monument. Coral Reefs 28, 635–650. http://dx.doi.org/10.1007/s00338-009-0490-z.

Shanks, A.L., Grantham, B.A., Carr, M.H., 2003. Propagule dispersal distance and the size and spacing of marine reserves. Ecol. Appl. 13, 159–169. http://dx.doi.org/10.1890/1051-0761(2003) 013[0159:PDDATS]2.0.CO;2.

Shomura, R.S., 1987. Hawaii's Marine Fishery Resources: Yesterday (1900) and Today (1986). Southwest Fisheries Center, National Marine Fisheries Service, Honolulu, HI.

Shomura, R.S., 2004. A historical perspective of Hawai'i's marine resources, fisheries, and management issues over the past 100 years. In: Friedlander, A.M. (Ed.), The Status of Hawaii's Coastal Fisheries in the New Millenium. American Fisheries Society, Hawaii Chapter, Honolulu, HI, pp. 6–11.

Sladek Nowlis, J., Friedlander, A.M., 2005. Marine reserve function and design for fisheries management. In: Norse, E.A., Crowder, L.B. (Eds.), Marine Conservation Biology: The Science of Maintaining the Sea's Biodiversity. Island Press, Washington, DC, pp. 280–301.

Sladek Nowlis, J., Roberts, C.M., 1999. Fisheries benefits and optimal design of marine reserves. Fish. Bull. 97, 604–616.

Smith, M.K., 1993. An ecological perspective on inshore fisheries in the main Hawaiian Islands. Mar. Fish. Rev. 55, 34–49.

Stamoulis, K.A., Friedlander, A.M., 2013. A seascape approach to investigating fish spillover across a marine protected area boundary in Hawai'i. Fish. Res. 144, 2–14. http://dx.doi.org/10.1016/j.fishres.2012.09.016.

Stevenson, T.C., Tissot, B.N., 2013. Evaluating marine protected areas for managing marine resource conflict in Hawaii. Mar. Policy 39, 215–223. http://dx.doi.org/10.1016/j.marpol.2012.11.003.

Stevenson, T.C., Tissot, B.N., Dierking, J., 2011. Fisher behaviour influences catch productivity and selectivity in West Hawaii's aquarium fishery. ICES J. Mar. Sci. 68, 813–822. http://dx.doi.org/10.1093/icesjms/fsr020.

Stevenson, T.C., Tissot, B.N., Walsh, W.J., 2013. Socioeconomic consequences of fishing displacement from marine protected areas in Hawaii. Biol. Conserv. 160, 50–58. http://dx.doi.org/10.1016/j.biocon.2012.11.031.

Stewart, R.E., Desai, A., Walters, L.C., 2011. Wicked Environmental Problems: Managing Uncertainty and Conflict. Island Press, Washington, DC.

Szuster, B.W., Needham, M.D., 2010. Marine recreation at the Molokini Shoal MLCD. (Final project report for Hawaii Division of aquatic resources). Department of Geography, University of Hawaii at Manoa, Honolulu, HI.

Tanaka, W., 2008. Ho ohana aku, Ho ola aku: first steps to averting the tragedy of the commons in Hawaii's nearshore fisheries. Asian Pac. L. Pol'y J. 10, 235.

Tanaka, W.C., Miyashiro, M.A., Kaulukukui, K.L., 2012. Enforcement chain analysis of aquatic resource enforcement on Oahu Island and north shore Maui. Oahu Resource Conservation and Development Council, Honolulu, HI.

Tava, R., Keale, M.K., 1990. Niihau: The Traditions of a Hawaiian Island. Mutual Publishing Company, Honolulu, HI.

Tissot, B.N., 2005. Integral marine ecology: community-based fishery management in Hawaiʻi. World Futures 61, 79–95. http://dx.doi.org/10.1080/02604020590902371.

Tissot, B.N., Hallacher, L.E., 2003. Effects of aquarium collectors on coral reef fishes in Kona, Hawaii. Conserv. Biol. 17, 1759–1768.

Tissot, B.N., Walsh, W.J., Hallacher, L.E., 2004a. Evaluating effectiveness of a marine protected area network in West Hawaiʻi to increase productivity of an aquarium fishery. Pac. Sci. 58, 175–188.

Tissot, B.N., Walsh, W.J., Hallacher, L.E., 2004b. Evaluating the Effectiveness of a marine reserve network in West Hawaii to improve management of the aquarium fishery. NOAA Coral Reef Conservation Program, Technical Report, Honolulu, HI.

Tissot, B.N., Walsh, W.J., Hixon, M.A., 2009. Hawaiian islands marine ecosystem case study: ecosystem- and community-based management in Hawaii. Coast. Manag. 37, 255–273. http://dx.doi.org/10.1080/08920750902851096.

Toonen, R.J., Wilhelm, T.ʻ.A., Maxwell, S.M., Wagner, D., Bowen, B.W., Sheppard, C.R.C., Taei, S.M., Teroroko, T., Moffitt, R., Gaymer, C.F., Morgan, L., Lewis, N., Sheppard, A.L.S., Parks, J., Friedlander, A.M., 2013. One size does not fit all: the emerging frontier in large-scale marine conservation. Mar. Pollut. Bull. 77, 7–10. http://dx.doi.org/10.1016/j.marpolbul.2013.10.039.

Townsend, R.E., Pooley, S.G., 1995. Corporate management of the Northwestern Hawaiian Islands lobster fishery. Ocean Coast. Manage. 28, 63–83. http://dx.doi.org/10.1016/0964-5691(95)00043-7.

Uchida, R.N., Tagami, D.T., 1984. Groundfish fisheries and research in the vicinity of seamounts in the North Pacific Ocean. Mar. Fish. Rev. 46, 1–17.

van Beukering, P., Cesar, H., 2004. Economic Analysis of Marine Managed Areas in the Main Hawaiian Islands. University of Hawaii, Honolulu, HI.

Vaughan, M.B., Vitousek, P.M., 2013. Mahele: sustaining communities through small-scale inshore fishery catch and sharing network. Pac. Sci. 67, 329–344. http://dx.doi.org/10.2984/67.3.3.

Walsh, W., 1987. Patterns of recruitment and spawning in Hawaiian reef fishes. Environ. Biol. Fish 18, 257–276. http://dx.doi.org/10.1007/BF00004879.

Walsh, W.J., Cotton, S.S., Dierking, J., Williams, I.D., 2003. The commercial marine aquarium fishery in Hawaiʻi. In: Friedlander, A.M. (Ed.), Status of Hawaiʻi's Coastal Fisheries in the New Millennium. American Fisheries Society, Hawaii Chapter, Honolulu, HI, pp. 132–159.

Walsh, W., Cotton, S., Barnett, C., Couch, C., Preskitt, L., Tissot, B., Osada-D'Avella, K., 2013. Long-term monitoring of coral reefs of the main Hawaiian Islands (Technical report). Hawaii Division of Aquatic Resources, Honolulu, HI.

Williams, I.D., Walsh, W.J., Miyasaka, A., Friedlander, A.M., 2006. Effects of rotational closure on coral reef fishes in Waikiki-Diamond head fishery management area, Oahu, Hawaii. Mar. Ecol. Prog. Ser. 310, 139–149.

Williams, I.D., Walsh, W.J., Schroeder, R.E., Friedlander, A.M., Richards, B.L., Stamoulis, K.A., 2008. Assessing the importance of fishing impacts on Hawaiian coral reef fish assemblages along regional-scale human population gradients. Environ. Conserv. 35, 261–272. http://dx.doi.org/10.1017/S0376892908004876.

Williams, I.D., Walsh, W.J., Claisse, J.T., Tissot, B.N., Stamoulis, K.A., 2009. Impacts of a Hawaiian marine protected area network on the abundance and fishery sustainability of the yellow tang, *Zebrasoma flavescens*. Biol. Conserv. 142, 1066–1073. http://dx.doi.org/10.1016/j.biocon.2008.12.029.

Wilson, J., Yan, L., Wilson, C., 2007. The precursors of governance in the Maine lobster fishery. Proc. Natl. Acad. Sci. 104, 15212–15217. http://dx.doi.org/10.1073/pnas.0702241104.

Woodroffe, R., Ginsberg, J.R., 1998. Edge effects and the extinction of populations inside protected areas. Science 280, 2126–2128. http://dx.doi.org/10.1126/science.280.5372.2126.

Yeh, J., Drazen, J.C., 2009. Depth zonation and bathymetric trends of deep-sea megafaunal scavengers of the Hawaiian Islands. Deep-Sea Res. I Oceanogr. Res. Pap. 56, 251–266. http://dx.doi.org/10.1016/j.dsr.2008.08.005.

Zeller, D., Darcy, M., Booth, S., Lowe, M.K., Martell, S., 2008. What about recreational catch? Potential impact on stock assessment for Hawaii's bottomfish fisheries. Fish. Res. 91, 88–97. http://dx.doi.org/10.1016/j.fishres.2007.11.010.

CHAPTER SIX

Marine Protected Area Networks in California, USA

Louis W. Botsford[1],*, J. Wilson White[†], Mark H. Carr[‡], Jennifer E. Caselle[§]

*Department of Wildlife, Fish, and Conservation Biology, University of California, Davis, California, USA
[†]Department of Biology and Marine Biology, University of North Carolina Wilmington, Wilmington, North Carolina, USA
[‡]Department of Ecology and Evolution, University of California, Santa Cruz, California, USA
[§]Marine Science Institute, University of California, Santa Barbara, California, USA
[1]Corresponding author: e-mail address: lwbotsford@ucdavis.edu

Contents

Abstract

California responded to concerns about overfishing in the 1990s by implementing a network of marine protected areas (MPAs) through two science-based decision-making processes. The first process focused on the Channel Islands, and the second addressed California's entire coastline, pursuant to the state's Marine Life Protection Act (MLPA). We review the interaction between science and policy in both processes, and lessons learned. For the Channel Islands, scientists controversially recommended setting aside 30–50% of coastline to protect marine ecosystems. For the MLPA, MPAs were intended to be ecologically connected in a network, so design guidelines included minimum size and maximum spacing of MPAs (based roughly on fish movement rates), an approach that also implicitly specified a minimum fraction of the coastline to be protected. As MPA science developed during the California processes, spatial population models were constructed to quantify how MPAs were affected by adult fish movement and larval dispersal, i.e., how population persistence within MPA networks depended on fishing outside the MPAs, and how fishery yields could either increase or decrease with MPA implementation, depending on fishery management. These newer quantitative methods added to, but did not supplant, the initial rule-of-thumb guidelines. In the future, similar spatial population models will allow more comprehensive evaluation of the integrated effects of MPAs and conventional fisheries management. By 2011, California had implemented 132 MPAs covering more than 15% of its coastline, and now stands on the threshold of the most challenging step in this effort: monitoring and adaptive management to ensure ecosystem sustainability.

Keywords: California, MPA, Channel Islands, Population models, Science, Process, Planning

1. INTRODUCTION

California responded to rising global concerns regarding the effects of overfishing on marine ecosystems in the 1990s by implementing a network of marine protected areas (MPAs). Here, we describe that effort in terms of the ecological setting, the initial concerns, the enabling legislation, the planning process, and the concurrent development of the science of MPAs. We synthesize the various kinds of success achieved, the challenges in the process, and the potential for the future. Our intent is to provide an example for other future MPA processes of how science interacted with the legal, social, ecological, and economic aspects throughout the implementation process based on our experiences as scientists involved in this process. We base our exposition on the relevant scientific data, as well as on the century-long history of the science of marine resource management. In particular, we take note of the scientific developments taking place over the lifetime of the implementation process in California and how the structure of that process influenced the degree to which science informed the MPA design. The California process was groundbreaking in many ways, not least of which was the goal of developing a functional *network* of ecologically connected MPAs, as opposed to a collection of multiple MPAs designed independently of one another. As such our summary of the process pays particular attention to the science of MPA network design.

1.1. Physical and biological context

The marine environment of California is defined by the contrast between the warm–temperate/subtropical southern region (from the Mexican border to Point Conception, with biota derived from the San Diegan biogeographical province) and the cold-temperate northern region (north of Point Conception, with biota belonging to the Oregonian region; Horn et al., 2006; Figure 6.1A). The northern region is heavily influenced by the equatorward–flowing California Current, a highly productive Eastern Boundary Current. High productivity is driven by spring upwelling winds, which are more prominent to the north of Point Conception (Checkley and

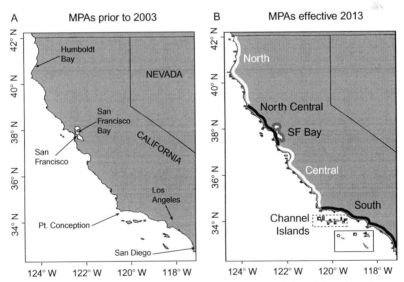

Figure 6.1 Map showing MPAs in California prior to 2003 (A) and those in place as of 2013 (B). MPAs administered by the state of California are red (black in the print version), and MPAs in U.S. territorial waters that are administered federally are outlined in blue (light grey in the print version). Estuarine MPAs are not shown. Coloured highlighting in (B) indicates the different Study Regions utilized in the design process that implemented the Marine Life Protection Act (MLPA). The dashed blue line (dashed light grey line in the print version) indicates the Channel Islands; MPAs were designed for those islands as part of a design process separate from the MLPA, and were not modified during the MLPA process for the South Coast region. The Channel Islands MPAs became effective in 2003, and MPAs designed under the Marine Life Protection Act (MLPA) became effective region-by-region between 2007 and 2013.

Barth, 2009). These winds vary from year to year, and productivity is also modulated on longer time scales by atmospheric and oceanographic conditions throughout the eastern North Pacific (Di Lorenzo et al., 2013). The continental shelf is narrower north of Point Conception and the shallow (<30 m depth) seafloor is primarily rocky reefs with kelp forests (comprises *Macrocystis pyrifera* and *Nereocystis leutkeana*) subject to frequent disturbance (Carr and Reed, 2015; Graham et al., 2008; Reed et al., 2011). In contrast, the California Current remains further offshore south of Point Conception, and nearshore surface waters are influenced more by warmer recirculating flow from the California Current and the poleward-flowing Davidson Current (the latter also extends northward past San Francisco Bay as a surface current in the winter; Hickey, 1998; Bray et al., 1999). In addition to warmer water, the southern region has a shallow, broad continental shelf and several large offshore islands and supports more persistent kelp (only *M. pyrifera*) forests than are found in the north (Carr and Reed, 2015; Graham et al., 2008; Reed et al., 2011). There are key ecological differences among the islands and the mainland (Ebeling et al., 1980). In particular, the northwestern most Channel Islands (San Miguel, Santa Rosa, and San Nicolas Islands) lie at the boundary between the bioregions, with cooler waters, more frequent disturbances, and a mix of San Diegan and Oregonian species (Hamilton et al., 2010; Pondella et al., 2005). Further south and east, the islands experience warmer waters and less frequent disturbances. The mainland coast south of Point Conception is more heavily influenced by human activities (e.g. ports, hardened coastlines, intake and discharge of power plants, recreational fishing, and urban runoff from the Los Angeles and San Diego metropolitan areas). In general, mainland south coasts are sandy with interspersed low relief rocky reefs, whereas the offshore islands contain primarily high relief rocky habitat and less turbid water (Pondella et al., 2005).

California has a Mediterranean climate, with wet winters and dry summers. Freshwater flow into the ocean is greater in the north, with several rivers forming large estuaries (e.g. San Francisco Bay, Humboldt Bay), although river damming has reduced both overall river outflow and variability during the twentieth century (Hanak et al., 2011; Hundley, 2001).

1.2. Context: History of fisheries management and conservation in California

The move to MPAs in California was influenced by the state of marine resource management from local to global levels in the late 1990s. Globally, there was growing concern for the high fraction of global fisheries that were reported overfished (e.g. Botsford et al., 1997; Hutchings, 2000; more recently reviewed by Worm et al., 2009). Reports of this fraction ranged from about 20% to more than 60%, depending on whether fully exploited fisheries were included in the overfished category (Mace, 2001). A second, related growing global concern was that marine resource management was falling short because of its focus on single species, ignoring the more extensive *ecosystem* effects of fishing (e.g. Botsford et al., 1997; Pikitch et al., 2004). The proposed solution was a more holistic, *ecosystem-based* approach that included the effects of (a) interactions among multiple species, (b) incidental take of nontarget species, (c) impacts on essential fish habitat, (d) the changing physical environment, and (e) the socioeconomic consequences of ecosystem status and marine ecosystem services. MPAs were considered to be an ecosystem-based management tool because they can protect both the physical (geomorphological, water quality) and biotic components of ecosystems from fishing and other anthropogenic impacts (Murray et al., 1999).

The local context was influenced by historical events of the previous several decades, as far back as the dramatic collapse of the California fishery for Pacific sardine (*Sardinops sagax*) around 1950 (Ueber and MacCall, 2005). A second fishery collapse occurred later in that decade with the decline of the central California Dungeness crab (*Metacarcinus magister*) fishery in 1958 (Botsford, 1981; Wild and Tasto, 1983). Other management crises followed in subsequent decades. In the 1980s, scientists and managers became aware that the history of California's abalone (*Haliotis* spp.) fishery was a prime example of serial depletion, leading to the near extinction of several species (Karpov et al., 2000). The many rockfish (*Sebastes* spp., Scorpaenidae) species off the California coast, ranging from nearshore reefs to the continental slope, went from being a concern as an under-utilized resource in the 1970s to having several species declared overfished in the 1990s (Love et al., 1998; Ralston, 1998). Interspersed among these declines was the rapid development in the 1980s of a fishery for the red sea urchin (*Strongylocentrotus franciscanus*) in northern California followed by a dramatic decline in catch (Botsford et al., 2004) as well as large increases in live-finfish fisheries (CDFG, 2002; Starr et al., 2002). Leet et al. (2001) provide a comprehensive survey of the status of California marine resources at that time.

This awareness of the vulnerability of California's marine resources set the context for improved management. It was coupled with an increasing conservation sentiment among California citizens, initiated in part by the effects of an oil spill in 1969 in the Santa Barbara region. These sentiments operated in the economic context of California's diverse modern economy (at least the eighth largest in the world since 1970s), with dominant entertainment, information technology, tourism and agricultural sectors, in contrast to a relatively small commercial fishing sector, and an economically more significant recreational fishing sector (Kildow and Colgan, 2005). Prior to the recent new MPAs (the subject of this chapter), there were only scattered, small, single-purpose MPAs in the state (McArdle, 1997; Figure 6.1A), accompanied by areas of excluded public use near military bases.

California fisheries are managed either by (a) the state of California (for species occurring only out to 3 nautical miles (nm; 5.56 km) offshore, the boundary of state waters within the United States), (b) the regional council of the federal management system, the Pacific Fishery Management Council (for species occurring from 3 to 200 nm, the U.S. territorial waters within the Exclusive Economic Zone), or (c) jointly by state and federal authorities.

1.3. Context: The state of fisheries and conservation science

By the late 1990s, the science of fisheries management around the globe had developed from concerns over declines in fishery catch in the early part of the twentieth century, to a standard procedure of calculating maximum sustainable yield (MSY) for a number of fisheries beginning in the 1950s, on to a gradual realization that simply seeking MSY would not be sufficient (Botsford, 2013). Concerns over the ineffectiveness of a simple MSY approach began to arise in the 1970s (Larkin, 1977), which ultimately led to development of a precautionary approach to fishery management in the early 1990s (FAO, 1996; Garcia, 1996; Mangel et al., 1996). The precautionary approach emphasized frequent observation of fisheries (e.g. biomass, age structure or catch), and comparison of these to reference points (i.e. predetermined values of those variables), with consequent responses by management, such as changes in allowable catch. These reference points included target reference points, which were essentially management goals similar to the earlier maximization of yield, and limit reference points, which were intended as critical limits to guard directly against overfishing and population collapse. Federal fisheries management in the U.S. operated under the Fisheries Conservation and Management Act (1976), which included specific attention to the potential for overfishing in its 1996 reauthorization as the Magnuson–Stevens Fisheries Conservation and Management Act (Restrepo and Powers, 1999; Rosenberg et al., 1994).

Parallel to the development of the reference point concept, a better understanding of the critical features of fish population dynamics emerged and largely supplanted the earlier approaches (e.g. logistic models, surplus production models) originally used to develop the MSY concept (Botsford, 2013). This new understanding centred on the realization that the key to persistence in marine populations is the maintenance of sufficient lifetime spawning to allow each adult to replace itself with a new recruit within its lifetime (i.e. remaining above a critical replacement threshold). Initial comparisons to empirical information on population collapses suggested that preserving 35% of unfished lifetime spawning would be a safe hedge against collapse (Clark, 1991; Mace and Sissenwine, 1993). Unfortunately, this 35% replacement level was too low for Pacific coast rockfishes, leading to overfishing (Clark, 2002; Ralston, 2002), and management has subsequently used more conservative replacement limits. For many fisheries, this limit is 40%. If the fishing mortality rate is high enough to cause lifetime reproduction to fall below the critical replacement limit (e.g. 35% or 40%) in the United States, the stock is declared to be undergoing overfishing. If the spawning stock biomass falls below a certain fraction of the unfished biomass (usually 40%), the stock is also declared to be overfished (Restrepo et al., 1998).

By the late 1990s, the federal fisheries management process in the United States had evolved to its current form (Fluharty, 2000). It generally involves a decision-making process in regional councils (e.g. http://www.pcouncil.com/), based on stock assessments involving population models fit to fishery data and fishery independent data, to determine periodically (annually in many cases) the amount of catch that should be taken. The stock assessments and technical aspects of decisions made by these councils are reviewed by a group of scientists called the 'Scientific and Statistical Committee'.

As the science of fishery management was maturing, conservation advocates and some fisheries biologists began to argue that fisheries could be managed more cautiously, and ecosystems could be better protected by reducing fishing effort to zero in designated protected areas, rather than attempting to control the overall level of fishing (Murray et al., 1999). These recommendations called for single protected areas, as well as 'networks' of protected areas; collections of protected areas linked by larval dispersal that replenish one another and the fished populations between them. There was also a growing realization among scientists that a decision-making process for management by MPAs would require new scientific understanding to predict their benefits and costs. For the most part, the models being used in conventional fisheries management did not consider how populations varied over space; they were concerned with temporal variability only. To manage populations using networks of MPAs, there would be a need to know (1) how many MPAs are required, how large they should be, and where they should be placed to ensure the persistence of multiple species and (2) how does fishery yield in management by MPAs compare to yield with conventional control of effort? These questions were only beginning to be addressed when the decision-making process for California's MPAs began in the late 1990s.

The effort to develop the science of marine reserve design and assessment was kick-started by a scientific working group at the National Center for Ecological Analysis and Synthesis (NCEAS) in Santa Barbara in 1998, and many of the seminal papers on the topic emerged from that group (Lubchenco et al., 2003 and references therein). With regard to the first questions (how many, how large, and where?), earlier population models had suggested that it was best to place an MPA in a 'source' location (e.g. an upstream reef in an archipelago) so that planktonic larvae spawned inside the MPA could seed populations in other patches (e.g. Crowder et al., 2000; but see Gaines et al., 2003; Hastings and Botsford, 2006 for potential draw-backs to this approach). Botsford et al. (2001) approached the question from a perspective more relevant to the California coast: a long, linear coastline with a network of evenly spaced MPAs, and relatively sedentary fish or invertebrate species that disperse widely as larvae. Analysis of their simple, strategic model (as opposed to a more detailed 'tactical' model of a specific location) showed that populations could persist in one of two ways: (1) in single MPAs that were at least as wide as the average dispersal distance of larvae (termed *self-persistence*) or (2) in a network of smaller MPAs covering an adequate fraction of the coastline. This mode of persistence was termed *network persistence* because even when individual MPAs within the network are too small to sustain themselves independently, larval connectivity among them allows the population distributed across the entire network to be sustained (White et al., 2010a). The minimum fraction of the coastline that must be protected to achieve network persistence was determined to be bio-logically related to the critical replacement threshold described above in a single-population context under conventional, non-spatial fishery manage-ment. Under the idealized assumption that fishing removed all reproduction outside MPAs (i.e. the 'scorched earth' assumption), the minimum fraction in MPAs necessary for network persistence would be equal to the critical replacement threshold from non-spatial population dynamics, presumed generally to be equal to 35% or 40%. When the amount of fishing outside the protected areas did not reduce reproduction to zero, the minimum fraction of coastline required for population persistence would be less. Also, the presence of alongshore flows transporting larvae would require higher fractions in reserves (Botsford et al., 2001). Later research would build on these basic results, further examining their sensitivity to such factors as alongshore currents, retention zones, and adult movement (Gaines et al., 2003; Kaplan, 2006; Moffitt et al., 2009; White et al., 2010a), but the central concept has proven highly influential. In particular, the second way of achieving the population persistence requirement was in part the inspiration for the idea that one could formulate general guidelines for the *size and spacing* of MPAs, and the idea that 35% of the coastline must be protected for MPAs to be effective (see Gaines et al., 2010). That percentage has been cited frequently as a theoretical requirement, while in reality the threshold actually would be less with less than scorched earth fishing outside the MPAs, and would depend on the settler–recruit relationship of a particular species, adult movement, and alongshore currents. It is not a general rule (Botsford et al., 2001; Kaplan and Botsford, 2005; Moffitt et al., 2009, 2011; White, 2010; White et al., 2010a).

With regard to the second question of differences in yield between MPAs and conventional management, analyses of simple, strategic models had shown that management by MPAs and conventional management by limiting catch or effort were essentially equivalent in the sense that under particular conditions, the potential yields from each would be equal (Hastings and Botsford, 1999; Mangel, 1998). These results implied that if a fishery were well managed (e.g. at MSY), adding MPAs would diminish yield because fishable area would be diminished (Holland and Brazee, 1996). However, if the fishery were overharvested beyond MPA boundaries, then MPAs could actually enhance fishery yields (Holland and Brazee, 1996; Sladek Nowlis and Roberts, 1999) and the enhancement would be greatest for networks of many small reserves (essentially maximizing the number of boundaries across which fish could spill over; Hastings and Botsford, 2003; Neubert, 2003).

2. ESTABLISHMENT OF MPAs IN CALIFORNIA

With the exception of the few individual MPAs established in *ad hoc* ways over the decades preceding the 1990s, two primary efforts in California led to implementation of science-guided networks of MPAs in California. The first effort was focused on the Channel Islands off southern

California (Figure 6.1), and the second concerned a statewide network of protected areas. We describe these processes here.

2.1. Channel Islands marine protected areas

In 1998, a group of fishermen, managers and other citizens who were concerned about declining fishery resources such as abalone, lobsters, and near-shore rockfishes, approached the California Fish and Game Commission with a proposal to a set aside areas for protection in the northern Channel Islands, bounding the Santa Barbara channel (CDFG, 2003; Osmond et al., 2010; Figures 6.1B and 6.2D). The Channel Islands region is complex from a planning perspective because of overlapping management and political jurisdictions as well as variable environmental and ecological conditions. Eleven federal, state, and local agencies have some jurisdiction in the planning region (Airamé et al., 2003). While both the Channel Islands National Marine Sanctuary (CINMS) and the Channel Islands National Park (CINP) overlap around the northern Channel Islands, neither agency regulates commercial or recreational fishing. The California Department of Fish and Wildlife (CDFW; previously the California Department of Fish and Game, CDFG, prior to 2013) manages all fisheries in state waters (within 3 nm (5.6 km) of shore), while the California Fish and Game Commission (an appointed body) has authority to set all state fishery regulations, including the creation of MPAs.

At the same time, the CINMS was beginning the process of updating its management plan and consideration of marine reserves was included as part of this plan. Rather than address the issue separately, the CDFG and the Channel Islands Sanctuary Advisory Council joined efforts in 1999 to create the Marine Reserves Working Group (MRWG), which included federal and state agencies, commercial and recreational fishermen, environmentalists, and other members of the Santa Barbara community (Bergen and Carr, 2003; CDFG, 2003; Figure 6.3). Additionally, two advisory panels were created to assist the work of the MRWG. A Science Advisory Panel (SAP) was tasked with assembling and evaluating ecological, physical and environmental data and a Socioeconomic Panel was formed to evaluate both recreational and commercial industries in the Channel Islands (Airamé et al., 2003). The MRWG developed several goals for marine reserves in the Channel Islands (Table 6.1; Airamé et al., 2003).

The MRWG, together with professional facilitators and the advisory panels, planned and debated for 3 years. While the MRWG was able to agree

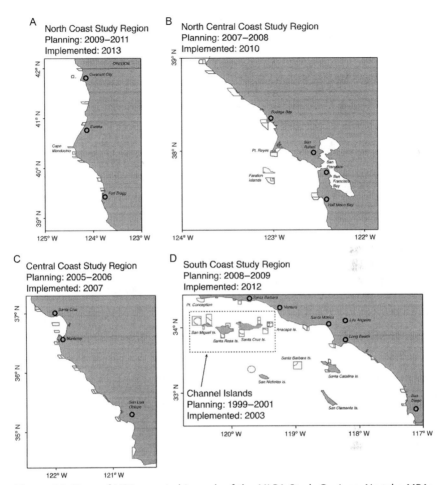

Figure 6.2 Maps of MPAs created in each of the MLPA Study Regions. No-take MPAs (most are State Marine Reserves) are outlined in red (light grey in the print version); limited-take MPAs (most are State Marine Conservation Areas) are outlined in blue (dark grey in the print version). Small dots, particularly in (A), are small special closures surrounding marine mammal haulout locations.

on overarching goals for the MPA network, the group dissolved in 2001 without reaching a consensus on the design of a potential MPA network, essentially ending the public process at that point (Helvey, 2004; Osmond et al., 2010). Following this, the superintendent of the CINMS and the Marine Region Manager of the CDFG developed a compromise solution that reflected the work of the MRWG and the advisory panels. This compromise plan, along with five other plans, was submitted to the

Channel Islands MPAs   Marine Life Protection Act Initiative

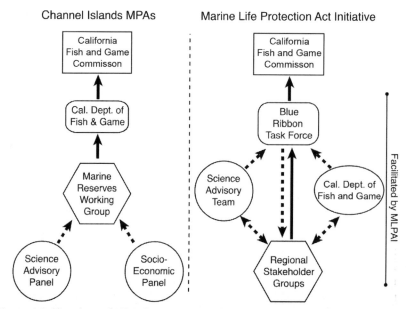

Figure 6.3 Flowchart of MPA design and decision making in the Channel Islands MPA process and the Marine Life Protection Act Initiative. Dashed arrows indicate flows of information (e.g. scientific guidelines); solid arrows indicate flows of MPA network proposals. Circles and ovals enclose groups providing scientific or regulatory guidance; hexagons enclose groups that originated MPA proposals based on guidelines; rounded rectangles enclose intermediate decision-making groups that refined and recommended proposals; rectangles enclose the final decision making and regulatory body.

California Fish and Game Commission. Ultimately, the compromise plan was approved by the Commission in 2002 and a network of MPAs (primarily marine reserves that allowed no commercial or recreational fishing) in state waters was implemented in April 2003. The compromise plan did include reserves which extended into federal waters but since the CINMS had no authority to manage fishing or other activities, formal protection was not extended until a separate, federal regulatory process was completed in 2007 (Osmond et al., 2010).

The science–based guidelines for reserve network design in the Channel Islands are detailed in Airamé et al. (2003) and briefly described here (Table 6.2). Taking both conservation and fisheries goals into account, the SAP recommended that 30–50% of the CINMS should be protected. Values this high were controversial. They were a collective professional judgement based on consideration of marine reserve literature, federal

Table 6.1 Summary of goals for marine protected areas established in the California Channel Islands (goals developed by the Marine Reserves Working Group; see Airamé et al., 2003) and along the entire California coastline (goals specified in the Marine Life Protection Act; see Kirlin et al., 2013)

Goal category	Channel Islands	Marine Life Protection Act
Ecosystem biodiversity	• Protect representative and unique marine habitats, ecological processes and populations of interest in the CINMS[a]	1. Protect the natural diversity and abundance of marine life and the structure, function and integrity of marine ecosystems
Sustainable fisheries	• Achieve sustainable fisheries by integrating marine reserves into fisheries management	2. Help sustain, conserve and protect marine life populations, including those of economic value, and rebuild those that are depleted
Economic viability	• Maintain long-term socioeconomic viability while minimizing short-term socioeconomic losses to all users and dependent parties	
Education	• Foster stewardship of the marine environment by providing educational opportunities to increase awareness and encourage responsible use of resources	3. Improve recreational, educational and study opportunities provided by marine ecosystems that are subject to minimal human disturbance, and manage those uses in a manner consistent with protecting biodiversity
Natural and cultural heritage	• Maintain areas of visitor, spiritual and recreational opportunities which includes cultural and ecological features and their associated values	4. Protect marine natural heritage, including protection of representative and unique marine life habitats in California waters for their intrinsic value
Management		5. Ensure that California's MPAs have clearly defined objectives, effective management measures and adequate enforcement, and are based on sound scientific guidelines
Network design		6. Ensure that the MPAs are designed and managed, to the extent possible, as a component of a statewide network

[a]Channel Islands National Marine Sanctuary.

Table 6.2 Science guidelines developed by the MLPA Science Advisory Teams for the design of MPA networks

	MPA design guideline	Design objective	Scientific rationale
Habitat representation	Every 'key' marine habitat should be represented in the MPA network	Protect the diversity of species that live in different habitats	Based on observed relationships between habitat type and marine community composition
Habitat replication	'Key' marine habitats should be replicated in multiple MPAs across large environmental gradients or geographic divisions	Protect the diversity of species that live in different ecological regions and geographical areas	Based on observed transitions in community composition across environmental gradients and geographic divisions
MPA size	• MPAs should extend from the intertidal zone to the offshore limit of state jurisdiction (5.56 km) • MPAs should have an alongshore span of 5–10 km (minimum) or 10–20 km (preferred)	• Accommodate the movements of individuals across depth zones • Protect populations of mobile organisms	Based on the reported movement scale of marine organisms, particularly adult fishes
MPA spacing	MPAs should be placed within 50–100 km (or less) of each other	Facilitate dispersal and connectedness among MPAs by benthic fish and invertebrates	Based on the reported movement scales of the larval stages of fish and invertebrates

Modified from Saarman et al. (2013).

fisheries management, dispersal rates and emerging fisheries in a qualitative way (PFMC, 2001). This differed from the population dynamic analyses described in this chapter in that 'No systematic assessments of populations within the CINMS were completed by the science panel' (PFMC, 2001).

The northern Channel Islands are situated in a complex geographical region with a strong environmental gradient across a relatively short geographic distance (see Section 1.1; also see Hamilton et al., 2010).

Consequently, the SAP defined three 'bioregions' and recommended that at least one, but preferably four, reserves be located in each bioregion.

The SAP combined all available information on substrate type, bathymetry and dominant macroalgal communities to characterize the habitats in order to ensure protections of each habitat type (Airamé et al., 2003). The SAP used information on species of concern or commercial importance to weight the importance of particular habitats. With this information, potential reserve configurations were generated using Sites v. 1, an analytical tool for planning regional-scale reserve networks (Andelman et al., 1999; Possingham et al., 2000). This program was precursor to the now widely used Marxan program (Ball et al., 2009) which identifies an efficient set of sites that collectively represent specified amounts of habitats, populations, or other features identified by the SAP (Airamé et al., 2003). These programs differ from the population dynamic, bioeconomic models in the MLPA process in that they do not calculate where populations of different species will actually persist based on spatial population dynamics (White et al., 2014; also see Section 2.3).

The network of MPAs finally implemented in the Channel Islands including Federal waters contained 21% of the CINMS waters in 11 state marine reserves (no commercial or recreational fishing allowed) and two conservation areas (where some types of fishing were allowed; Figures 6.1B and 6.2D).

2.2. Marine Life Protection Act

The second MPA effort in California applied to the whole state, and was initiated by conservation groups lobbying the legislature to obtain passage of legislation called the Marine Life Protection Act (MLPA) in 1999 (Osmond et al., 2010). This law directed the state to redesign its tiny collection of MPAs (0.2% of state waters) to meet six goals (summarized in Table 6.1). These goals were quite general, and even though the law was the enabling legislation for the MPAs, they contained few specific operational metrics. The goals were concerned with protection at the *ecosystem* level, but they did require the state to help sustain, conserve and protect marine life *populations* (Goal 2). They contained considerable ambiguities (e.g. what does it mean exactly 'to protect natural diversity', and what is a 'statewide network' of MPAs?). The MLPA had two other important requirements: (1) that it makes use of the best readily available science and (2) that after implementation, the MPAs be monitored and subject to adaptive management.

Enactment of the MLPA was not accompanied by sufficient funding to implement such a far-reaching decision-making process, one that would change marine fishery management throughout the State. This limited funding led to problems in early implementation efforts (Weible, 2008). In the first attempt, the CDFG formed a committee of marine scientists to suggest locations, configurations, and boundaries for MPAs throughout state waters. These proposed maps were presented at public meetings in 2001 as a starting point for discussion of the implementation of MPAs, but a strong negative reaction by stakeholders to already-developed maps led to the immediate failure of this approach. A second attempt a year later added statewide regional stakeholder groups (RSGs) and paid facilitators to the volunteer scientists. That attempt was also deemed inadequately funded, and was halted in the spring of 2003 (Gleason et al., 2010; Kirlin et al., 2013), although it foreshadowed some of the components of the later process that eventually succeeded.

In 2004, an agreement was struck between the state government agencies and a private foundation, the Resource Legacy Fund Foundation (funded by conservation-minded philanthropic foundations[1]), to fund a decision-making process to implement the MLPA. This process was to be controlled by an organization known as the MLPA Initiative (MLPAI). MLPAI staff included some state agency personnel and contractors with expertise in facilitation, spatial planning, geographic information systems, and policy analysis.

The planning process initiated and managed by the MPLAI divided the California coast into five Study Regions (Figures 6.1B and 6.2), and conducted the design process sequentially in each region, converting a statewide design problem into a sequence of regional-scale processes. Within each Study Region, the MLPAI appointed a RSG, and a Science Advisory Team (SAT), both based in part on nominations by interested citizens within each region. The RSGs comprises representatives of various constituencies (e.g. commercial and recreational fishing sectors, conservation groups, education and research sectors, interested state and federal agencies, tribal governments, and others[2]). There was also a Blue Ribbon Task Force (BRTF) appointed by the state Secretary of Resources in consultation with the Governor's office. The BRTF comprises four to five individuals with highly regarded experience in policymaking processes, although not necessarily in marine or fisheries conservation. The BRTF was responsible for overseeing

---

[1] http//www.resourcesllegacyfund.org/.

[2] https://www.dfg.ca.gov/marine/mpa/centralcoast_rsg.asp.

the integrity of the process to ensure it moved forward in a timely manner and was true to the goals of the MLPA (e.g. pushing for consensus among stakeholders, ensuring the RSGs strove to meet the science guidelines while recognizing the socioeconomic trade-offs in each region). The BRTF was responsible for winnowing lists of proposed plans emerging from the RSG, eventually submitting a short list of potential plans (usually including a 'consensus plan' preferred by the BRTF) to the California Fish and Game Commission. In addition, the CDFG provided feedback to the BRTF, SAT, and RSG on the regulatory and logistical feasibility of networks proposed by the RSGs, eventually submitting their recommended network proposal to the Commission in parallel with the BRTF (Figure 6.3). The Commission made the final decision on all MPA designs as the controlling authority for fishery regulations in state waters.

During the planning process for the first Study Region (Central Coast), the MLPAI and SAT developed the MLPA Master Plan (CDFG, 2008). This document dictated the detailed procedures of the MPA design process led by the MLPAI, and translated the somewhat vague policy goals of the MLPA into more specific, ecologically based objectives and design guidelines. The Master Plan was also approved by the Fish and Game Commission, and was used by the RSG and SAT in the development and evaluation of MPA network proposals in the first, and subsequent, Study Regions.

Within each Study Region, planning began with the MLPAI and CDFG preparing a Regional Profile that described the ecology, human uses and economics of the particular marine Study Region. Based on that profile and general MPA design principles, the SAT developed a series of region-specific scientific guidelines, presumably consistent with the Master Plan (Table 6.2). The RSG then began the process of developing a range of alternative, proposed spatial configurations of MPAs (Figure 6.4). Various subgroups of the RSG, with specific perspectives (e.g. favouring either conservation, recreation, commercial fishing, tribal, or other considerations) and staff support, were encouraged to develop collaborative, consensus proposals. External groups were also allowed to submit plans for consideration. Draft MPA plans were submitted to the SAT, who evaluated how well each plan met the scientific guidelines codified in the MLPA Master Plan. An iterative process followed, with the BRTF providing advice on the SAT-evaluated draft plans, the RSG then revising those plans and resubmitting them to the SAT. After three to four such rounds the BRFT submitted its recommendations to the Fish and Game Commission.

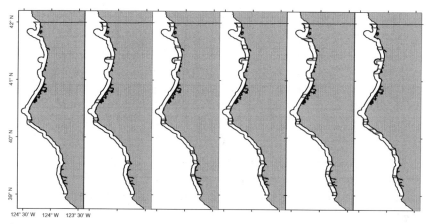

Figure 6.4 As an example, a representative range of proposed MPA networks during the first round of the North Coast design process. Proposed MPA boundaries are indicated in red (light grey in the print version); boundaries of California state waters are indicated in black. The leftmost panel shows the MPAs that existed in the region prior to the MLPAI process. Each proposed network contains a mixture of no-take and limited-take MPAs but these regulatory differences are not indicated on the figure.

This planning process was implemented first in the Central Coast Study Region beginning in 2003, and the MPAs in that region took effect in 2007. The planning process then moved to the North Central Coast (planning initiated 2007, implementation of MPAs 2010), the South Coast (initiated 2008, implementation 2012), and finally the North Coast (initiated 2009, implementation 2013; Figures 6.1B and 6.2). Across all four Study Regions, the MLPAI process created or expanded 124 MPAs, covering 16% of state waters; of these 61 (9.4% of state waters) were no-take State Marine Reserves, no-take State Marine Conservation Areas, or no-take State Marine Recreational Management Areas (in the latter fishing is prohibited but waterfowl hunting is permitted). The remaining MPAs were designated limited-take State Marine Conservation Areas or limited-take State Marine Parks (Gleason et al., 2013b).

As the planning process proceeded across the Study Regions, the MLPAI gradually improved outreach and interactions with stakeholders (Fox et al., 2013a,b; Sayce et al., 2013) by broadening the range of scientific expertise included on the SAT. In addition, new scientific tools were brought to bear on the process, including economic analyses, increasing consideration of spatially explicit, mathematical population models (Kaplan et al., 2006, 2009; Moffitt et al., 2009; White et al., 2010b, 2013a), and a Web-based spatial planning interface ('MarineMap', which later evolved into 'SeaSketch', Merrifield et al., 2013).

There were considerable differences among Study Regions in the stakeholder community, ranging from large groups of recreational fishermen and recreational water-users (kayakers, surfers, etc.) in the South to predominantly commercial fishing interests in the North Central and North Study Regions, with a large presence of Native American tribal stakeholders in the North (Fox et al., 2013a; Sayce et al., 2013). The stakeholders also became more involved in the process and more organized in their opposition or support, particularly after fishers for spot-prawn (*Pandalus platyceros*) abstained from the planning process in the Central Coast, resulting in some fishermen having all of their fishing grounds included in no-take MPAs.

Planning for the fifth Study Region (San Francisco Bay) had not yet begun fully when Governor Arnold Schwarzenegger left office in 2011, and as a new administration took office the political will and funding for the MLPAI process dissipated (particularly given the number and diversity of regulatory institutions and complicated stakeholder relationships in that bay). As of this writing, an MPA planning process has not begun for San Francisco Bay beyond an initial science review and considerations for the application of the network design guidelines for that region.

2.3. Scientific guidelines in the MLPA planning process

In each Study Region, as region-specific or additional science considerations emerged, regional science advisory teams (SATs) developed design guidelines in addition to those codified in the Master Plan. The intent of the guidelines was to ensure that MPAs would meet the statutory requirements of the MLPA, which required translating vague statutory language (e.g. 'preserve biological diversity') into an ecological and operational context (e.g. 'ensure that all habitat types were represented inside at least two MPAs in each Study Region'). The guidelines included recommendations for local habitat representation (what area of each key habitat should be included across the network of MPAs), habitat replication (how many MPAs in a Study Region should include each habitat type), and the minimum size and maximum spacing between MPAs (Saarman et al., 2013; Table 6.2). Eventually, guidelines were also developed for the minimum area of a habitat represented within an MPA that is required for an MPA to contribute to the spacing guidelines (i.e. network) for that habitat. There were also non-specific guidelines that each MPA should extend from the shore all the way to the 3 nm boundary of state waters (in order to accommodate cross-shore movements of fishes), and nonscientific guidelines promulgated by the CDFG enforcement division suggesting that MPAs have straight-line boundaries and be aligned with natural landmarks. These guidelines were formalized in the MLPA Master Plan document and used by the RSG in drafting proposed MPA networks. These draft network plans were then evaluated by the SAT as to how well each proposal met the guidelines.

In addition to evaluating how well MPA network proposals met these design guidelines above, the SAT also assessed the degree to which each proposed MPA intersected with locations relevant to other types of marine spatial planning. These included seabird foraging areas and rookeries, marine mammal haulouts, and regions affected by discharge from streams with high contaminant loads or wastewater outfalls. There was some debate among the SAT as to whether MPAs should be designed to avoid locations impacted by contamination, so they are more 'pristine' or whether they should target impacted locations in order to leverage improvements in water quality in the future. Similarly it was unclear how relevant seabird and marine mammal habitats were to MPA planning because those species were largely already protected by separate federal and state statues (e.g. the Marine Mammal Protection Act) and potentially move large distances. In general, these assessments had little bearing on the final configuration of MPA networks.

An additional aspect of the SAT's evaluation of MPA network proposals was characterizing the impact of specific extractive activities permitted in limited-take MPAs. Proposed networks typically included both no-take reserves and multiple types of limited-take MPAs (Figure 6.2). To evaluate limited-take areas, the SAT developed a protocol for characterizing the level of protection (LOP) afforded by each specific permitted activity, depending on the gear type used, ecosystem role of the targeted species, and other considerations (Saarman et al., 2013; Figure 6.5). In the SAT evaluations, the degree to which an MPA network proposal satisfied the scientific design guidelines was reported in terms of those LOPs; for example, a proposal might satisfy the size and spacing requirements if all MPAs with at least a 'Moderate–Low' LOP (some activities that will alter community structure are permitted) were counted along with MPAs with higher LOPs, but not if only MPAs with a 'High' LOP (no or very little extraction) were counted, for example, of two proposed networks that similarly met the size and spacing guidelines, the proposal comprises MPAs with higher levels of protection was considered to better meet the science guidelines and goals of the MLPA.

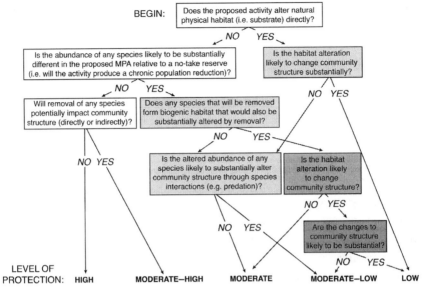

Figure 6.5 Decision tree flowchart used by the Science Advisory Team to determine the 'Level of Protection' afforded by an activity (e.g. fishing using a specific gear and target species) proposed to be allowed in a limited-take MPA. *Adapted from Saarman et al. (2013).*

2.3.1 Size and spacing guidelines

The size and spacing guidelines formulated by the SAT during the decision-making process for the first Study Region (Central Coast) were perhaps the most influential action taken by the scientists involved in the MLPA. The guidelines stated that MPAs should span at least 5–10 km in extent along the coastline, but that spanning 10–20 km along the coastline would be preferred (CDFG, 2008). This size guideline was based on qualitative examination of available information on home range sizes of California species (see CDFG, 2008 for references). The spacing guideline was that MPAs should be separated by no more than 50–100 km. This spacing guideline was based on information regarding larval dispersal distances (see CDFG, 2008 for references). Thus, although the guidelines did not specify a fraction of the coastline or habitat to be placed in reserves, the initial SAT at the outset effectively specified that between 5% (5 km MPAs spaced 100 km apart) and 28% (20 km MPAs spaced 50 km apart) of the coastline should be placed in MPAs. Recall from above that the percentage of coastline in no-take MPAs was 9.4%, while the percentage including limited-take MPAs was 16.0% (Gleason et al., 2013b). Note that these percentages refer to the entire coastline in a Study Region, not the percentages of particular habitats or species ranges. The size and spacing guidelines were not linked to particular habitat types or distributions; habitat-specific protection was instead addressed by separate guidelines for habitat representation and replication (Table 6.2).

There was not uniform agreement among the scientists on the Central Coast SAT that specifying *a priori* size and spacing guidelines was the best approach. Some members argued that a more comprehensive evaluation of all of the relevant factors affecting persistence of the fish populations present would lead to better performance of MPAs in the end. The state of the science at the time of the initial discussion (2003) was that the spatial configuration of MPAs required to ensure persistence of a species was known to depend on the larval dispersal distance of the species as well as how heavily that species was being fished outside the MPAs (Botsford et al., 2001). Moreover, population modeling tools that could calculate how specific MPA configurations, larval dispersal distances, and different levels of fishing would affect the spatial pattern of species abundance were under development, and would soon allow more direct evaluation of the effects of proposed spatial configurations of MPAs on fish populations (Kaplan et al., 2006, 2009; Moffitt et al., 2009; White et al., 2010b, 2013a). Some members of that initial SAT agreed to support the size and spacing guidelines as only a first step representing the best available science at that time, and the guidelines were incorporated into the Master Plan. Although the Master Plan was proposed as a 'living document' that could change as the best available science evolved (Kirlin et al., 2013), in practice it was deemed not possible to remove or fundamentally alter the primacy of size and spacing guidelines included in the original Master Plan as the 'best available science' improved. This was because (1) formally updating the Master Plan would require action by the Fish and Game Commission and (2) the MLPAI was concerned about components of the evaluation for some Study Regions differing from those used in the other Study Regions. Additionally, some SAT members argued that because spatially explicit population model evaluations could only be performed for certain species with adequate information, adopting size, and spacing guidelines without explicit calculations for any species was a conservative buffer against uncertainty about the response of the full suite of affected species. Consequently, the size and spacing guidelines remained the key component of the evaluation of proposed MPAs, even as more comprehensive modeling evaluations became available and were also used by the SAT in evaluating MPA network proposals (see Section 2.3.2).

Another characteristic of size and spacing guidelines noted by scientists familiar with decision making in natural resource problems was that by effectively specifying how much of the coastline to set aside in MPAs at the beginning of the process, they limited the scope for later decision making (Osmond et al., 2010). The size and spacing guidelines effectively limited the 'decision-making space' being considered by the SAT, the BRTF, and ultimately by the California Fish and Game Commission. However, even in the final designs implemented by the Fish and Game Commission not all of the MPAs in the network met the 'preferred' size and spacing requirements (i.e. 20 km MPAs spaced 50 km apart).

The size and spacing guidelines were not the only factor that constrained the design of MPA configurations by the stakeholders. Requirements for representation of minimum areas of key habitats, particularly rare habitats, and that those habitats be replicated in multiple MPAs throughout at Study Region (Saarman et al., 2013) effectively ensured that MPAs would be placed in certain key locations. Together, these requirements led to RSG groups proposing alternative MPA proposals that were largely quite similar to each other, particularly once plans that failed to meet the scientific guidelines were winnowed out in early evaluation stages (see modeling results below).

2.3.2 Population models and fisheries

Another characteristic of the scientific evaluation of proposed MPA networks that raised questions among the SAT members was the decision made by the MLPAI in the first region to ignore the relationship of the proposed MPAs to fisheries and their management outside the MPAs. This went so far as MLPAI staff directing the population modelers not to use the word 'sustainability' to describe population status, because it implied the MLPA decision making was related to sustainable fishery management. Scientists knew by that time that including the level of fishing outside the MPAs was necessary to predict the effects of the proposed MPAs on persistence of fish populations of various species (e.g. as noted above in Botsford et al., 2001). From the point of view of the MLPAI, however, this was a legal issue involving whether the implementation of the MLPA was required to interact with the implementation of a new law changing the way that California fisheries were managed, the Marine Life Management Act (MLMA) (Fox et al., 2013c).

The views of the MLPAI on the value of including fishery information changed in response to the publication of a population modeling study that addressed proposed MPAs in the Central Coast Region (Walters et al., 2007). These authors concluded that (1) movement of adult fishes could lead to lower fish abundance in MPAs, (2) population persistence in MPAs depended critically on fishery management outside the MPAs, (3) the size and configuration of MPAs had little impact on population dynamics, and (4) the MPAs were unlikely to benefit key fish species. While other modelers associated with the MLPA process noted potential flaws in that paper (see comments in Moffitt et al., 2009), the MLPAI began to support population modeling more formally after that. It was decided at that time that two groups, one at the University of California Davis and one at the University of California Santa Barbara should each formulate population models and report the effects of fishing on MPA performance and the effects of MPAs on fishery catch. These two modeling efforts converged on similar model structures and assumptions and produced similar results and were eventually folded into a single joint effort (White et al., 2013a). Many stakeholders and SAT members initially resisted inclusion of the models in the decision-making process, in large part because of debate about whether the model should assume that future fishing outside MPAs should be assumed to be at sustainable levels or unsustainable levels (White et al., 2013a). In the end, the models did not supplant the primary role of the size and spacing guidelines in the decision making. These guidelines were based on the assumption that current fishery management provided little protection against overfishing (Gaines et al., 2010; MRWG SAP, 2001).

The population models developed under the MLPA initially were extensions of the original modeling approach taken by Botsford et al. (2001) in the sense that they assumed the California coastline was essentially linear and that larval dispersal could be approximated by a symmetrical, spatially homogenous dispersal kernel (e.g. Figures 6.6A–D). That type of model was used to advise the design process in the Central Coast and North Central Coast regions, but in the South Coast and North Coast regions, the modeling groups developed two-dimensional models with finer (1 km^2) spatial resolution, and used results from Lagrangian simulations of larval dispersal in ocean circulation models (Drake et al., 2011; Mitarai et al., 2009) to obtain connectivity matrices for the population models (e.g. Figures 6.6E–H; White et al., 2013a). These models afforded much finer-scale assessments of the likely performance of individual MPAs (White et al., 2013a), and later analysis showed that they could have guided the planning process to network designs with higher fish biomass and higher fishery yields than those obtained by following the SAT's more general guidelines (Costello et al.,

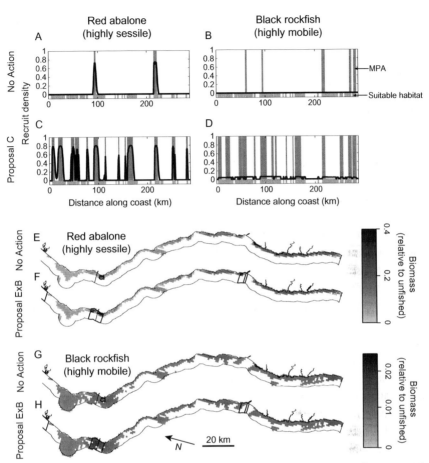

Figure 6.6 Representative results showing differences in the response of species with different larval dispersal distances to alternative MPA network proposals. In (A–D), a population model that approximated the North Central Coast Study Region as a linear coastline predicted that red abalone (*Haliotis rufescens*) would have self-persistent populations within MPAs in either the sparse 'No Action' proposal (with only previously existing MPAs; (A) or the conservation-oriented Proposal C (C). By contrast, black rockfish (*Sebastes melanops*) were predicted to have network persistence only in Proposal C (D). In (E–H), similar results for the same two species are displayed for the higher-resolution two-dimensional model used in the North Coast Study Region, for either the No Action alternative or the conservation-oriented Proposal ExB. In the North Coast, black rockfish populations were sustained by network persistence by the North Central Coast MPAs to the south of the Study Region, even in the No Action scenario (G). For model details, see White et al. (2010b, 2013a). In all of these examples, the populations were presumed to be overfished (i.e. lifetime reproduction was below the critical replacement level).

2010; Rassweiler et al., 2014). However, even these models had a key limitation: in the absence of estimates of present-day population density of species along the coastline, it was not possible to initialize the models to make short-term predictions, and only long-term equilibrium abundances could be forecast. This would prove to be an obstacle to using the models to guide short-term assessment and adaptive management of the MPAs (White et al., 2011; see Section 3).

Although the models became more sophisticated over the course of the MLPA process, the basic way that the modeled populations responded to MPA network designs and fishing did not change. The models developed to predict population responses to network designs evolved in complexity, from early modeling with an assumed straight coastline, and an assumed shape of a larval dispersal kernel for the Central Coast Study Region, along with results of later modeling with real coastlines and bottom topography, and larval transport from a circulation model for the North Coast Study Region (Figure 6.6). In the former models (Figures 6.6A–D), a short distance disperser, red abalone (*Haliotis rufescens*), persists in some locations even without additional MPAs while under Proposal C, this species persists wherever there is both suitable habitat and an MPA, but not elsewhere. The black rockfish (*Sebastes melanops*), a species with long larval dispersal distances and a large home range, does not persist anywhere under the assumed level of overfishing, and even with substantial area in MPAs, does not persist at a very high level (however, this highly mobile species was predicted to persist at higher biomass under lower levels of fishing; White et al., 2010b). Using the more sophisticated circulation model (Figures 6.6E–H) in a different region, the results are not as dramatically different between species and proposed plans, but the benefits of MPAs for both species can be clearly seen.

Because this form of graphical results (e.g., Figure 6.6) from the population models involved too much detail for MLPA decision-making groups, the results for a number of different proposals were summarized (Figure 6.7). Stakeholders and decision-makers could see how the conservation value (total biomass of all model species) and the economic value (total fishery yield) of each proposed MPA network varied with the different fractions of coastline in MPAs, at different levels of fishing outside the MPAs. As the total area in MPAs increased from the plan with the lowest fraction to the highest fraction, conservation value never decreased, and often increased. However, fishery yield increased with MPA area only in the case when it was assumed that overfishing was occurring outside MPAs. When it was assumed that fishing levels outside were at the level producing

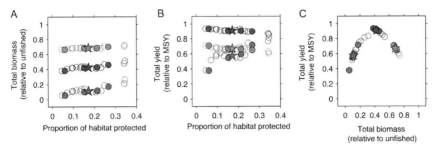

Figure 6.7 Example of the summary analysis of proposed MPA networks by the UC Davis bioeconomic model (White et al., 2013a). Each point is the result for one proposed South Coast Study Region MPA network, evaluated for one model species (the California sheephead, *Semicossyphus pulcher*). Panels show (A) equilibrium biomass in the Study Region as a function of the proportion of shallow rocky reef habitat (<30 m depth) in the Study Region protected by MPAs where sheep head fishing would be prohibited; (B) equilibrium fishery yield as a function of the proportion of habitat protected; and (C) equilibrium biomass as a function of equilibrium fishery yield. Biomass (kg) is expressed relative to the equilibrium unfished biomass estimated by the model, and yield (kg) is expressed relative to the maximum sustainable yield (MSY) for that species in the absence of MPAs, as estimated by the model. Because future fishing levels are unknown, results are shown for three different possible levels of harvest outside of MPA boundaries: unsustainable (the fishery would collapse without MPAs; red symbols (grey in the print version)), MSY-type (the fishery would be at or near MSY without MPAs; blue symbols (dark grey in the print version)), and conservative (the fishery would be below MSY without MPA (because of low fishing); green symbols (light grey in the print version)). Open symbols indicate proposals generated in the first two rounds of the design process, closed symbols indicate proposals from the final round, and the star indicates the 'preferred' proposal selected by the Blue Ribbon Task force. See White et al. (2013a), for model details.

MSY or less, fishery yield declined with increasing area in MPAs (White et al., 2013a).

These results for the South Coast Study Region also provide an example of how the range of proportions of habitat protected by proposed MPAs contracted as the deliberation among proposals proceeded (Figure 6.7). The original proposals (Figure 6.7, open circles) ranged from 0.05 to 0.35, then contracted to a range of 0.07 to 0.28 (Figure 6.7, solid circles) by the final round of decisions for the BRTF. The plan eventually chosen by the California Fish and Game Commission (Figure 6.7, star) was midway between these.

Viewing the conservation and economic values together (Figure 6.7C) indicates whether there are trade-offs involved or a win–win situation. When the species are overfished, total yield and total biomass increase together. When the fishing level is at or below that causing MSY, fishing yield declines as total biomass increases.

The science of population modeling for MPAs developed rapidly during the period of the MLPA process. Kaplan et al. (2006, 2009) showed how differences in larval dispersal distances among species, and spatial heterogeneity in the distribution of habitat would affect the persistence and spatial distribution of different species with alternative MPA network proposals for the Central Coast Study Region. Moffitt et al. (2009) then showed how accounting for the additional effect of movement of adult fish within home ranges reduces the effectiveness of MPAs. This model achieved the capability to quantitatively evaluate population persistence considering the combination of adult and larval movement, as the SAT had done qualitatively in developing the size and spacing guidelines. Accordingly, Moffitt et al. (2011) next analyzed the effects of size and spacing guidelines in a way that also accounts for the level of fishing and the spatial configuration of the MPAs. Because the effects of proposed MPA networks on populations depends on the intensity of fishing outside MPA boundaries (which is highly uncertain at the local scale of MPAs), White et al. (2010b) used population models to perform a decision analysis, evaluating likely MPA performance over a probability distribution of different 'states of nature' (Hilborn and Walters, 1992) representing different levels of fishing. The results illustrated how recommendations for MPA design could depend on decision-makers' assumptions about the effectiveness of conventional fisheries management in the future (fewer and smaller MPAs would be recommended under optimistic assumptions about conventional management), but this type of analysis was not adopted by the MLPAI.

2.3.3 Economic assessments

Although economic considerations were not among the statutory goals of the MLPA (Table 6.1), the MLPAI and BRTF recognized that the fishing communities in each Study Region faced potentially substantial economic costs from new MPAs. Consequently, the SAT in each region also considered assessments of the economic costs of lost fishing grounds in each MPA network proposal. Scholz et al. (2004, 2011) describe the details of the analysis used. Essentially, in each Study Region they surveyed a stratified sample of participants in important commercial and recreational fisheries as to the extent and relative stated importance of their fishing grounds, as well as a number of demographic and operating cost variables. The aggregated data were then used to calculate the percentage of fishing grounds closed under each MPA network proposal, as well as the minimum first order economic losses due to that closure. These assessments were highly valued by RSG members and the BRTF because it was one of the few SAT analyses (along with the bioeconomic population models) that centred on the economic costs of MPAs rather than the potential ecological benefits. An important limitation to the economic analysis was its static nature: it implicitly assumed that fishing grounds inside MPAs were a complete loss; i.e., there was no way to account for the potential increase in biomass inside MPAs that could eventually 'spill over' and sustain fishery yields. Thus the analysis reflected only the initial short-term costs of closing fishing grounds; this was a mirror image of the limitation of the population modeling, which could project long-term equilibrium outcomes but not short-term trajectories (White et al., 2013a).

2.3.4 What species were likely to benefit?

The science guidelines codified in the Master Plan (Table 6.2) operated on the assumption that by setting aside a certain fraction of habitat area, species would persist at higher levels within those areas, and ecosystems would be preserved in a more natural state. Consequently, the list of 'species likely to benefit' from the MPAs assembled by the SAT in each region, as required by the MLPA, typically included any species that could be taken in a fishery, or that might benefit from reduced disturbances or habitat improvements inside MPAs. Such lists did not reflect species differences in harvest pressure present, larval dispersal patterns, adult movement rates, or other life history characteristics that were known or predicted to affect the response of species to MPAs (Botsford et al., 2001; White et al., 2010b, 2011). In general, it should be reasonable to expect any fished species to increase in abundance to some degree after protection in an MPA (assuming that fishing mortality exceeds any negative effects of the MPA on predator–prey and competitive interactions among fished species), but the compilation of a broad, unranked list contributed to the expectation that there should be -the-board increases in fish abundance after MPA implementation. This turned out not to be the case: preliminary *post hoc* assessments show that not all fished species have increased, and some have increased much more than others (Hamilton et al., 2010; see Section 3.1).

3. IMPACTS OF THE MPAs

Because the Channel Island MPAS were implemented in 2003, and the first region of the MLPA was implemented in 2007, there has been a relatively short time for impacts to occur, be observed and be interpreted through analysis, especially as regard potential network benefits. Moreover, there have been very few studies conducted to evaluate social or economic impacts for either of these networks.

3.1. Ecological impacts

The ecological impacts of the Channel Islands MPAs are more apparent because of the longer time since implementation. One key development has been the realization that removing the confounding effects of biogeo-graphic and physical factors is key to detecting effects (Hamilton et al., 2010). Results after 5 years showed large increases for several fished finfish species, but curiously no net change in abundance for other fished species (e.g. the recreationally fished kelp bass, *Paralabrax clathratus*) possibly due to environmentally driven failures in larval recruitment in the years after implementation and the short time scale involved. Kay et al. (2012a) also documented an increase in catch-per-unit-effort (CPUE) and the size of spiny lobsters (*Panulirus interruptus*) inside of MPAs, but not outside. More-over, inside the MPAs, lobster CPUE increases with distance from the MPA boundary, which could imply spillover, but could also be due to poaching (Kay et al., 2012b).

For the MLPA MPAs, there is a mandated periodic 5-year review and evaluation process. This has been completed for the Central Coast Study Region, and is underway for the North Coast Study Region. The monitor-ing effort is managed jointly by the California Ocean Science Trust (CalOST) and CDFW (OST, 2013). It began early enough to be considered baseline monitoring and continuing monitoring is planned. The monitoring effort is directed at measuring ecosystem-level effects, but it includes single-species population outcomes. Initial evaluations for the Central Coast Study Region have been mixed, with some species showing increases and others showing decreases; potentially a result of high variability in environmental conditions and larval recruitment (OST, 2013). This initial report is of lim-ited utility in that reported results of monitoring are not accompanied by associated measures of uncertainty, such as confidence limits.

3.2. Fishery impacts

As noted above, a method for assessing the loss of preferred fishing grounds based on interview data was developed during the implementation of the MLPA MPAs (Scholz et al., 2011; White et al., 2013a). These are short-term, worst case cost projections that do not account for long-term trends (positive or negative) and they have not been tested since the implementation of the MPAs.

Currently, several economic studies of the response of fishing are in place to track the effects of the MPAs on fishing, but it is too early to describe extensive results (OST, 2013). The results of a survey of fishermen reporting how many were affected by the implementation of the MPAs are reported in OST (2013).

3.3. Interface with fisheries organizations

There has been some coordination with fisheries management regarding the Channel Islands MPAs. The Pacific Fisheries Management Council, the regional federal management body, reviewed the Channel Islands implementation process in 2001 (PFMC, 2001). The Channel Island MPAs were originally implemented in State waters only (i.e. out to 3 nm from shore), and later extended into federal waters (i.e. out to 200 nm from shore).

Although California's MLMA, passed near the same time as the MLPA specifically recognized the MLPA as a means by which the state could move toward a more ecosystem-based approach to fisheries management, there has been no formal consideration for integration of the MPA networks into the state's approach to fishery management (CDFG, 2001, 2002). As noted above (e.g. Figure 6.7), the population models indicated that fishery yields would decline with increasing area in MPAs if the fishing effort was that producing MSY, or less. However, the MLPA process concluded that this should not be accounted for because future fishing levels were highly uncertain.

However, two studies have explored the application, or potential application, of these reserves for informing stock assessments. Schroeter et al. (2001) demonstrated the application of reserves in evaluating the fishery status of the warty sea cucumber (*Parastichopus parvimensis*) in the northern Channel Islands. Similarly, Babcock and MacCall (2011) explored the application of reserves for stock assessments for a suite of nearshore California fishes.

3.4. Social impacts

There were some strong negative responses to MPAs by fishermen. The strongest was the response by fishermen in northern California to the early attempt to implement MPAs (see Section 2). Later, there was strong resistance by fisherman in the Southern California Study Region (including lawsuits seeking to enjoin the implementation of the MPAs; Fox et al., 2013a), and resistance by Native Americans in the northern California Region (see Fox et al., 2013c for details). Nonetheless, the MLPAI process was purposefully inclusionary and iterative, and strove to ensure that all stakeholder groups had opportunities to voice their views (Fox et al., 2013c; Sayce et al., 2013).

3.5. Enforcement and its effectiveness

The importance of enforcement to management with MPAs is widely appreciated, but the deleterious effect of violations of MPAs on monitoring and adaptive management may not be as well appreciated. The presence of poaching in MPAs can render the task of assessing the protective effects of MPAs almost impossible. One remedy that can reduce that effect is carefully keeping records of violations, and, if possible, their biological effects. This point is underscored in the history of California's MPAs. Recorded levels of poaching was one of the potential reasons for the lack of a significant difference in fish density between reserve and non-reserve sites Hopkins Marine Life Refuge (one decade old) and Pt. Lobos Marine Reserve (two decades old), both in Central California (Paddack and Estes, 2000).

CDFW is the agency responsible for enforcement of the MPAs. They patrol by boat, and can respond to poaching in progress. Records of violations are kept, and presumably will be available for analyses associated with adaptive management of the MPAs. Between 2008 and 2011 (4 years) between 3 and 16 violations of MPAs occurred per year in the central coast region (OST, 2013).

4. OVERVIEW: LOOKING AHEAD

4.1. What was achieved?

In the Channel Islands a contentious, early decision-making process led to the implementation of 13 MPAs in state waters, which were eventually extended to federal waters (Figures 6.1B and 6.2D). These covered 21% of the CINMS waters.

The Channel Islands process likely influenced the development of the MLPA by calling attention to the effects of stakeholder involvement and a strong role for science-based guidelines. One important difference between the Channel Islands process and the MLPA is that in the Channel Islands, local community members initiated an *ad hoc* process that grew into a joint state and federal partnership, but without overarching legislation to drive the process (Osmond et al., 2010). Though there exists no formal state-sponsored monitoring to evaluate the impacts of the Channel Island reserves separately from the South Coast Study Region, independent academic (Partnership for Interdisciplinary Studies of Coastal Oceans) and federal (Channel Islands National Park Service) studies continue.

A number of achievements were accomplished under the MLPA. The law was passed, and 124 MPAs were implemented through a public decision-making process. A number of publications are now available focused on how implementation of the law was successfully accomplished (Gleason et al., 2013a and references therein). It has not yet been demonstrated that the central goal of the MLPA, i.e., improvement of the sustainability of California's coastal ecosystem, has been accomplished. That will require implementation of the adaptive management of the MPAs, which includes, as a first step, evaluation of monitoring data to determine whether they are 'working'. As we have noted in this chapter, adaptive management following implementation is a requirement of the MLPA. Baseline monitoring has been accomplished and a monitoring framework is under development. So far monitoring of abundance and size distributions of key species, both inside and outside of the MPAs has been accomplished over the 7 years since implementation in the Central Coast Region. A meeting organized by the CalOST in February 2013 celebrated proposed indications of success (OST, 2013). However, the results presented at that meeting did not include an account of uncertainty in estimates (e.g. confidence limits). Additional time and analysis will be required to assess the performance of these MPAs more definitively. Work in progress by the authors on direct assessment of potential increase of abundance and mean size of three species, inside and outside of three MPAs in the Central Coast Region indicate abundance and sizes have not increased. Population modeling of expected population responses of these three species, accounting for observed levels of recruitment variability and local estimates of fishing mortality, indicate that it is too early to expect to detect positive indications that these MPAs have had the desired effect.

4.1.1 Resistance to global change

California faces a number of specific, identified threats from climate change, and the predicted responses of species indicate that California's MPAs will provide some resilience to their effects. Two key design traits of the MLPA network underpin the potential for the network to buffer the effects of climate change on species and communities; the depth range of individual

MPAs and the spacing between MPAs scaled to larval dispersal distances (Carr et al., 2010). Increases in sea surface temperatures and thermal stratification cause increased vulnerability of species to thermal stress at shallow depths. Whether stress-related or reflecting thermal preferences, species populations find thermal refuge in deeper cooler waters (e.g. Dulvy et al., 2008). California MPAs that extend from the intertidal to the outer edges of the continental shelf provide protection for species as populations shift to deeper depths. Another phenomenon associated with climate change is an overall warming of ocean temperatures and a concomitant latitudinal (poleward) shift in species ranges (e.g. Perry et al., 2005; Pinsky et al., 2013; Poloczanska et al., 2013), including a poleward shift in intertidal species over a 30-year period documented in California (Barry et al., 1995). Larval transport is an important mechanism by which species shift distributions along the coast (Gaylord and Gaines, 2000). Spacing and larval connectivity among MPAs can allow species to track changes in water temperature while maintaining protection afforded by MPAs by shifting from one protected area to another as their ranges shift along the coast.

The adaptive management scheme specified in the MLPA also contributes to the resilience to global change. The MPAs will be sampled for monitoring every 5 years, tendencies for species to shift distribution can be detected and accounted for by moving boundaries if desired.

There are also indications that both the magnitude and seasonal timing of upwelling in the California Current large marine ecosystem have changed and will continue to change (Bakun, 1990; Garcia-Reyes and Largier, 2010; Snyder et al., 2003). In addition to the proposed long-term, gradual changes, there have been a number of episodic changes in physical conditions that have affected marine populations, including a period during which upwelling began later in the year near 2005 (Barth et al., 2007), and occasional periods of anoxia at various locations (Chan et al., 2008; Grantham et al., 2004). The most recently identified effect of increasing CO_2 is the observation that upwelled waters are becoming increasingly acidic (Feely et al., 2008).

The evaluations of proposed MPAs throughout the MLPA implementation indicate there will likely be some amelioration of the effects of climate change and ocean acidification through the increase in lifetime reproduction implied by the increase in biomass with area in MPAs in virtually all of the population model results. The results of population modeling during the MLPA process showed that increasing coverage in MPAs would not cause a decline in biomass, and in many examples it would cause an increase in biomass. This increase in biomass was due to an increase in lifetime repro-duction in the affected species provided by the proposed MPAs, lifting them further above their critical replacement level. This increase in replacement provides a buffer against both long-term decline in population productivity and occasional episodic low rates of survival or productivity. A MPA implemented nearby in Mexico provides a clear empirical example of this increase in resilience provided by MPAs (Micheli et al., 2013). There the greater potential for reproduction in the protected population was sampled and the consequences for sustained settlement during a hypoxic period were directly observed.

5. FUTURE REQUIREMENTS

The task remaining in the management of California's new MPAs (i.e. implementing and executing their adaptive management) is arguably the most important part of the Channel Islands MPA and the MLPA efforts, both from the perspective of local resource management and the global need for information regarding the performance of MPAs. One of the initial steps required was to reanalyze the population model responses to MPAs, focusing on the short-term, transient population response, rather than the long-term effects used in the decision making for the MLPA. The general expectations of fish population transient responses to a removal of fishing mortality have been described (White et al., 2013b). In addition, how these responses would be detected from sampling over a range of temporal and spatial scales have been compared (Moffitt et al., 2013). Both of these results depend, of course, on the level of fishing to which the populations have been subjected prior to the implementation of the MPAs. Because fishery management commonly resolves variability in fishing only with coarse spatial resolution, the local fish-ing mortality rates affecting specific MPAs will need to be estimated from local size distributions. The observed population responses to implementa-tion will also depend on annual variability in past recruitment, as well as measurement errors. We are in the process of evaluating the combined effects of these for a number of species and locations in the central coast region. The results indicate that it is currently too early to detect expected increases in population abundance or individual size, given the time scales and stochasticity inherent in the dynamics of these populations.

The fact that physical oceanographic conditions influence recruitment, individual growth, fecundity and mortality rates, as well as larval dispersal of species in a network of MPAs suggests that physical observations will be essential for interpretation of the monitoring of the biological status of fish species in a network of MPAs. Carr et al. (2011) explained how monitoring of the effects of MPAs on fish size and abundance, as well as the relative make-up of the species composition in the fish community could benefit from monitoring of specific physical oceanographic variables by recently developed ocean observation systems (e.g. in the United States, the various OOSs).

A framework plan for future monitoring in support of adaptive management of MPAs in California is currently under development through a collaboration between the CDFW and California's Ocean Science Trust. This plan is aimed at the ecosystem level, as were the MPA implementation efforts described herein. However, it correctly seeks operational information at the population level, i.e., species densities and size distributions. From the material presented in this chapter, it appears that if this planned monitoring occurs, it would provide the information needed for adaptive management only if it (a) made use of the population results regarding transient responses to link MPA effects to monitoring observations and (b) provided the information necessary to allow the adaptive management program to account for uncertainty. The former would be required to connect life histories and MPA designs to the observations, as required for adaptive management (i.e. for asking whether the observations match the 'predictions'?). The latter would be vital, simply put, to guarantee that the adaptive management would be based on statistically significant results. The need for both of these is especially acute in this case because of the complexity of this kind of resource management, and the nascent nature of our understanding of it.

5.1. Could it have been achieved differently/more effectively?

Not surprisingly, there is a range of opinions regarding whether implementation of these MPAs should have been done differently. These opinions depend largely on one's view of the ultimate goals of the MLPA, and more generally, the role of science in resource decision making.

Whether the size and spacing guidelines should have played such a dominant role is a central question. They were formulated on a qualitative basis as 'rules of thumb', statements formulated to facilitate the formulation of initial spatial configurations for proposed MPAs (Carr et al., 2010). Such rules serve a useful purpose in expediting broadly based decision making, and these certainly played that role in the MLPA, as did the similar specification of the fraction to be placed in MPAs in the Channel Islands implementation.

The problem with these guidelines was that they came to be very influential in the design process, and were treated as rigorous scientific requirements, which they were not. That is, there is not a quantitative basis for saying that the MPAs would be optimal in any sense, if they satisfy those guidelines. As can be seen in Figure 6.8, greater area in MPAs will protect more species, and even a figure such as that one depends critically on the amount of fishing assumed. Analyses such as these with population dynamic models could have been used to create more realistic lists of species expected to benefit.

The size and spacing guidelines were based on the intuitive ideas that (1) MPAs should be big enough for some species home ranges to be contained

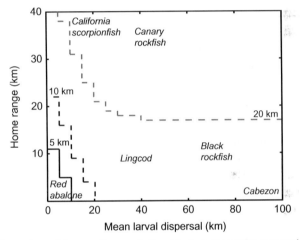

Figure 6.8 Illustration of the nonlinear relationship between MPA size and spacing and the persistence of marine populations. Results are derived from a population model of idealized species living on an infinite linear coastline with homogenous habitat. Model species had different combinations of larval dispersal distance (expressed as the standard deviation of a symmetrical dispersal kernel) and adult home range size (expressed as the diameter of the home range). The coastline had no-take MPAs spaced 50 km apart, and with sizes of 5, 10, or 20 km wide (alongshore dimension). Movement scale combinations interior of each contour lines indicates the range of species that would persist within different size and spacing configurations. Approximate estimates of adult and larval movement scales for several California species are shown for comparison. For 5 or 10 km MPAs, only 10% or 20% of the coastline respectively, in is MPAs, and species only persist if movement is low enough to allow self-persistence. For 20 km MPAs, the total MPA area (40% of coastline) is sufficient for network persistence, protecting a much greater range of movement combinations. Thus gradual increases in MPA size (or decreases in MPA spacing) can yield abrupt jumps in the protection afforded by the MPAs. *Adapted from Moffitt et al. (2011).*

within them and (2) MPAS should be close enough together for the larvae from one MPA to reach other MPAs. When implementation of MPAs under the MLPA began, these rules were probably the best way to begin drawing lines on maps. But at about that same time, scientists were discovering that (a) the dependence of persistence of populations on size and spacing change completely with different amounts of fishing outside the MPAs and (b) the dependence of persistence on size and spacing was not linear, as presumed by these rules of thumb (Botsford et al., 2001; Hastings and Botsford, 1999; Mangel, 1998; Moffitt et al., 2011; Figure 6.8).

Use of the size and spacing guidelines narrowed the range of decisions that could be made by the RSGs, the SATs, the BRTFs, and the California Fish and Game Commission (Osmond et al., 2010). The question of whether scientific advice should have played that role in the decisions is certainly a reasonable one. Should the role of scientific advice in resource decision making be merely to provide the best possible estimate of the consequences of various policy actions, or should scientists be specifying the policy actions? Such specification of scientific advice would be regarded by some as bordering on advocacy of a specific policy, rather than just the provision of objective scientific advice regarding the consequences of a specific policy.

Adoption of the size and spacing guidelines led to the early misunderstanding that the effects of MPAs on populations and ecosystems did not depend on the level of fishing outside of MPAs, hence fishing could be ignored in the MLPA process. This changed with the publication of Walters et al. (2007), after which there was greater acceptance of population modeling that accounted for the controlling effects of fishing and individual movement rates. However, the effects of fishing levels on MPA responses were never specifically accounted for in the decision-making process, even though a decision analysis based on the uncertainty in fishing rates was developed (White et al., 2010b). The effects of MPAs on fishing were only minimally accounted for through the economic analysis of Scholz et al. (2011). Moreover, thus far the monitoring of fishing activities outside MPAs at relevant spatial scales across California's network has not been implemented.

Additional support for the broad application of the size and spacing guidelines was recently proposed by Gaines et al. (2010), who presented the same qualitative argument that MPA size should depend on home range and MPA spacing should depend on larval dispersal distances, but also related the resulting suggested fraction in MPAs to the results from Botsford et al. (2001). They noted that their conclusion that one-third of the coastline should be in MPAs was close to the 35% result in Botsford et al. (2001). That reference by Gaines et al. (2010) was appropriately qualified by noting that it was the result obtained for the case in which fishing removes all of the fish outside the MPAs (i.e. the 'scorched earth assumption'). However, use of the value of 35% was justified nonetheless by stating that we needed to assume complete fishing removal to guard against that eventuality since there was so much overfishing in the world. Making such an assumption eliminates consideration of how much fishing there is (or should be), and supplants policy decisions regarding the level of precaution that should taken, narrowing the scope for decision to be made by the BRTF and the California Fish and Game Commission.

Ignoring the effects of MPAs on fishing (i.e. possibly reduced fishery yield), and the effects of fishing on MPAs (possibly lower MPA fractions needed) are not recommendations that we would make for future MPA efforts. In fact, an increase in fishery yield is often the promise associated with implementation of MPAs. Our recommendation would be to attempt to determine which situation applies in Figure 6.7B: overfished (red symbols), fished at MSY (green symbols) or fished less than MSY (blue symbols).

Adopting size and spacing guidelines with primacy over explicit consideration of species movement rates and level of fishing through population modeling in the implementation of the MPAs of the MLPA limited efforts to meet two requirements of the act: (1) the requirement for the use of the best available science and (2) the requirement for adaptive management. As the science of quantitative assessment of MPAs developed, new scientific advances could not be incorporated into the MLPA decision making because that would be perceived as changing the rules between Study Regions. Lack of quantitative, population-specific results of the expected benefits of the MPAs limited the ability to perform an essential element of adaptive management: comparison with the predicted benefits (and costs) of the management action. This removed explicit consideration of major uncertainties such as those in fishing levels and larval dispersal, as well as consideration of another societal input, fishing outside the MPAs. There is broad appreciation of the value of adaptive management in combating uncertainty to avoid management failures in biological resources (Parma and NCEAS, 1998). California now has the opportunity to move forward with the adaptive management of these newly implemented MPAs in a way that accounts for uncertainty and fishing through population modeling and monitoring (e.g. Moffitt et al., 2013; White et al., 2013b).

6. SUMMARY

In summary, California implemented a network of 132 MPAs, covering more than 15% of its coastline, in a science-based, stakeholder-inclusive process. The success of this outcome was driven by the passage of a law, substantial funding by philanthropy, qualitatively justified size and spacing guidelines, and paying minimal attention to fishery interactions (Fox et al., 2013c; Osmond et al., 2010). The goals of the Channel Islands and the MLPA processes go beyond mere implementation of MPAs, but rather concern the ultimate effects of those MPAs on California's coastal ecosystem. Whether these goals are met will depend strongly on the outcome of future monitoring and adaptive management.

In spite of the fact that future attempts to implement MPAs will likely not have the same financial resources as California (see Gleason et al., 2013b for an accounting of costs), and may not have the same fishery infrastructure, California's experience may be valuable. It seems that some level of evaluation of the interactions of proposed MPAs with locally fished species will be possible and worthwhile, rather than simply choosing a fraction of coastline to be placed in MPAs. This more comprehensive approach will enable (a) direct interaction with management of the fishing outside the MPAs and (b) direct integration of MPA design into an MPA monitoring and evaluation program, achieving true adaptive management. That type of integrated management approach is necessary for the science of MPAs to proceed.

REFERENCES

Airamé, S., Dugan, J.E., Lafferty, K.D., Leslie, H., Mcardle, D.A., Warner, R.R., 2003. Applying ecological criteria to marine reserve design: a case study from the California Channel Islands. Ecol. Appl. 13 (1), S170–S184, Supplement.

Andelman, S., Ball, I., Davis, F., Stoms, D., 1999. Sites V 1.0. An Analytical Toolbox for Designing Ecoregional Conservation Portfolios. Manual Prepared for The Nature Conservancy. National Center for Ecological Analysis and Synthesis, Santa Barbara, CA, USA.

Babcock, E.A., MacCall, A.D., 2011. How useful is the ratio of fish density outside versus inside no-take marine reserves as a metric for fishery management control rules? Can. J. Fish. Aquat. Sci. 68, 343–359.

Bakun, A., 1990. Global climate change and the intensification of coastal ocean upwelling. Science 247, 198–201.

Ball, I.R., Possingham, H.P., Watts, M., 2009. Marxan and relatives: software for spatial conservation prioritisation. In: Moilanen, A., Wilson, K.A., Possingham, H.P. (Eds.), Spatial Conservation Prioritisation: Quantitative Methods and Computational Tools. Oxford University Press, Oxford, UK, pp. 185–195.

Barry, J.P., Baxter, C.H., Sagarin, R.D., Gilman, S.E., 1995. Climate-related, long-term faunal changes in a California rocky intertidal community. Science 267, 672–675.

Barth, J.A., Menge, B.A., Lubchenco, J., Chan, F., Bane, J.M., Kirincich, A.R., McManus, M.A., Nielsen, K.J., Pierce, S.D., Washburn, L., 2007. Delayed upwelling alters nearshore coastal ocean ecosystems in the northern California current. Proc. Natl. Acad. Sci. U.S.A 104, 3719–3724.

Bergen, L.K., Carr, M.H., 2003. Marine reserves: how can science best inform policy? Environment 45 (2), 8–19.

Botsford, L.W., 1981. The effects of increased individual growth rates on depressed population size. Am. Nat. 117 (1), 38–63.

Botsford, L.W., 2013. Maximum Sustainable Yield. Oxford Bibiolographies Online, http://dx.doi.org/10.1093/OBO/9780199830060-0071.

Botsford, L.W., Castilla, J.C., Peterson, C.H., 1997. The management of fisheries and marine ecosystems. Science 277, 509–515.

Botsford, L.W., Hastings, A., Gaines, S.D., 2001. Dependence of sustainability on the configuration of marine reserves and larval dispersal distance. Ecol. Lett. 4, 144–150.

Botsford, L.W., Campbell, A., Miller, R., 2004. Biological reference points in the management of North American sea urchin fisheries. Can. J. Fish. Aquat. Sci. 61, 1325–1337.

Bray, N.A., Keyes, A., Morawitz, W.M.L., 1999. The California current system in the Southern California bight and the Santa Barbara Channel. J. Geophys. Res. 104, 7695–7714.

California Department of Fish and Game (CDFG), 2001. The Master Plan: A Ruide for the Developmnst of Fishery Management Plans. http://www.dfg.ca.gov/marine/masterplan.asp.

California Department of Fish and Game (CDFG), 2002. Nearshore Fisheries Management Plan, Chapter 2: Background. http://www.dfg.ca.gov/marine/nfmp/.

California Department of Fish and Game (CDFG), 2003. History of the Channel Islands Marine Reserve Working Group Process. https://nrm.dfg.ca.gov/FileHandler.ashx?DocumentID=31320.

California Department of Fish and Game (CDFG), 2008. Marine Life Protection Act Master Plan for Marine Protected Areas. http://www.dfg.ca.gov/marine/mpa/masterplan.asp.

Carr, M.H., Reed, D.C., 2015. Shallow rocky reefs and kelp forests. In: Mooney, H., Zavaleta, E. (Eds.), Ecosystems of California. UC Press, Berkeley.

Carr, M.H., Saarman, E., Caldwell, M.R., 2010. The role of 'rules of thumb' in science-based environmental policy: California's Marine Life Protection Act as a case study. Stanf. J. Law Sci. Policy 2, 1–17.

Carr, M.H., Woodson, C.B., Cheriton, O.M., Malone, D., McManus, M.A., Raimondi, P.T., 2011. Knowledge through partnerships: integrating marine protected area monitoring and ocean observing systems. Front. Ecol. Environ. 9, 342–350.

Chan, F., Barth, J.A., Lubchenco, J., Kirincich, A., Weeks, H., Peterson, W.T., Menge, B.A., 2008. Emergence of anoxia in the California current large marine ecosystem. Science 319, 920.

Checkley, D.M., Barth, J.A., 2009. Patterns and processes in the California current system. Prog. Oceanogr. 83, 49–64.

Clark, W.G., 1991. Groundfish exploitation rates based on life history parameters. Can. J. Fish. Aquat. Sci. 48, 734–750.

Clark, W.G., 2002. $F_{35\%}$ revisited ten years later. N. Am. J. Fish. Manag. 22, 251–257.

Costello, C., Rassweiler, A., Siegel, D., De Leo, G., Micheli, F., Rosenberg, A., 2010. The value of spatial information in MPA network design. Proc. Natl. Acad. Sci. U.S.A 107, 18294–18299.

Crowder, L.B., Lyman, S.J., Figueira, W.F., Priddy, J., 2000. Source-sink population dynamics and the problem of siting marine reserves. Bull. Mar. Sci. 66, 799–820.

Di Lorenzo, E., Combes, V., Keister, J.E., Strub, P.T., Thomas, A.C., Franks, P.J.S., Ohman, M.D., Furtado, J.C., Bracco, A., Bograd, S.J., Peterson, W.T., Schwing, F.B., Chiba, S., Taguchi, B., Hormazabal, S., Parada, C., 2013. Synthesis of Pacific Ocean climate and ecosystem dynamics. Oceanography 26, 68–81.

Drake, P.T., Edwards, C.A., Barth, J.A., 2011. Dispersion and connectivity estimates along the U.S. west coast from a realistic numerical model. J. Mar. Res. 69, 1e37.

Dulvy, N.K., Rogers, S.I., Jennings, S., Steizenmuller, V., Dye, S.R., Skjoldal, H.R., 2008. Climate change and deepening of the North Sea fish assemblage: a biotic indicator of warming seas. J. Appl. Ecol. 45, 1029–1039.

Ebeling, A.W., Larsen, R.J., Alevzion, W.S., 1980. Habitat groups and island-mainland distribution of kelp-bed fishes off Santa Barbara, California. In: Power, D.M. (Ed.), Multidisciplinary Symposium on the California Islands, Santa Barbara Museum of Natural History.

Feely, R.A., Sabine, C.L., Hernandez-Ayon, J.M., Ianson, D., Hales, B., 2008. Evidence for upwelling of corrosive 'acidified' water onto the continental shelf. Science 320, 1490–1492.

Fluharty, D., 2000. Habitat protection, ecological issues, and implementation of the sustainable fisheries act. Ecol. Appl. 10, 325–337.

Food and Agriculture Organization (FAO), 1996. Precautionary approach to capture fisheries and species introductions. Elaborated by the technical consultation on the Precautionary Approach to capture fisheries (including species introductions), 6–13 June 1995, Lysekil, Sweden. FAO Tech. Guidelines Responsible Fish. No 2.

Fox, E., Miller-Henson, M., Ugoretz, J., Weber, M., Gleason, M., Kirlin, J., Caldwell, M., Mastrup, S., 2013a. Enabling conditions to support marine protected area network planning: California's Marine Life Protection Act as a case study. Ocean Coast. Manag. 74, 14e23.

Fox, E., Poncelet, E., Connor, D., Vasques, J., Ugoretz, J., McCreary, S., Monie, D., Harty, M., Gleason, M., 2013b. Adapting stakeholder processes to region-specific challenges in marine protected area network planning. Ocean Coast. Manag. 74, 24e33.

Fox, E., Hastings, S., Miller-Henson, M., Monie, D., Ugoretz, J., Frimodig, A., Shuman, C., Owens, B., Garwood, R., Connor, D., Serpa, P., Gleason, M., 2013c. Addressing policy issues in a stakeholder-based and science-driven marine protected area network planning process. Ocean Coast. Manag. 74, 34e44.

Gaines, S.D., Gaylord, B., Largier, J.L., 2003. Avoiding current oversights in marine reserve design. Ecol. Appl. 13, S32–S46.

Gaines, S.D., White, C., Carr, M.H., Palumbi, S., 2010. Designing marine reserve networks for both conservation and fisheries management. Proc. Natl. Acad. Sci. U.S.A 107, 18286–18293.

Garcia, S., 1996. The precautionary approach to fisheries and its implications for fishery research, technology and management: an updated review. FAO Fish. Tech. Pap. No. 350/2.

Garcia-Reyes, M., Largier, J., 2010. Observations of increased wind-driven coastal upwelling off central California. J. Geophys. Res. 115, C04011.

Gaylord, B., Gaines, S.D., 2000. Temperature or transport? Range limits in marine species mediated solely by flow. Am. Nat. 155, 769–789.

Gleason, M., McCreary, S., Miller-Henson, M., Ugoretz, J., Fox, E., Merrifield, M., McClintock, W., Serpa, P., Hoffman, K., 2010. Science-based and stakeholder-driven marine protected area network planning: a successful case study from north central California. Ocean Coast. Manag. 53, 52–68.

Gleason, M.G., Kirlin, J., Fox, E., 2013a. California's marine protected area network planning process: introduction to the special issue. Ocean Coast. Manag. 74, 1–2.

Gleason, M.G., Fox, E., Vasques, J., Whiteman, E., Ashcraft, S., Frimodig, A., Serpa, P., Saarman, E., Miller-Henson, M., Kirlin, J., Weber, M., Caldwell, M., Ota, B., Pope, E., Wiseman, K., 2013b. Designing a network of marine protected areas in California: achievements, costs, lessons learned, and challenges ahead. Ocean Coast. Manag. 74, 90–101.

Graham, M.B., Halpern, B.S., Carr, M.H., 2008. Diversity and dynamics of California subtidal kelp forests. In: McClanahan, T.R., Branch, G.M. (Eds.), Food Webs and the Dynamics of Marine Reefs. Oxford University Press, New York, NY, pp. 103–134.

Grantham, B.A., Chan, F., Nielsen, K.J., Fox, D.S., Barth, J.A., Huyer, A., Lubchenco, J., Menge, B.A., 2004. Upwelling-driven nearshore hypoxia signals ecosystem and oceanographic changes in the northeast Pacific. Nature 429, 740–754.

Hamilton, S.L., Caselle, J.E., Malone, D., Carr, M.H., 2010. Incorporating biogeography into evaluations of the Channel Islands marine reserve network. Proc. Natl. Acad. Sci. U.S.A 107, 18272–18277.

Hanak, E., Lund, J., Dinar, A., Gray, B., Howitt, R., Mount, J., Moyle, P., Thompson, B., 2011. Managing California's Water from Conflict to Reconciliation. Public Policy Institute of California, San Francisco, CA.

Hastings, A., Botsford, L.W., 1999. Equivalence in yield from marine reserves and traditional fisheries management. Science 284, 1537–1538.

Hastings, A., Botsford, L.W., 2003. Comparing designs of marine reserves for fisheries and for biodiversity. Ecol. Appl. 13, S65–S70.

Hastings, A., Botsford, L.W., 2006. Persistence of spatial populations depends on returning home. Proc. Natl. Acad. Sci. U.S.A 103, 6067–6072.

Helvey, M., 2004. Seeking consensus on designing marine protected areas: keeping the fishing community engaged. Coast. Manag. 32, 173–190.

Hickey, B.M., 1998. Coastal oceanography of western North America from the tip of Baja California to Vancouver Island. In: Robinson, A.R., Brink, K.H. (Eds.), In: The Sea: The Global Coastal Ocean: Regional Studies and Syntheses, vol. 11. John Wiley, New York, pp. 345–393.

Hilborn, R., Walters, C.J., 1992. Quantitative Fisheries Stock Assessment and Management: Choice, Dynamics and Uncertainty. Chapman and Hall, New York, USA.

Holland, D.S., Brazee, R.J., 1996. Marine reserves for fisheries management. Mar. Resour. Econ. 11, 157–171.

Horn, M.H., Allen, L.A., Lea, R.N., 2006. Biogeography. In: Allen, L.G., Horn, M.H. (Eds.), Ecology of Marine Fishes: California and Adjacent Waters. University of California Press, Berkeley, CA, pp. 3–25.

Hundley, N., 2001. The Great Thirst: Californians and Water, a History. University of California Press, Berkeley, CA.

Hutchings, J.A., 2000. Collapse and recovery of marine fishes. Nature 406, 882–885.

Kaplan, D.M., 2006. Alongshore advection and marine reserves: consequences for modeling and management. Mar. Ecol. Prog. Ser. 309, 11–24.

Kaplan, D.M., Botsford, L.W., 2005. Effects of variability in spacing of coastal marine reserves on fisheries yield and sustainability. Can. J. Fish. Aquat. Sci. 62, 905–912.

Kaplan, D.M., Botsford, L.W., Jorgensen, S., 2006. Dispersal per recruit: an efficient method for assessing sustainability in marine reserve networks. Ecol. Appl. 16, 2248–2263.

Kaplan, D.M., Botsford, L.W., O'Farrell, M.R., Gaines, S.D., Jorgensen, S., 2009. Model-based assessment of persistence in proposed marine protected area designs. Ecol. Appl. 19, 433–448.

Karpov, K., Haaker, P.L., Taniguchi, I., Rogers-Bennett, L., 2000. Serial depletion and the collapse of the California abalone fishery. Can. Spec. Publ. Fish. Aquat. Sci. 130, 11–24.

Kay, M.C., Lenihan, H.S., Guenther, G.M., Wilson, J.R., Miller, C.J., Shrout, S.W., 2012a. Collaborative assessment of California spiny lobster population and fishery responses to a marine reserve network. Ecol. Appl. 22, 322–335.

Kay, M.C., Lenihan, H.S., Kotchen, M.J., Miller, C.J., 2012b. Effects of marine reserves on California spiny lobster are robust and modified by fine-scale habitat features and distance from reserve borders. Mar. Ecol. Prog. Ser. 451, 137–150.

Kildow, J., Colgan, C.S., 2005. California's ocean economy: Report to the Resources Agency. National Ocean Economics Program, State of California.

Kirlin, J., Gleason, M., Ashcraft, S., Caldwell, M., Fox, E., Harty, M., Miller-Henson, M., Ota, B., Weber, M., Wiseman, K., 2013. California's Marine Life Protection Act Initiative: supporting implementation of legislation establishing a statewide network of marine protected areas. Ocean Coast. Manag. 74, 3e13.

Larkin, P.A., 1977. An epitaph for the concept of maximum sustainable yield. Trans. Am. Fish. Soc. 106, 1–11.

Leet, W.S., Dewees, C.M., Klingbeil, R., Larson, E.J., 2001. California's Living Marine Resources: A Status Report. California Resources Agency, California Department of Fish and Game, Sacramento, CA, USA.

Love, M.S., Caselle, J.E., van Buskirk, W., 1998. A severe decline in the commercial passenger fishing vessel rockfish (*Sebastes* spp.) catch in the southern California Bight 1980–1996. CalCOFI Rep. 39, 180–195.

Lubchenco, J., Palumbi, S.R., Gaines, S.D., Andelman, S., 2003. Plugging a hole in the ocean: the emerging science of marine reserves. Ecol. Appl. 13, S3–S7.

Mace, P.M., 2001. A new role for MSY in single-species and ecosystem approaches to fisheries stock assessment and management. Fish Fish. 2 (1), 2–32.

Mace, P.M., Sissenwine, M.P., 1993. How much spawning per recruit is enough? Can. Spec. Publ. Fish. Aquat. Sci. 120, 110–118.

Mangel, M., 1998. No-take areas for sustainability of harvested species and a conservation invariant for marine reserves. Ecol. Lett. 1, 87–90.

Mangel, M., Talbot, L.M., Meffe, G.K., Agardy, M.T., Alverson, D.L., Barlow, J., Botkin, D.B., Budowski, G., Clark, T., Cooke, J., et al., 1996. Principles for the conservation of wild living resources. Ecol. Appl. 6, 338–362.

Marine Reserves Working Group Science Advisory Panel, 2001. Draft Summary of the Joint Meeting of the Science Advisory Panel of the Marine Reserves Working Group/Science and Statistical Committee ad hoc Marine Reserve Committee. http://www.pcouncil.org/bb/2001/1101/Ex_F.1.c_Supp_MRWG_Sci_PanRep_Nov2001BB.pdf.

McArdle, D.A., 1997. California Marine Protected Areas. California Sea Grant College System, La Jolla, California, Publication No. T-039.

Merrifield, M., McClintock, W., Burt, C., Fox, E., Gleason, M., Serpa, P., Steinback, C., 2013. MarineMap: a web-based platform for collaborative marine protected area planning. Ocean Coast. Manag. 74, 67e76.

Micheli, F., Saenz-Arroyo, A., Greenley, A., Vazquez, L., Espinoza Montes, J.A., Rossetto, M., De Leo, G.A., 2013. Evidence that marine reserves enhance resilience to climatic impacts. PLoS One 7, e40832.

Mitarai, S., Siegel, D.A., Watson, J.R., Dong, C., McWilliams, J.C., 2009. Quantifying connectivity in the coastal ocean with application to the Southern California Bight. J. Geophys. Res. 114, C10026.

Moffitt, E.A., Botsford, L.W., Kaplan, D.M., O'Farrell, M.R., 2009. Marine reserve networks for species that move within a home range. Ecol. Appl. 19, 1835–1847.

Moffitt, E.A., White, J.W., Botsford, L.W., 2011. The utility and limitations of size and spacing guidelines for designing marine protected area (MPA) networks. Biol. Conserv. 144, 306–318.

Moffitt, E.A., White, J.W., Botsford, L.W., 2013. Accurate assessment of marine protected area success depends on metric and spatiotemporal scale of monitoring. Mar. Ecol. Prog. Ser. 489, 17–28.

Murray, S.N., Ambrose, R.F., Bohnsack, J.A., Botsford, L.W., Carr, M.H., Davis, G.E., Dayton, P.K., Gotshall, D., Gunderson, D.R., Hixon, M.A., Lubchenco, J., Mangel, M., MacCall, A., McArdle, D.A., Ogden, J.C., Roughgarden, J., Starr, R.M., Tegner, M.J., Yoklavich, M.M., 1999. No-take reserve networks: sustaining fishery populations and marine ecosystems. Fisheries 24, 11–25.

Neubert, M.G., 2003. Marine reserves and optimal harvesting. Ecol. Lett. 6, 843–849.

Ocean Science Trust (OST), 2013. State of the Central California Coast: Results from Baseline Monitoring of Marine Protected Areas from 2007–2012. http://oceanspaces.org/sites/default/files/cc_results_report.pdf.

Osmond, M., Airame, S., Caldwell, M., Day, J., 2010. Lessons for marine conservation planning: a comparison of three marine protected area planning processes. Ocean Coast. Manag. 53, 41–51.

Pacific Fishery Management Council, 2001. Exhibit F.1: Status of Marine Reserve Proposals for the Channel Island National Marine Sanctuary. http://www.pcouncil.org/resources/archives/briefing-books/november-2001-briefing-book/#marine.

Paddack, M.J., Estes, J.A., 2000. Kelp forest fish populations in marine reserves and adjacent exploited areas of central California. Ecol. Appl. 10, 855–870.

Parma, A.M., NCEAS Working Group on Population Management, 1998. What can adaptive management do for our fish, forests, food, and biodiversity? Integr. Biol. 1, 16–26.

Perry, A.L., Low, P.J., Ellis, J.R., Reynolds, J.D., 2005. Climate change and distribution shifts in marine fishes. Science 308, 1912–1915.

Pikitch, E.K., Santora, C., Babcock, E.A., Bakun, A., Bonfil, R., Conover, D.O., Dayton, P., Doukakis, P., Fluharty, D., Heneman, B., Houde, E.D., Link, J., Livingston, P.A., Mangel, M., McAllister, M.K., Pope, J., Sainsbury, K.J., 2004. Ecosystem-based fishery management. Science 305, 346–347.

Pinsky, M.L., Worm, B., Fogarty, M.J., Sarmiento, J.L., Levin, S.A., 2013. Marine taxa track local climate velocities. Science 341, 1239–1242.

Poloczanska, E.S., Brown, C.J., Sydeman, W.J., Kiessling, W., Schoeman, D.S., Moore, P.J., Brander, K., Bruno, J.F., Buckley, L.B., et al., 2013. Global imprint of climate change on marine life. Nat. Clim. Chang. 1, 919–925.

Pondella, D.J., Gintert, B.E., Cobb, J.R., Allen, L.G., 2005. Biogeography of the nearshore rocky-reef fishes at the southern and Baja California islands. J. Biogeogr. 32, 187–201. http://dx.doi.org/10.1111/j.1365-2699.2004.01180.x.

Possingham, H., Ball, I., Andelman, S., 2000. Mathematical methods for identifying representative reserve networks. In: Ferson, S., Burgman, M. (Eds.), Quantitative Methods for Conservation Biology. Springer-Verlag, New York, pp. 291–305.

Ralston, S., 1998. The status of federally managed rockfish on the US West Coast. In: Yoklavich, M.M. (Ed.), Marine Harvest Refugia for West Coast Rockfish: A Workshop, pp. 6–16, NOAA-TM-NMFS-SWFSC-255, La Jolla, CA.

Ralston, S., 2002. West coast groundfish harvest policy. N. Am. J. Fish Manag. 22, 249–250.

Rassweiler, A., Costello, C., Hilborn, R., Siegel, D.A., 2014. Integrating scientific guidance into marine spatial planning. Proc. R. Soc. B281, 20132252.

Reed, D.C., Rassweiler, A., Carr, M.H., Cavanaugh, K.C., Malone, D.P., Siegel, D.A., 2011. Wave disturbance overwhelms top-down and bottom-up control of primary production in California kelp forests. Ecology 92, 2108–2116.

Restrepo, V.R., Powers, J.E., 1999. Precautionary control rules in US fisheries management: specification and performance. ICES J. Mar. Sci. 56, 846–852.

Restrepo, V.R., Thompson, G.G., Mace, P.M., Gabriel, W.L., Low, L.L., MacCall, A.D., Methot, R.D., Powers, J.E., Taylor, B.L., Wade, P.R., Witzig, J.F., 1998. Technical Guidance on the Use of Precautionary Approaches to Implementing National Standard 1 of the Magnuson-Stevens Fishery Conservation and Management Act. NOAA-TM-NMFS-F/SPO-31.

Rosenberg, A., Mace, P., Thompson, G., Darcy, G., Clark, W., Collie, J., Gabriel, W., MacCall, A., Methot, R., Powers, J., Restrepo, V., Wainwright, T., Botsford, L., Hoenig, J., Stokes, K., 1994. Scientific review of Definitions of Overfishing in U.S. Fishery Management Plans, NOAA Tech. Memo. NMFS-F/SPO-17, 205 p.

Saarman, E., Gleason, M., Ugoretz, J., Airamé, S., Carr, M., Fox, E., Frimodig, A., Mason, T., Vasques, J., 2013. The role of science in supporting marine protected area network planning and design in California. Ocean Coast. Manag. 74, 45–56.

Sayce, K., Shuman, C., Connor, D., Reisewitz, A., Pope, E., Miller-Henson, M., Poncelet, E., Monie, D., Owens, B., 2013. Beyond traditional stakeholder engagement: public participation roles in California's statewide marine protected area planning process. Ocean Coast. Manag. 74, 57e66.

Scholz, A.J., Bonzon, K., Fujita, R., Benjamin, N., Woodling, N., Black, P., Steinback, C., 2004. Participatory socioeconomic analysis: drawing on fisher men's knowledge for marine protected area planning in California. Mar. Policy 28, 335e349.

Scholz, A., Steinback, C., Kruse, S., Mertens, M., Silverman, H., 2011. Incorporation of spatial and economic analysis of human-use data in the design of marine protected areas. Conserv. Biol. 25, 485e492.

Schroeter, S.C., Reed, D.C., Kusher, D.J., Estes, J.A., Ono, D.S., 2001. The use of marine reserves in evaluating the dive fishery for the warty sea cucumber (Parastichopus parimensis) in California, U.S.A. Can. J. Fish. Aquat. Sci. 58, 1773–1781.

Sladek Nowlis, J., Roberts, C., 1999. Fisheries benefits and the optimal design of marine reserves. Fish. Bull. 97, 604–616.

Snyder, M.A., Sloan, L.C., Diffenbaugh, N.S., Bell, J.L., 2003. Future climate change and upwelling in the California current. Geophys. Res. Lett. 30, 1823.

Starr, R.M., Cope, J.M., Kerr, L.A., 2002. Trends in Fisheries and Fishery Resources Associated with the Monterey Bay National Marine Sanctuary from 1981–2000. California Sea Grant College Program, University of California, San Diego, La Jolla, CA, p. 156. www.montereybay.noaa.gov/research/techreports/fisherytrends.pdf.

Ueber, E., MacCall, A., 2005. The rise and fall of the California sardine empire. In: Glantz, M. (Ed.), Climate Variability, Climate Change, and Fisheries. Cambridge University Press, Cambridge UK, pp. 31–48.

Walters, C., Hilborn, R., Parrish, R., 2007. An equilibrium model for predicting the efficacy of marine protected areas in coastal environments. Can. J. Fish. Aquat. Sci. 64, 1009–1018.

Weible, C.M., 2008. Caught in a Maelstrom: implementing California marine protected areas. Coast. Manag. 36, 350–373.

White, J.W., 2010. Adapting the steepness parameter from stock–recruit curves for use in spatially explicit models. Fish. Res. 102, 330–334.

White, J.W., Botsford, L.W., Hastings, A., Largier, J.L., 2010a. Population persistence in marine reserve networks: incorporating spatial heterogeneities in larval dispersal. Mar. Ecol. Prog. Ser. 398, 49–67.

White, J.W., Botsford, L.W., Moffitt, E.A., Fischer, D.T., 2010b. Decision analysis for designing marine protected areas for multiple species with uncertain fishery status. Ecol. Appl. 20, 1523–1541.

White, J.W., Botsford, L.W., Baskett, M.L., Barnett, L.A.K., Barr, R.J., Hastings, A., 2011. Linking models and monitoring data for assessing performance of no-take marine reserves. Front. Ecol. Environ. 9, 390–399.

White, J.W., Scholz, A.J., Rassweiler, A., Steinback, C., Botsford, L.W., Kruse, S., Costello, C., Mitarai, S., Siegel, D., Drake, P.T., Edwards, C., 2013a. A comparison of approaches used for economic analysis in marine protected area planning in California. Ocean Coast. Manag. 74, 77–89.

White, J.W., Botsford, L.W., Hastings, A., Baskett, M.L., Kaplan, D.M., Barnett, L.A.K., 2013b. Transient responses of fished populations to marine reserve establishment. Conserv. Lett. 6, 180–191.

White, J.W., Schroeger, J., Drake, P.T., Edwards, C.A., 2014. The value of larval connectivity information in the static optimization of marine reserve design. Conserv. Lett. http://dx.doi.org/10.1111/conl.12097.

Wild, P.W., Tasto, R.N. (Eds.), 1983. Life History, Environment, and Mariculture Studies of the Dungeness Crab, Cancer Magister, with Emphasis on the Central California Fishery Resource. California Department of Fish and Game Fish Bulletin, Sacramento, CA, USA, p. 172.

Worm, B., Hilborn, R., Baum, J.K., Branch, T.A., Collie, J.S., Costello, C., Fogarty, M.J., Fulton, E.A., Hutchings, J.A., Jennings, S., Jensen, O.P., Lotze, H.K., Mace, P.M., McClanahan, T.R., Minto, C., Palumbi, S.R., Parma, A.M., Ricard, D., Rosenberg, A.A., Watson, R., Zeller, D., 2009. Rebuilding global fisheries. Science 325, 578–585.

CHAPTER SEVEN

Inadequate Evaluation and Management of Threats in Australia's Marine Parks, Including the Great Barrier Reef, Misdirect Marine Conservation

Bob Kearney[*,1], Graham Farebrother[†]
*Emeritus Professor in Fisheries Management, Institute for Applied Ecology, University of Canberra, Bruce ACT, Australia
†Sydney Fish Market, Pyrmont, NSW, Australia and Fisheries Aquaculture and Coasts Centre, Institute for Marine and Antarctic Studies, University of Tasmania, Hobart, Tasmania, Australia
[1]Corresponding author: e-mail address: bob.kearney@canberra.edu.au

Contents

Abstract

The magnificence of the Great Barrier Reef and its worthiness of extraordinary efforts to protect it from whatever threats may arise are unquestioned. Yet almost four decades after the establishment of the Great Barrier Reef Marine Park, Australia's most expensive and intensely researched Marine Protected Area, the health of the Reef is reported to be declining alarmingly. The management of the suite of threats to the health of the reef has clearly been inadequate, even though there have been several notable successes. It is argued that the failure to prioritise correctly all major threats to the reef, coupled with the exaggeration of the benefits of calling the park a protected area and zoning subsets of areas as 'no-take', has distracted attention from adequately addressing the real causes of impact. Australia's marine conservation efforts have been dominated by commitment to a National Representative System of Marine Protected Areas. In so doing, Australia has displaced the internationally accepted primary priority for pursuing effective protection of marine environments with inadequately critical adherence to the principle of having more and bigger marine parks. The continuing decline in the health of the Great Barrier Reef and other Australian coastal areas confirms the limitations of current area management for combating threats to marine ecosystems. There is great need for more critical evaluation of how marine environments can be protected effectively and managed efficiently.

Keywords: Great Barrier Reef Management, Evidence-based marine conservation, Adaptive marine management, Poorly justified restriction of fishing in MPAs, Australian marine conservation strategy, Uncritical science used to underpin marine policy

1. INTRODUCTION

By the late 1960s, growing appreciation within Australia of the magnificence of the Great Barrier Reef (GBR) and the need to protect it was boosted greatly by highly publicised proposals for mining of coral and exploration for oil within the Reef. The Royal Commission into oil exploration, 1970–1974 (as detailed by Queensland State Archives, 2013) galvanised public pressure for the protection of the Reef to the extent that action was accepted increasingly to be a political imperative. Acceptance that mining of coral was an obvious and unquestioned threat to coral reefs anywhere and that oil spills were potentially damaging across huge but undetermined areas made a management approach that included the whole of the Reef an obvious strategy.

Appreciation of the terrestrial model of national parks on land as a means of conserving large areas supported acceptance that the park approach would be appropriate; the GBR Marine Park Act was enacted in 1975 (Australian Government, 1975). Designating the Reef as a special management area was, and remains, an appropriate means of confirming that the area is of great importance and warrants exceptional efforts dedicated to its protection.

However, it is now apparent that the failure from the beginning of management of the Park to give adequate priority in the management process to critical evaluation of what threats the GBR needed protection from and how effective protection could be provided has contributed to the contents of the Park continuing to degrade as described by Brodie and Waterhouse (2012) and De'ath et al. (2012).

Successive Australian governments have continued to claim that the terrestrial park model is appropriate for marine conservation across all types of environments, for example, the Federal Environment Minister's announcement in 2013 that '*We're for protecting our oceans in the same way that we have protected precious areas on land with national parks*' (Burke, 2013). This has been coupled with fostering of the public perception that areas classified as 'marine protected areas' (MPAs) will indeed be protected. Acceptance that marine areas need protection has been distorted to support the assertion that more marine parks, as implemented in Australia, will provide the required protection.

Recent evidence of declines in the health of the GBR (De'ath et al., 2012; UNESCO, 2013), Australia's biggest and most intensely researched and managed park, has, however, exposed the limitations of the marine parks approach as a strategy for protecting the Reef against numerous significant threats. It also suggests that even though there have been benefits from giving the GBR elevated priority for management, encompassing it in a park has not resulted in the level of protection anticipated for what is Australia's most iconic MPA.

This chapter elaborates on why the GBR remains inadequately protected (Kearney, 2013a) and the impact the GBRMP has had on the development of Australia's approach to marine conservation, in particular the system of marine parks, the National Representative System of Marine Protected Areas (NRSMPA). It elaborates on five suggested inadequacies in the national approach:

1. The failure to investigate adequately the causes of impacts (threats) to different marine environments and ecosystems and to prioritise appropriately their management;

2. The failure to acknowledge the fundamental differences between terrestrial and marine environments and to accommodate adequately the peculiarities of the marine realm that necessitate management that is tailored to the individuality of the components of it;

3. Excessive priority being given to description of what is in various parts of areas that make them more worthy of protection, to the exclusion of determination of how best to provide protection;

4. Inaccurate evaluation of the impacts of fishing and exaggeration of the benefits of closing areas to all forms of fishing; and

5. Insufficient evaluation and attribution of management outputs and outcomes and the resulting inadequate and/or inappropriate adaptive management.

2. THE BACKGROUND TO AUSTRALIA'S USE OF THE MARINE PARK CONCEPT TO PROTECT THE ICONIC GBR

In the latter half of the twentieth century, Australian opinion had begun to mirror growing international appreciation that the total global environment was an entity and not just a series of independent national domains or a boundless abstraction. It had also become evident that there were definite limits to how humans could use the planet; cracks were appearing. There was growing realisation that many ecosystems, including those in marine areas that were particularly important, even vital, for human well-being, were threatened and therefore needed unusual and mostly unprecedented effort to ensure their protection (Carson, 1962; Ehrlich, 1968; Hardin, 1968; Taylor, 1970). This became particularly publicised for large iconic and particularly valued areas, such as Australia's GBR (as reported in AMCS (QLS), 2014b).

Exactly what human activities selected areas needed protection from remained obscured by the thickening haze of impacts and potential threats. These ranged from sudden catastrophes, such as nuclear disasters or major oil or other chemical spills, through damaging extraction and overexploitation in numerous forms, to other largely invisible but insidious, diffuse and relentless side effects of human population growth, technology use and affluence epitomised by increasing air and water pollution. It was in this environment of increased awareness coupled to a growing sense of responsibility for issues beyond local and present confines that broad-ranging, including international, environmental activism flourished.

One of Australia's prominent early products of this activism was the 'Save the Barrier Reef' campaign (AMCS (QLS), 2014a). This campaign was formalised in Queensland in 1967 and fuelled by the concern over coral mining and oil exploration in the Reef (discussed above) quickly generated intense public interest nationally and subsequently internationally. The degree to which public support for saving the Barrier Reef was eventually able to transcend the boundaries of party politics in Australia is exemplified in the creation of the GBRMP under the joint sponsorship of two famously antagonistic Australian governments, the Australian Commonwealth, Whitlam Labor Party Government and the Queensland, Bjelke-Petersen National Party Government. The Australian Commonwealth Act to create the Park was passed with the support of all political parties in 1975 (Australian Government, 1975).

Throughout Australia growing acceptance of the importance of the Reef was matched by recognition of the need to protect it. Apprehension was allayed by the popular belief that the issue was being more than adequately addressed by encompassing the Reef in a huge marine park, an unprecedented action. The declaration of the GBRMP greatly increased attention on marine conservation in Australia. The immediate success in addressing the specific issues that had catalysed the creation of the Park, curtailing coral extraction and oil exploration, that were actioned in the Act that underpins the Park (Australian Government, 1975), bolstered the marine park concept. The existence of the Park has continued to catalyse the creation of more marine parks under both Commonwealth and state jurisdictions as part of the NRSMPA. The extent of the continued momentum has been such that by early 2014 Australia had a total of 3,244,100 km^2 of its territorial waters in marine parks (Australian Government, 2014), approximately 34.6% of the global total reported by the IUCN in late 2013 (IUCN, 2013).

It is now apparent that the declaration of the GBRMP has resulted in considerable effort to protect the Reef with some notable successes. These include curtailing coral extraction and oil exploration (Australian Government, 1975) and controlling resort development (GBRMPA, 1990, 1992, 1994). However, the total effort to protect the Reef has been much less than completely effective. Nonetheless, Australia has continued to greatly expand its system of marine parks without critical evaluation of the benefits and limitations of existing parks and how protection of the many different types of marine environments can be provided most cost-effectively. Australia's approach to marine conservation has been distracted by a fundamental failure to prioritise management on an evidence-based approach to addressing cause and effect in different situations. The substitute has been overly optimistic projection of a seductively simplistic area management paradigm in the form of the NRSMPA. This has been confounded by the failure to differentiate between the specific management responses necessary in different types of areas and against different types of threats. The unfortunate reality is that almost four decades after its creation, the condition of the contents of Australia's most iconic and intensively managed marine park, the GBRMP, continues to deteriorate (Brodie and Waterhouse, 2012; De'ath et al., 2012; UNESCO, 2013) at an obviously unacceptable rate (UNESCO, 2013, 2014). Many other marine parks continue to be created with even less specific purpose and less commitment to identification and management of what threatens the areas in question than was the case for the GBRMP.

3. THE MANAGEMENT SHORTCOMINGS

3.1. The failure to investigate adequately the causes of impacts (threats) to different marine environments and ecosystems and to prioritise appropriately their management

The wisdom of making exceptional efforts to protect the Barrier Reef was, and remains, beyond reproach. The Reef is iconic and its immense importance is not restricted to Australia; it has World Heritage status, although because of its declining condition the continuation of that status has been questioned (UNESCO, 2013, 2014). Its worthiness of enhanced protection against whatever human-induced threats have arisen, or may arise in the future, is obvious and the expectation that governments would make special efforts to provide that protection is logical. It is also logical to expect that the efforts made to address the problem, how to protect the Reef, would be based on first identifying and then addressing in priority order at least the major causes of the problem, that is the threats against which protection was required. While some of the most obvious threats were addressed in early actions of the Great Barrier Reef Marine Park Authority (GBRMPA), when notable successes were achieved, the commitment throughout Australia to a threat-based approach has been progressively diminished as the NRSMPA developed.

The selection of the term 'marine park' to designate the whole of the Reef as a special area worthy of protection was not unreasonable, particularly at the time. In hindsight, however, it is apparent that the word 'park' fostered unsubstantiated expectations of compatibility with terrestrial management paradigms. This was largely at the expense of identification of the management measures that would have been more specific for addressing key threats and more appropriate for marine environments. Unrealistic expectations of the level of protection of the entity that would come from 'setting (parts of) it aside' (discussed in Section 3.2) distracted efforts from targeted and effective management of threats to the entity, or even parts of it, no matter where they originated.

Other prominent expectations or assumptions that have not been adequately questioned include firstly that declaring the area of the Reef as a marine park would ensure that adequate and appropriate protection would be provided, and secondly, that the designation of the area as a marine 'protected' area was descriptive and not merely wishful. Unfortunately, the nuance of the titles 'marine park' and 'MPA', and in particular the absolute and past-tense implication of 'protected', has been used to support public perception of 'mission accomplished'. This misconception has been continuously nurtured by governments in incorrectly definitive statements such as the Ministerial pronouncement, '*Australia's precious marine environments have been permanently protected*' by marine parks (Burke, 2012). The continued overstatement of what has actually been achieved has diminished public scrutiny of the adequacy and appropriateness of the management action taken and what has been proposed. The resulting lack of scrutiny has continued to help conceal the limitations of what management has been provided.

The groundswell of enthusiasm for saving the splendour of the Barrier Reef has since the 1960s been such that the question from a minority of the many who accepted the need for extra efforts to ensure its preservation, and from many of those who did not, of exactly what it needed saving from was smothered by the comfort afforded its proponents by the extreme imprecision, but righteous implication, of the popular answer: 'Everything'. An absolute and uncompromising response was justified at the time by the assertion, in hindsight prophetic, that the Reef must be protected against all threats and that many threats to the Reef had not yet been identified and new ones would arise as human populations grew and technology developed. Decades later, even though the GBRMP has been the subject of considerable management effort, detail of exactly what actions would be necessary, and how they would be most efficiently provided to save the Reef from everything, or even the threats that are now known to be priorities, has unfortunately continued to remain obscure. Recent reports of declines in the health of the Reef (Brodie and Waterhouse, 2012; Brodie et al., 2013; De'ath et al., 2012; UNESCO, 2012) confirm that Government desire to take advantage of public support for marine parks has not been matched by adequate support for, or direction to, managers, such as GBRMPA, to empower them to provide effective management of the increasingly obvious threats.

As 'The Reef' in total, or at least the component of it that constituted the 'barrier' off the east coast of northern Australia, has been accepted as an entity, it was, and remains, logical that efforts to protect it should be based on consideration of the area in its entirety. The call has always been to 'save the Barrier Reef', not just some parts, or some fraction, of it. Acceptance that the Reef as an entity is worthy of protection necessitates a holistic approach to its management. To this end, it is noteworthy that the first Annual Report of GBRMPA states '*the complexity of physical and biological systems in the seas is such that their long-term survival depends ultimately upon integrated planning of all human activities that have significant impacts upon them*' (GBRMPA, 1978a, p. 1). Having acknowledged the need to address '*all human activities that have significant impacts*', it is unfortunate that the causes of the known impacts were not further investigated at that early stage and appropriately prioritised in the planning and management processes.

In reality right up to the late 1990s, the primary impacts on the Reef were accepted to be direct use, i.e., resource extraction and tourism within the boundaries of the Park (GBRMPA, 1995, 1997), and efforts to identify the full suite of threats, including those arising outside the Park that are now accepted to be having great impact on the Reef, were minimal. Management efforts were stated by GBRMPA to be concentrated on surveying and cataloguing the contents of the Park (for example, GBRMPA, 1978a,b, 1979, 1980). Frustration with the lack of financial, technological and human resources to even describe the area of the Reef that needed to be managed is apparent throughout earlier reports of the GBRMPA (for example, GBRMPA, 1979, 1984, 1986, 1988, 1990, 1993). It was stated that the '*lack of long-term funding has restricted the design and implementation of many other projects and will likely limit the meaningfulness of the results*' (GBRMPA, 1994, p. 37). One annual report went so far as to contrast the GBRMPA annual budget with the cost of a few parts of military hardware, '*People may be surprised to learn that the Authority's annual budget is equivalent to about two and a half times the cost of a wing assembly of an FA-18 fighter aircraft*' (GBRMPA, 1993, p. 1). Thus, the need to manage all threats to the Reef, including those that arise outside the GBRMP, had been initially recognised, but not translated into management priorities that, if appropriately resourced, may have led to the actions that were necessary to ensure the area was indeed protected.

The failure to protect the Reef adequately is to a large degree because the origins of several of the major threats, such as climate change and ocean acidification, are global in nature and beyond the control of Reef managers (Atkins et al., 2011; Elliott, 2011). But several of the major causes of the problems, including those identified by UNESCO as the ones urgently needing management (UNESCO, 2012, 2013, 2014), for example, pollution from agricultural runoff, inappropriate coastal development and dumping of dredge spoils, are the specific responsibility of Australian governments, federal and state. It is apparent from the most recent reports of GBRMPA that at least some levels of management now accept the need for greater priority to manage threats from outside the Park, particularly those arising in adjacent, and even somewhat distant, terrestrial environments (GBRMPA, 2010, 2011b, 2012, 2013). In recent years, GBRMPA has been giving increasing attention to addressing threats that originate in terrestrial locations.

Acceptance from the beginning that the Reef was worthy of protection against all threats inadvertently helped conceal the need to describe individual threats adequately and to determine what management would be necessary to address each of them effectively. The uncritical and imprecise use of the expression 'protected area' (discussed above), coupled with the encouraged, but unrealistic, expectation that the contents of a marine park will automatically be endowed with the same level of protection as those in a terrestrial park, further diminished public expectation of urgency for additional action, at least until very recently. The use of the 'area' concept to facilitate protection also tended to focus attention on actions within the area and not more holistically on all threats that warranted management. The absence of priority to specifically address all major threats also meant that the need for determination of where and when the necessary action would be best applied was diminished greatly. In effect, the lack of *a priori* determination of exactly what needed to be done facilitated avoidance of determination of how, where and when protective action would most cost-effectively be applied, and of course, the cost of actually doing it effectively.

At the time of creation of the GBRMP, priority was given to addressing at least some of the clearly identified threats, e.g., coral mining and oil exploration, and these were effectively addressed across the whole of the Park. Notably, the Park is still vulnerable to these activities, particularly oil extraction, if they occur in close proximity to the boundaries of the Park. Other major threats, especially pollution from coastal activities, were identified, as discussed above, but not effectively addressed, presumably because of the cost and political difficulty of doing so. The failure to maintain priority on management of major threats across their whole area of influence facilitated increased use of a management strategy, zoning within the Park, that was not appropriate for amelioration of the impacts of the major threats, even within the designated zones. Effective management of the dominant threats to marine systems including the Reef, such as pollution in its many forms, necessitates that they be addressed at their source, not by the zoning of their destinations. By as early as 1989, GBRMPA had reported that the threat from pollution was increasing and was by then an acknowledged priority (GBRMPA, 1989), yet the next major change to the management of the GBRMP was when zoned no-take areas were increased from 4.6% to 33% in 2004. This sevenfold increase in no-take zoned areas was not an effective response to the major threats; it did little more than greatly increase the area closed to fishing.

Acknowledgement by Australian governments of the different types of threats to marine environments is increasing, but the reluctance to adequately address threats persists throughout the country. In the recent Commonwealth Government Regulatory Impact Statement (RIS)[1] (*Completing the Commonwealth Marine Reserves Network*) (RIS, 2012), which cites the 2011 Australian *State of the Environment Report* (Hatton et al., 2011), the overall threats to marine systems are introduced as follows: '*The main risks to the future of the marine environment are from the impacts of climate change*' (RIS, 2012, p. 6). The 2009–2010 Year Book Australia (ABS, 2010) from which this RIS also takes information selectively states, '*the Great Barrier Reef is experiencing significant damage from a number of factors, including agricultural run-off and rapid changes in climate. The Intergovernmental Panel on Climate Change has warned that the Great Barrier Reef faces 'functional extinction' within decades*' (ABS, 2011, p. 10). Furthermore, the recent UNESCO review (UNESCO, 2012) of the status of the GBRMP confirms that there are very serious threats to the contents of the Park that are not being adequately managed. These poorly managed threats are stated to be dominated by pollution in many forms and inappropriate coastal development (UNESCO, 2012). Yet, in spite of clear recognition of these major threats that could lead to 'functional extinction within decades', the need to address them in the NRSMPA is ignored in the actions supported by the 2012 RIS.

The '*Goals and Principles underpinning the development of the Marine Reserves Network*', as repeated in the 2012 RIS (RIS, 2012, p. 73), all relate to what is within the areas declared. What the areas to be selected need protection from (the threats) or how that protection is to be provided to '*maximise conservation*

---

[1] "*A RIS is a document prepared by the department, agency, statutory authority or board responsible for a regulatory proposal, following consultation with affected parties. It formalises and provides evidence of the key steps taken during the development of the proposal, and includes an assessment of the costs and benefits of each option*" (Australian Government, 2013).

outcomes' (RIS, 2012, p. 73) are not described. Not surprisingly then, 13 of the 14 'selection and design principles' described for the NRSMPA to meet these goals relate to the number of reserves, what is in them and how the boundaries are determined (RIS, 2012), not how to protect even these areas, let alone Australia's marine environments more generally. The preoccupation with having more comprehensive and representative reserves is to the exclusion of determination of how effective protection is to be provided even within the selected areas and against the threats that were known to be priorities for management (further discussed in Section 3.3).

3.2. The failure to acknowledge the fundamental differences between terrestrial and marine environments and to accommodate adequately the peculiarities of the marine realm that necessitate management that is tailored to the individuality of the components of it

At the time of the declaration of the GBRMP in 1975, Australians had long been supportive of the need for terrestrial reserves to prevent the increasingly obvious environmental devastation of unconstrained human development linked to population growth, increasing affluence and increased use of technology. The direct effects of development in terrestrial environments were readily visible in the forms of urban and industrial sprawl and inadequately regulated agriculture, all of which represented forms of destruction of habitats and visible despoiling of natural ecosystems. Australia had had a succession of national parks since the declaration of the Royal National Park south of Sydney in 1879 and the resulting benefits were obvious and publicly acclaimed. Terrestrial national parks are justifiably, hugely popular in Australia. They continue to represent the country's most obvious expression of environmental conservation.

The commitment to terrestrial parks was founded on acceptance that the total area of Australia was not going to be fully protected against human pressures, so management based on protecting selected parts of key environments represented an obvious compromise; no realistic alternatives were apparent. The management of national parks on land is predominantly in the form of exclusion, from areas with obviously defined boundaries, of the visible forms of intensive human activity, including agriculture. This approach has provided a relatively high degree of protection from associated threats of development to the areas involved. Furthermore, the results of this protection against extraction have been clearly visible and, as a result, readily accepted and progressively acclaimed by the general public. It is only in the last few decades that the limitations of terrestrial parks for providing comprehensive protection, particularly against mobile and pervasive threats such as air and water pollution, and their common ineffectiveness in the control of introduced or translocated species and organisms and in mitigating global-scale impacts, such as climate change, have been more openly appreciated (Beeton and Lynch, 2012; Bradshaw et al., 2007; IUCN WCPA, 2005; Woinarski et al., 2010).

The 'setting it aside' approach is being increasingly questioned (Flannery, 2012; IUCN WCPA, 2010; Margules and Pressey, 2000; Penn and Fletcher, 2010). Acceptance of the benefit of the alternative strategy of actively managing the full extent of ecosystems, including by actively addressing threats across their full range, is growing (CBD, 2006; Kareiva et al., 2011). This appreciation is, however, largely limited to those with detailed knowledge of park management; civil society's acceptance of the outwardly obvious conservation benefits from national parks continues to greatly outweigh acknowledgement of their far less obvious limitations (for example, Flannery, 2012). However, the increasing acceptance of the inadequacies of the total effort to conserve planet Earth, which has for more than a century included national parks in terrestrial environments, has created a form of collective guilt. A primary outcome of this guilt has been international promotion by governments of the need for more holistic (comprehensive) and appropriate (effectively managed) approaches to conservation (Kearney et al., 2013).

The emergence of recognition of the increasing degradation of marine environments, the guilt over acknowledged inadequacies in the management of the global environment and the lack of evidence-based evaluation of what was necessary to provide effective management of oceans, fuelled more idealistic intentions and associated optimistic expectations. Unfortunately, however, in the absence of adequate identification of threats and appropriate risk analyses, idealistic objectives were, and remain, largely tied to inadequately tested assumptions about the effectiveness of the protection provided to various types of marine environments by area management techniques. These techniques had been developed for terrestrial environments that are very different to their marine counterparts and commonly have different vulnerabilities to transmission of the major threats (Kearney et al., 2013).

Even though the inherent and fundamental differences between marine and terrestrial environments have been described (for example, Carr et al., 2003; Steele, 1985, 1991), there has remained an unfortunate and serious lack of critical appraisal of whether management systems which produced obvious (highly visible), and generally highly desirable, even if less than ideal, outcomes in terrestrial environments could be transposed effectively to marine systems (Kearney et al., 2013). Furthermore, most components of marine systems, including even prominent structural foundations, are invisible to the general public. The lack of readily visible cause and effect relationships for many of the major threats in the marine realm means that the uncertainty created by the lack of accuracy and precision in identifying individual threats, where they originated and how they could be effectively managed, is actually much greater than it is in terrestrial domains. Unfortunately, accuracy and precision in the evaluation of threats and how best to address them have not been features of the development of the NRSMPA in Australia. In reality, Australia has, as detailed in Section 3.1, progressively distanced itself from basing marine management on accurately addressing precisely described threats.

Terrestrial parks provide considerable protection against the most obvious threats (the direct effects of industrial development, urbanisation and agriculture) which are in effect different forms of extraction of habitats and their contents. Management within marine parks has uncritically continued a similar preoccupation with regulating extraction in the form of fishing and mining, even though the major threats to most marine environments are not extraction but various forms of injection, primarily pollution, introduced organisms and runoff from coastal land usage. Many of these threats are common to terrestrial environments where management by parks is increasingly acknowledged to not represent an effective approach. They are much less amenable to area management approaches in marine environments. Marine environments are largely opaque and have extremely high connectivity and mobility of key components compared to their terrestrial counterparts. As a result, control of invasive threats, such as pollution, once they enter the marine system is extremely difficult or even impossible. In contrast, extraction is relatively easy to eliminate in marine environments, particularly if it is already tightly regulated. Governments have been seduced by this simplicity and the associated claim that effective, even total (Burke, 2012) protection can be or even has been provided and at comparatively small cost.

The prevention, or at least slowing, of visible change is primary to public appreciation of area management in terrestrial parks. Most change is not readily visible in marine environments. However, preoccupation with the limited change that is visibly obvious has flowed over into marine management. As a result, inadequate attention is being paid to less obvious but more insidious manifestations of threats to broader systems.

The impacts of unjustified priority being given to visible change in only the most obvious components of ecosystems as sentinels of ecosystem health are magnified greatly in marine environments. Many of the agents of major change to marine systems, such as pollution, are commonly invisible and few of the primary or initial changes they cause are readily detectable by the non-expert. These changes are also usually less intensely researched than their more obvious counterparts. This has contributed to seriously disproportionate use of the relative abundance of larger, more visible species in different areas as measures of ecosystem health and the effectiveness of its management. The many claims that marine parks work (e.g. Possingham, 2011 reports that there are more than 100 examples) are dominated by reports of the elevated relative abundance of large, visibly prominent, fish or crustacean species in 'protected' versus fished areas. Evidence of cause and effect on the health of even the major components of all trophic levels is seldom provided.

Undetected change is particularly a problem at lower trophic levels where the first, and often the greatest, primary impacts of invasive vectors such as pollution commonly occur. Unfortunately, few studies have used primary indicators of ecosystem health or biodiversity conservation from the trophic levels most impacted by pollution as descriptors of the effectiveness of marine parks. The use of fished species that quickly and directly respond to changes in fishing pressure as indicators exaggerates the relationship between fishing and true ecosystem well-being. Detecting higher abundance of selected species in an area closed to fishing primarily confirms that abundance of target and associated species can be regulated by changing catch. In the absence of evaluation of more direct measures of protection against the most influential threats, it is not a robust measure of conservation outcomes. Its preferential use disguises expression of the effectiveness, or lack thereof, of protection for the individual components of different levels of the system (i.e. biodiversity).

3.3. Excessive priority being given to description of what is in various parts of areas that make them more worthy of protection, to the exclusion of determination of how best to provide protection

Successive Australian Commonwealth governments between 2000 and 2013 demonstrated a strong desire to have more marine parks; under the nurtured perception of their conservation credentials, such ideas were extremely popular in the electorate. The public wanted marine areas protected and they were led to believe that 'MPAs' as implemented in Australia fulfilled that requirement (e.g. Minister Burke's statement that they were '*permanently protected*'; Burke, 2012). Pursuit by governments of more parks was not predicated on detailed assessment of the positive and negative attributes of those parks that were already in place (normally a prerequisite for efficient adaptive management; B.C. Government, 2014; Holling, 1978), including the GRBMP. This is exemplified in the most recent Government RIS that supports massive expansion of the NRSMPA (RIS, 2012). This formal Government description of the regulatory process relating to the establishment of marine parks assumes, rather than evaluates, benefits of having more areas zoned to regulate fishing. As outlined in Section 3.1, what is in areas has been used as the primary criterion for determination of the number, location and size of parks and closures within. The threats to each area are not described and how protection is to be provided is not discussed.

The determination of the outer boundaries of a marine park is in itself a form of zoning. When the GBRMP was first declared, it was huge by world marine reserve standards; the intention was clearly to protect the bulk of the Barrier Reef as an entity. Indeed, the concept of protecting the entire GBRMP was portrayed from the beginning by prevention of oil and gas exploration and mining in the whole area of the Park and a comprehensive approach to the regulation of tourism development and shipping (Australian Government, 1975; GBRMPA, 1978a). Almost three decades later, in 2004, in by far the greatest single change to the management of the Park, the prominence in the management process afforded to identification of what was in areas and the closure of at least parts of all of them to fishing was greatly increased under the Representative Areas Program (RAP) (GBRMPA, 2011e). Under this Program, 70 separate bioregions (30 reef and 40 non-reef bioregions) within the GBRMP were identified and prioritised for greater 'protection'. The additional primary management action provided was the closure of a minimum of 20% of each bioregion to all fishing (GBRMPA, 2011e). Other more limited restrictions, including on selected types of fishing, were included in additional zones.

The RAP within the GBRMP was one of Australia's earliest formal enactments of 'bioregional planning' (originally a terrestrial concept; Lambert et al., 1995) as the justification for subdividing marine areas and basing increased priority for management on the assignment of greater

worthiness for protection to some parts of some of them. It is further encap-
sulated in the Comprehensive, Adequate and Representative (CAR) principle, another terrestrial construct (Commonwealth of Australia, 1992b)
translocated without evidence of relevance to marine environments
(Kearney et al., 2013) as the guiding paradigm for the whole of Australia's
NRSMPA (ANZECC TFMPA, 1999). It has underpinned the RAP and
most other recent marine park initiatives in Australia.

The concept of having a CAR fraction of every type of environment
effectively protected is extremely seductive, but the achievement of such
a lofty objective is dependent on management that is effective in each
and every type of environment. Australia has deliberately distanced itself
from a commitment to effectively manage marine environments or even
subsets of them by removing the words 'effectively managed' from the internationally agreed (Convention on Biological Diversity) principle for MPA
management (CBD, 2005) and replacing them with 'adequate' in the CAR
principle. Adequacy is then not defined, but the design and selection criteria
used to support it are restricted to the number and size of marine parks and
fishing closures, not the delivery of effective protection or even the implementation of adequate management strategies (Kearney et al., 2013).

Australia's avoidance of international principles in the interests of
declaring more marine parks has not been limited to diminution of the
commitment to effectively manage that is implicit in the CBD (CBD,
1992). Australia repeatedly states that it takes a precautionary approach
to marine conservation (for example, Australian Government, 1975;
Commonwealth of Australia, 1992a), yet the internationally accepted definition of the Precautionary Principle has been redefined by Australia specifically for marine parks (ANZECC TFMPA, 1999). The Australia-specific
definition mandates that declaration of parks must not be constrained by scientific uncertainty (ANZECC TFMPA, 1999), thus removing the need for
precaution over uncertainty in the effectiveness of the parks themselves
(Kearney et al., 2012b). Not only has Australia diminished its commitment
to basing its marine conservation on addressing properly identified threats
(Section 3.1), but also it has formalised processes that reduce the requirements to assess the delivery of effective protection and of precautionary
management.

Adoption of the RAP in the GBRMP resulted in a strategy of combining
bioregional planning and area management that is driven and evaluated by the
number and size of areas and their content and not the effectiveness of management of the marine environment. This is more questionable when it is noted

that the subdivisions in the RAP were made on the basis of the individuality of each of the multitude of bioregions. In spite of the fundamental priority given to the individuality of each type of area (bioregion), a homogeneous management approach of closing approximately the same percentage of each area to fishing was the primary management strategy determined for each and every type of area. There is further irony in that management based on subdivision and dominated by restriction of a single type of activity rose to prominence in Australia in the GBRMP where the founding concept was to provide effective protection of the whole of an iconic entity against everything.

The limitations of the utility of zoning for managing threats to the GBRMP are actually evident in the way GBRMPA manages those threats it has specifically and relatively successfully addressed. Three such major threats to the area of the GBRMP are mining within the Park, tourism development and shipping throughout the Park. The management of all three by GBRMPA is across the entire Park and is not constrained by the zones within the Park. Equally noteworthy is that the two forms of fishing given special attention, trawling and long–lining for pelagic species such as tuna, were excluded from most (66%) of the Park in the case of trawling (Pears et al., 2012) and all of the park for long–lining (GBRMPA 2000), not from just the 33% of the Park's zoned 'no–take' areas. In reality, no threat has been identified in the GBRMP for which zoning as implemented in the Park is the most appropriate conservation measure. Within the GBRMP when clearly identified threats have been effectively managed the approach taken has not been reliant upon, or constrained by, zoning.

It is also noteworthy that the whole of the GBRMP and other areas of Queensland's waters are subject to fisheries management legislation that can be, and has been, used to specifically address identified threats from fishing by a variety of measures, including area closures. The utility of traditional fishery management, even for managing trawling, is recognised by GBRMPA; '*Fishery management tools that actively manage effort within sustainable levels for each of the key trawl fishery sectors could provide a mechanism to control risks and impacts on harvested species and the environment*' (Pears et al., 2012, p. iv). 'Fishery management tools' available under existing fisheries legislation include much more than effort controls; other available measures include area closures that can be specific for seasons and/or applied to individual gear types or any combination of types and can include no-take zones.

It is also obvious that effective protection of the Reef from the major threats that are now known to be impacting it (ocean acidification, climate change, pollution, agricultural runoff and poorly managed coastal development) will continue to require management of activities that are largely independent of the boundaries of the total Park, not just independent of the zones within it. This is increasingly recognised by GBRMPA and apparent in the attempts to manage more tightly the impacts of agriculture and coastal development in Queensland (GBRMPA, 2009). The limitations of area management for controlling the level of threat and the distribution of impact in marine environments can also be inferred from the geographic, environmental and economic extent of damage from the failure to adequately manage the dispersed impacts of point source pollution that has occurred in other locations, for example, from a recent oil spill in the Gulf of Mexico (The White House, 2010).

3.4. Inaccurate evaluation of the impacts of fishing and exaggeration of the benefits of closing areas to all forms of fishing

In spite of the requirement under the Inter-Governmental Agreement on the Environment (Commonwealth of Australia, 1992a) to address each threat in proportion to its impact (Section 3.1), Australian Commonwealth and state governments have continued to give disproportionate priority to the regulation of fishing in marine parks. This has been to the detriment of efforts to adequately identify and manage all threats. The justification for the singular priority given to further restricting fishing in Australian marine parks has not been matched by detailed evaluation of the net benefit that can be anticipated in each type of environment. Rather it has been fuelled by campaigns that do not differentiate between the threats from well-managed fishing and those from inadequately managed fishing, or how different threats would be best addressed. A threat from fishing is assumed, leading to the uncritically evaluated conclusion that area closures are the appropriate way to address it: for example, 'fishing kills fish' and it is argued therefore that more marine parks are needed (Possingham, 2011). When the CAR principle is then applied, the need for more and bigger fishing closures becomes unavoidable. Unfortunately, the less precise and efficient the management and the more abstract the measures of outcomes, the bigger the area that can be justified to meet an open-ended commitment to 'adequate'!

The case for large fishing closures has been underpinned by uncritical exaggeration of the impacts of fishing, including imprecise projections of total global fisheries collapses (for example, those suggested by Pauly et al., 1998, 2002; Worm et al., 2006). It has been further biased by distortion of the relevance to Australia (for example, by GBRMPA, 2002) of overseas examples of the benefits of preventing destructive fishing practices and/or gross overfishing by the management measure of last resort, closing areas to all fishing (for example, those provided by Goñi et al., 2010; McClanahan and Kaunda-Arar, 1996; Russ et al., 2003), as further discussed in Kearney et al. (2012a).

That inadequately managed fishing and destructive fishing practices are serious problems in numerous parts of the world is not questioned. In many developing countries that do not have Australia's fisheries management capability and in regions such as the European Union where international, cooperative management is essential but elusive, area closures can be the most efficient, or possibly the only immediately available, means of preventing gross overfishing in at least some areas. But destructive fishing practices and gross overfishing are neither major nor irreversible threats in modern Australia (Kearney, 2013b), and many areas, including the GBRMP, are very lightly fished (discussed below). Destructive practices and overfishing are certainly not threats of a sufficient magnitude that they justify the creation of extremely large, total 'no-take' zones.

There are numerous far greater threats to marine environments than well-managed fishing. More importantly where problems with fishing do arise in Australia, there are more efficient ways of addressing them than blanket closures to all forms of fishing regardless of which form may be causing the problem. Furthermore, as discussed above, most threats from fishing can be effectively and efficiently managed under existing fisheries legislation that is specifically designed for the purpose. Most importantly, this legislation is usually designed to cover the whole region where the problem arises and its results are manifest. After all, marine environments and fisheries resources outside zoned fishing closures must also be protected! Because of the strong interconnectivity in marine systems, even the environments and resources within closures will be threatened if they are not.

The degree to which Australia's preoccupation with further restricting fishing in marine parks detracts from balanced management is clearly demonstrated for Commonwealth waters, including the GBRMP, by the relevant Commonwealth Government RISs of 2003 (RIS, 2003) and 2012 (RIS, 2012). As outlined in Section 3.2, both of these assessments of the anticipated impacts of expanding marine parks deal virtually exclusively with fishing and the asserted benefits, including to fisheries, that are projected to come from further unselective restriction of all forms of it. In identification of '*The problem*' (RIS, 2003, p. 6) that must be addressed by the Representative Areas Program, the 2003 RIS states, '*Water pollution, global climate change and overfishing have contributed to the decline of coral reef health*' (RIS, 2003, p. 6). And yet the '*Risk Identification*' in this same RIS (2003, p. 8) completely ignores pollution and climate change. It is completely restricted to the threat of overfishing or inadequately managed fishing (RIS, 2003); no other threats are even mentioned in the risk identification and no actions other than increased fishing closures are proposed. This is in spite of the failure to identify where overfishing was a threat or to describe how blanket closures to all types of fishing would represent efficient conservation, even if some form of overfishing were identified or anticipated.

The primary scientific advice to governments on the social impacts of the RAP as provided by the Federal Government's Bureau of Rural Sciences (BRS) suggested possible benefits to fisheries that could come from the closures although it was noted that data specific to commercially caught species in the region were limited. The advice included the projection that there would only be a relatively small reduction (10%) in value of total commercial catches and that this could be offset in as little as 3 years for short-lived species if evidence suggesting spillover benefits was correct (BRS, 2003). The RIS interpretation of available scientific data was that while the RAP was focused on biodiversity protection, '*[t]he implementation of a system of MNPZs* [Marine National Park Zones] *is likely to benefit fisheries through improved recruitment and 'spillover' effects*' (RIS, 2003, p. 21) while noting that benefits were species dependent but that they would increase over time (RIS, 2003). Clearly, the available data were interpreted to project the assumption that fisheries in the region would benefit, or at least not suffer.

The actual commercial fishery catch data from the GBR region show that as a result of the RAP closures, there was an initial reduction in commercial catches of approximately 26% (only slightly less than the extra 28.4% of the area that was closed) and that catch data for 7 years after the closure showed no evidence of a recovery (Fletcher et al., in press). Clearly, the scientific advice to governments by BRS, and as interpreted in the RIS, that was to the fore in justifying increased fishing closures, over-optimistically projected the outcome from the closures.

The ability of Australian governments to rectify overfishing under existing fisheries legislation has been clearly demonstrated in the last decade (Kearney, 2013b). Overfishing has not been completely eliminated, but Australian governments have demonstrated that they can effectively address specific problems with fishing (the results of threats) when they are accepted to be a priority. Fisheries management in Australia has not by any means been perfect. Over-capitalisation of individual fisheries has been a particularly common problem. It has, however, repeatedly demonstrated that relative abundances of overfished species have recovered when catches have been deliberately reduced (Kearney, 2013b).

In total, while Australia's fisheries management has had deficiencies, it has been internationally recognised for its conservatism (for example, Pitcher et al., 2009a,b). This conservatism is particularly evident in the GBRMP. The reported retained harvest from the whole of the 344,400 km^2 of the GBRMP is approximately 14,000 tonnes (t) per annum from commercial, recreational and indigenous fisheries combined (GBRMPA, 2011d). Significant parts of this catch come from trawl fisheries in non-reef areas and fisheries for pelagic species in non-reef areas which constitute more than 90% of the Park. There are more than 24,000 km^2 of reef in the GBRMP and approximately 16,000 km^2 of these are open to fishing. Estimates from numerous other countries of the harvest that can be taken sustainably from coral reefs vary considerably and include a range from 3 t/km^2/year (Bell et al., 2009), 5 t/year (Newton et al., 2007), and 10–20 t/year (Alcala, 2014, citing Russ, 1991) to a high in 1 year of 36.9 t/year (Alcala and Russ, 1990). Using a low estimate of 5 t/km^2/year, a sustainable harvest of 80,000 t per year could be taken from the 16,000 km^2 of open reef. The current combined harvest from the whole of the GBRMP of 14,000 mt/year is therefore <20% of the yield that international precedents suggest could be taken sustainably from only the reef area currently open to fishing and this fished reef area is <5% of the total area of the Park. Even at the most conservative of the above estimates of potential yield of 3 t/km^2/year, the Reef and its surrounding waters are still very lightly exploited. Yet the priority for management within the GBRMP remains the zoning of restrictions on fishing that have been justified by concern over unidentified overfishing and which are additional to conventional management measures which, on the basis of the available evidence, already appear to be extremely (excessively?) conservative!

Throughout Australia, the management of fishing under existing, conventional fisheries legislation has been demonstrated to be far more successful and conservative than the management of many other threats to marine resources or marine environments, such as pollution and introduced organisms: 429 introduced or cryptogenic species had been reported in Australia by 2008 (Hewitt and Campbell, 2010), but no species have been reported to have been fished to extinction. The preoccupation of the Australian Commonwealth Government with further regulating fishing in marine parks confirmed by the 2013 proclamation of many more parks dominated by fishing closures (Burke, 2013) not only fails to acknowledge its own capabilities and successes in fisheries management but is at the expense of addressing the major threats that are not being adequately managed, including those named in the Government's own RIS.

Even though zoning is not the most appropriate way to address over-fishing, it does have a role to play in the regulation of fishing in Australia, particularly in places such as the GBRMP where underwater observation is a priority. Its primary utility lies in the allocation of resource access. Unlike indirect human impacts that dominate impacts on the Reef, such as pollution from distant sources, regulation of direct human access can usually be based effectively on zoning, assuming that cost-effective compliance can be achieved. Closing previously fished reef areas in the GBRMP to fishing does, as anticipated, usually result in visibly obvious higher levels of relative abundance and availability of larger individuals of selected, relatively sedentary and commonly targeted species (for example, McCook et al., 2010; Russ et al., 2008). Most forms of management of fishing that reduce catches can be expected to lead to similar responses in previously affected species. Within selected zones, increases in numbers and size are particularly noticeable for those individuals or species that are relatively sedentary inhabitants of shallow structured habitats and become habituated to, or even welcoming of, human presence, as many do naturally when they are not challenged, or when they are fed by humans as they are in parts of the GBRMP (GBRMPA, 2011f). Zoning is commonly the most efficient means of allocating access to selected areas and their contents to the benefit of selected individuals or groups, such as researchers and non-extractive users.

That industries, groups or individuals, such as divers, that find advantage in the predictable presence of more and larger examples of selected species would benefit from such an outcome is to be expected and the direct and associated benefits should not be understated. That non-extractive use of the GBR is already of much more monetary value to the region and to Australia generally than extractive use of it is not questioned. The value-added contribution of tourism from the greater GBR and its catchment to the Australian economy was estimated to be $5.2 billion in 2011/2012 (the actual contribution of the Reef would be considerably less, but still considerable) compared to $0.16 billion for commercial fishing within the GBR area (Deloitte, 2013). But what must be questioned is the continued use throughout Australia of unsubstantiated claims of net conservation and fisheries benefits from closing large areas of all types to all fishing and/or the closing of even much smaller areas, particularly in riverine estuaries and ocean beaches, that are subject to agents of major change (e.g. floods and storms), are not overfished, have not been demonstrated to be critical to ecosystem services and are not of significant benefit to non-extractive users.

Within the GBRMP, the great bulk of tourism activity is concentrated in a small percentage of its total area. GBRMPA reports that 80% of tourism activity occurs in 7% of the region (GBRMPA, 2011a). As the total tourism activity includes considerable use of areas that are not closed to fishing, the percentage of the Park that is actively used primarily because it is closed to fishing would be considerably lower than the 7% figure. An official, more precise estimate of usage by area was not available (GBRMPA 15/04/2014), but Starck (2014) has suggested that non-fishing tourists may actually use considerably less than 1% of reef areas. The most cost-effective allocation and management of areas to support the non-fishing tourism industry would therefore likely require much less area than current extensive closures (33%). A smaller area could be more precisely tailored and intensely managed to meet the assessed needs of the tourism industry.

The management of Australia's many marine parks is based on multiple use, and it is important not to confuse the need for areas managed for tourism with the need for area management to protect against the identified impacts of fishing. It is, however, also important that the use of parts of areas by more lucrative activities is not incorrectly used to support the closure of much larger areas to other activities. If the justification for each area closure was adequately described, more appropriate performance indicators could then be developed. These would be influenced by the types of multiple use, for example, in reef versus non-reef areas. More specific goals and performance indicators could then replace the current commitment to 'adequate' that mandates more and bigger closures to meet predetermined but inadequately justified percentages in all types of areas.

3.5. Inadequate evaluation of outputs and inappropriate adaptive management

The absence, in the management strategy for the NRSMPA, including the GBRMP, of prioritisation of threats and timelines for monitoring the effectiveness of management of each threat has resulted in distraction of performance evaluation away from evidence-based assessment of outcomes. The failure to mandate assessment of the performance of management against properly prioritised outcomes, such as the continued health of the coral reefs of the greater GBRMP, has meant that the ability to improve future outcomes within the GBR based on adaptive management has been diminished. This shortcoming is common throughout the whole of the NRSMPA.

As outlined in Section 3.1, the selection and design principles in the national strategy for marine reserves, the NRSMPA (Australian Government, n.d.; RIS, 2012), are dominated (13 of 14) by measures of the number and size of parks and their contents to the exclusion of measures of the delivery of effective protection of marine environments, even those within the parks. The failure to give priority to measurement of conservation outcomes and to define appropriate performance indicators has in turn weakened the link between evaluations of the effectiveness of the actions taken and where subsequent management expenditure should be directed. Collectively, these failings have meant that governments' very public expressions of support for the grand concept of saving the GBR have not been constrained by the need to demonstrate cost–effective delivery of promised outcomes. The lack of appropriate performance assessment has facilitated concealment of the failure to arrest the continued deterioration in the state of the Reef as reported by Brodie and Waterhouse (2012), De'ath et al. (2012) and UNESCO (2013, 2014).

The deficiency in the provision of services necessary to deliver the desired outcome, the protection of the GBR, has been publicly acknowledged for many years by the manager of the Park, the GBRMPA; it has repeatedly expressed great concern over the inadequacy of funding in its Annual Reports (as discussed in Section 3.1) (GBRMPA, 1979, 1993, 1994) even though by comparison with other parks around the world expenditure and stated commitment to management have been high (Brodie and Waterhouse, 2012). The expressed commitment has included, for example, expenditure of more than 100 million dollars in 2012 on scientific research and management (Deloitte, 2013). The lack of regular assessment of performance against appropriate key performance indicators of the delivery of effective protection has greatly diminished the means of evaluating performance against this commitment. This has, in turn, meant that from the outset, failure to deliver would not be as obvious and therefore would be less likely to force budget review.

The most prominent change in management action taken since the declaration of the GBRMP was the expansion in 2004 under the RAP of the area of the Park closed to fishing, from 4.6% to 33% (GBRMPA, 2011c). This major change to management was not predicated by evidence-based assessment of cost–effective conservation outcomes from pre-existing fisheries closures and evaluation of how the threats to the Reef from the identified threats, including pollution from coastal runoff which had been prioritised as a threat by that time (GBRMPA, 1989), would be better ameliorated by the adaptation. The benefits outlined in the Government's Impact Statements and the scientific advice to Government (BRS, 2003) on these changes were restricted to those that were hypothesised to result from having more fishing closures to address unidentified overfishing and to provide benefits to fisheries. As discussed above, the predictions of benefits were unrealistic.

The continued deterioration in the condition of the GBR has more recently resulted in greater efforts by GBRMPA to manage threats arising outside the Park (for example, GBRMPA, 2013), such as pollution from agricultural runoff, that are now accepted to be primary to the decline in the health of the Reef. Notably, the measures being used for the management of these specific threats are again independent of the zoning of fishing closures within the Park.

4. THE BROADER IMPLICATIONS

The 1975 enactment of the GBRMP Act was pivotal in changing attitudes to marine conservation in Australia, and the resulting GBRMP continues to influence the global marine conservation debate. It has had unequalled influence on the development of Australia's extensive NRSMPA. The iconic nature of the Barrier Reef, its World Heritage status, the magnitude of the Park and the amount of effort and money spent on research are amongst the reasons for its global prominence in marine conservation. These same attributes contribute to the reluctance of governments to acknowledge the increasingly obvious lack of success in actually protecting the Reef. The lack of critical appraisal of how threats are being addressed and the effectiveness of what protection is actually being provided to marine systems throughout Australia has helped conceal the limitations of the conservation outcomes. Non-governmental organisations (NGOs) have also actively promoted the adoption of more MPAs through orchestrated, well-funded and effective campaigns. Such campaigns have been largely based on philosophies and ideals (Hilborn, 2006; Kearney, 2012) and exaggerated and sensationalised claims of benefits from restricting fishing rather than evidence-based assessment of conservation and resource use outcomes from marine parks.

Australian states have embraced the NRSMPA and have declared their own series of parks as part of the national initiative. These series of parks have followed the established process of giving priority to a presumption of adequacy based on the number and size of parks with 'comprehensive' contents (the CAR principle). This has followed the Australian Commonwealth Government's example and been at the expense of determination of how best to address priority threats to the total ecosystem, even though this is required under Australia's Inter-Governmental Agreement on the Environment (Commonwealth of Australia, 1992a).

A prominent defence of the NRSMPA is that it is necessary to satisfy Australia's international commitment (CBD, 1992) to having marine parks. This commitment was developed in a global (often developing country) context that involved compromise and is not fully cognisant of particular competencies of individual countries. The requirements of the relevant international conventions are more consistent with the use of MPAs as a fisheries management measure of last resort, not as a measure to protect against invasive threats such as pollution. They were developed in an ambience of aiding developing countries or guiding groups of countries that do not have Australia's specific management capabilities. Australia is a developed country with full competence as an island state to address most threats across the whole area of the influence of each threat. It is inefficient for it to rely on management measures of last resort or international guidance that is excessively generalised in an attempt to be applicable to a full range of global scenarios, or diluted to accommodate the lowest common denominator.

Australian states have unfortunately compounded the commitment to non-specific management by using the Federal Government's commitment to the NRSMPA to justify their own series of very large parks. This is exemplified in South Australia that has established MPAs over 45% of its marine area (Australian Government, 2014) under its 2007 Marine Parks Act. This Act has a primary object to '*protect and conserve marine biological diversity and marine habitats by declaring and providing for the management of a comprehensive, adequate and representative system of marine parks*' (Government of South Australia, 2007, p. 7). The presumption that the parks will 'protect' is to the fore. Furthermore it is acknowledged that the management of this system of parks is deliberately not threats based (Government of South Australia, 2011).

The system of marine parks in New South Wales (NSW) was initiated in 1998 (NSW Government, n.d.) on similar priorities to those later adopted in South Australia. Significantly, a recent NSW Government review has suggested that the whole of this State's marine waters, including marine parks, should be managed as a single marine estate and their management based on addressing identified threats and how best to ameliorate each throughout the estate (Beeton et al., 2012). The NSW Government has established a Marine Estate Management Authority to implement this process (NSW Government, 2013). It is noteworthy that NSW is Australia's most populous state where anthropogenic activities, such as recreational fishing, are denser and impacts on marine environments, particularly coastal ones, from terrestrial-based threats are high. The recent actions in NSW confirm that as the impacts of anthropogenic influence become manifest, it is more obvious that for management to be effective, it must be based on directly addressing threats no matter where they arise. The limitations of zoning for addressing impacts, particularly those from invasive threats such as pollution and introduced organisms that have no respect for boundaries of zones are progressively being more widely accepted (for example, Agardy et al., 2011; Jameson et al., 2002).

The weaknesses in policies based on setting marine areas aside and treating them as though they have boundaries that are vital for the provision of protection against threats are slowly being recognised. What is not yet being adequately recognised is the full impact of remote threats, including those that are terrestrial in origin, on marine environments and in particular on the lower trophic levels of marine ecosystems, and the limitations of area management approaches to addressing these threats.

While the concept of identifying an adequate sample of all types of habitats and their contents (CAR) and subsequently protecting it is appealing, the evidence is progressively challenging the possibility of cost-effectively achieving this outcome in marine environments. Nonetheless, Australian governments have continued to roll out marine parks that prioritise further restriction of fishing, usually to the exclusion of addressing other threats. This is primarily because it is relatively easy to do so and much cheaper than preventing the impacts on marine environments of industrial activity, agriculture, coastal development and introduced organisms. The electoral rewards of nurturing the perception that effective management of marine environments is being provided at limited expense are considerable.

5. CONCLUSIONS

The encapsulation of the GBR in a marine park provided appropriate confirmation of the iconic significance of the Reef and acknowledged that it was worthy of extraordinary efforts to protect it as an entity. The creation of the GBRMP not only provided public recognition of the importance of the Reef but also facilitated considerable extra government support, including funding, for its research and management. Successes in controlling the threats that were prominent at the time, particularly mining, and had been influential in the initial justification for the Park, strengthened public confidence in the marine park concept. The declaration of the GBRMP and its acceptance as a major component of the national CAR system of MPAs resulted in very favourable international comparisons (for example, Agardy et al., 2003). The early successes of the management of the Park catalysed the rapid expansion of systems of marine parks throughout Australia and internationally that was unfortunately not accompanied by adequate analysis of the limitations of the park approach and its applicability to the GBR or other types of marine environments. Recent declines in the health of the Reef within the GBRMP, Australia's most intensively managed marine park, have raised serious doubts over the level of protection that can be anticipated from marine parks as currently managed in Australia.

Detail of the actual effectiveness of the protection of Australia's total marine environment being provided throughout the NRSMPA has been concealed by the lack of appropriate performance indicators and a bias towards uncritical acceptance that the declaration of more and bigger parks is automatically a net benefit. However, the ineffectiveness of the approach has been exposed by the progressively obvious results of failing to address the suite of threats to the GBR (Brodie and Waterhouse, 2012; De'ath et al., 2012; UNESCO, 2013, 2014). It is apparent that the cause of this ineffectiveness is not unique to the GBRMPA; it is common to the whole of Australia's marine parks management process.

While it is reasonable to assume that the condition of the Reef would be considerably worse if those management actions taken under the auspices of the GBRMPA had not been initiated, the continued deterioration in the health of the Reef confirms that the creation of the Park has not resulted in the GBR being adequately protected. The limitations of the declaration of a marine park where disproportionate priority is given to closure of parts of it to fishing at the expense of managing more insidious threats are becoming obvious.

There have been considerable benefits from the GBRMP to non-extractive users resulting from the preferential access to extensive areas that has effectively been allocated to them by fishing closures. But these allocation benefits have not been matched by conservation outcomes; the health of the Reef continues to decline. The benefits to fisheries that were predicted to come from closure of extensive additional areas of the GBRMP in 2004 did not materialise; total fisheries production in the GBR region has been reported to have fallen in proportion to the amount of area closed. This result, in combination with the continued declines in the health of the Reef, has exposed overly optimistic scientific advice to governments. Similar uncritical optimism has been a feature of the NRSMPA process. The process has been further confused by Australia striving to surpass international commitments to marine parks and allowing international NGO campaigns to confuse and misdirect public appreciation of marine conservation and resource use issues.

The limitations of the process of zoning have been masked to a large degree by uncritical presumption that the elevated abundance and size of individuals of selected species in fishing closures represent conservation and fisheries outcomes that justify the use of zoning of fishing as the primary management process. The continued decline in the GBR confirms the need for more timely and direct indicators of the health of marine ecosystems and for management of marine environments that more directly addresses threats.

The lack of precisely defined objectives for marine parks and the zones within and the associated lack of appropriate and binding performance indicators have continued to mask the inadequacy of marine conservation outcomes throughout Australia. However, the declining health of the GBRMP, the flagship of Australia's marine parks, is increasingly exposing the weaknesses in the broader strategy. Hopefully, more thorough evaluation of protection of the GBR may once again see management of the Reef as the catalyst for change in marine conservation and resource use policies!

NOTE ADDED TO PROOF

More than three months after the completion of this chapter (27/5/2014) the Australian Commonwealth Government and the Queensland State Government have jointly announced (15/9/2014) the release, for comment, of a "35-year plan for a healthier and more resilient Great Barrier Reef" (http://www.environment.gov.au/minister/hunt/2014/mr20140915.html). Significant to the text and conclusions in this chapter is that this draft plan greatly increases the emphasis on the need to manage the suite of threats to the Reef that are now stated to be—"climate change, poor water quality from land-based run off, impacts from coastal development and some fishing activities". The prominence now given to land-based threats is most significant but the replacement of the previously stated need to combat generic 'overfishing' with the intention to address "some fishing activities" (presumably those identified as specific threats) must also be noted.

REFERENCES

ABS, 2010. 2009–10 Yearbook Australia. Australian Bureau of Statistics, Canberra. Available: http://www.ausstats.abs.gov.au/Ausstats/subscriber.nsf/0/AC72C92B23B6 DF6DCA257737001B2BAB/$File/13010_2009_10.pdf (accessed September 1, 2012).

ABS, 2011. 2009–10 Year Book Australia. Australian Bureau of Statistics, Commonwealth of Australia, Canberra. Available: http://www.ausstats.abs.gov.au/Ausstats/subscriber.nsf/0/AC72C92B23B6DF6DCA257737001B2BAB/$File/13010_2009_10.pdf (accessed September 3, 2012).

Agardy, T., Bridgewater, P., Crosby, M.P., Day, J., Dayton, P.K., Kenchington, R., Laffoley, D., Mcconney, P., Murray, P.A., Parks, J.E., Peau, L., 2003. Dangerous targets? Unresolved issues and ideological clashes around marine protected areas. Aquat. Conserv. Mar. Freshwat. Ecosyst. 13, 353–367.

Agardy, T., Notarbartolo Di Sciara, G., Christie, P., 2011. Mind the gap: addressing the shortcomings of marine protected areas through large scale marine spatial planning. Mar. Policy 35, 226–232.

Alcala, A.C., 2014. Marine Reserves as Tools for Fishery Management and Biodiversity Conservation: Natural Experiments in the Central Philippines, 1974–2000. CBD, Montreal. Available: https://www.cbd.int/doc/nbsap/fisheries/ALCALA.pdf (accessed April 28, 2014).

Alcala, A.C., Russ, G.R., 1990. A direct test of the effects of protective management on abundance and yield of tropical marine resources. ICES J. Mar. Sci. 47, 40–47.

AMCS (QLS), 2014a. Along the Way. Australian Marine Conservation Society (Queensland Littoral Society). Available: http://www.sustainableseafood.org.au/About-AMCS.asp?active_page_id=256 (accessed April 22, 2014).

AMCS (QLS), 2014b. In the Beginning. Australian Marine Conservation Society (Queensland Littoral Society). Available: http://www.sustainableseafood.org.au/About-AMCS.asp?active_page_id=207 (accessed March 25, 2014).

ANZECC TFMPA, 1999. Strategic Plan of Action for the National Representative System of Marine Protected Areas: A Guide for Action by Australian Governments. Australia and New Zealand Environment and Conservation Council Task Force on Marine Protected Areas, Canberra.

Atkins, J.P., Burdon, D., Elliot, M., Gregory, A.J., 2011. Management of the marine environment: integrating ecosystem services and societal benefits with the DPSIR framework in a systems approach. Mar. Pollut. Bull. 62, 215–226.

Australian Government, 1975. Great Barrier Reef Marine Park Act 1975. ComLaw, Commonwealth of Australia, Canberra. Available: http://www.comlaw.gov.au/Details/C2012C00109 (accessed March 26, 2014).

Australian Government, 2013. What is a Regulation Impact Statement (RIS)? Department of Prime Minister and Cabinet, Canberra. Available: http://www.dpmc.gov.au/deregulation/obpr/faq.cfm#what_ris (accessed April 23, 2014).

Australian Government, 2014. Collaborative Australian Protected Areas Database (CAPAD) 2012—Marine. (updated April 29, 2014)Department of the Environment, Canberra. Available: http://www.environment.gov.au/metadataexplorer/full_metadata.jsp?docId=%7BAB09E7E0-E6BC-47A7-B522-B426C8E572AE%7D&loggedIn=false (accessed April 29, 2014).

Australian Government, n.d. Goals and Principles for the Establishment of the National Representative System of Marine Protected Areas in Commonwealth waters [Online]. Canberra: Department of the Environment. Available: http://www.environment.gov.au/resource/goals-and-principles-establishment-national-representative-system-marine-protected-areas (Accessed April 29, 2014).

B.C. Government, 2014. An Introductory Guide to Adaptive Management for Project Leaders and Participants. Government of British Columbia, Victoria. Available: http://www.for.gov.bc.ca/hfp/amhome/Training/am-intro-guide.htm (accessed March 31, 2014).

Beeton, R.J.S., Lynch, A.J.J., 2012. Most of nature: a framework to resolve the twin dilemmas of the decline of nature and rural communities. Environ. Sci. Policy 23, 45–56.

Beeton, R.J.S., Buxton, C.D., Cutbush, G.C., Fairweather, P.G., Johnston, E.L., Ryan, R., 2012. Report of the Independent Scientific Audit of Marine Parks in New South Wales. NSW Department of Primary Industries and Office of Environment and Heritage, Sydney.

Bell, J.D., Kronen, M., Vunisea, A., Nash, W.J., Keeble, G., Demmke, A., Pontifex, S., Andréfouët, S., 2009. Planning the use of fish for food security in the Pacific. Mar. Policy 33, 64–76.

Bradshaw, C.J.A., Field, I.C., Bowman, D.M.J.S., Haynes, C., Brook, B.W., 2007. Current and future threats from non-indigenous animal species in northern Australia: a spotlight on World Heritage Area Kakadu National Park. Wildl. Res. 34, 419–436.

Brodie, J., Waterhouse, J., 2012. A critical review of environmental management of the 'not so Great' Barrier Reef. Estuar. Coast. Shelf Sci. 104–105, 1–22.

Brodie, J., Waterhouse, J., Schaffelke, B., Kroon, F., Thorburn, P., Rolfe, J., Johnson, J., Fabricius, K., Lewis, S., Devlin, M., Warne, M., Mckenzie, L., 2013. Scientific Consensus Statement: Land Use Impacts on Great Barrier Reef Water Quality and Eco-system Condition. The State of Queensland, Brisbane. Available: http://www.reefplan.qld.gov.au/about/assets/scientific-consensus-statement-2013.pdf (accessed March 26, 2014).

BRS, 2003. Implementing the Representative Areas Program in the Great Barrier Reef Marine Park: Assessment of Potential Social Impacts on Commercial Fishing and Associated Communities. Australian Government Bureau of Rural Sciences, Canberra. Available: http://data.daff.gov.au/brs/brsShop/data/12928_GBRMP.pdf (accessed May 2, 2014).

Burke, T., 2012. Gillard Government Proclaims the Final Network of Commonwealth Marine Reserves. Commonwealth of Australia, Canberra. Available: http://www.environment.gov.au/minister/archive/burke/2012/mr20121116.html (accessed November 24, 2012).

Burke, T., 2013. Marine Park Management Plans Finalised. Commonwealth of Australia, Canberra. Available: http://www.environment.gov.au/minister/archive/burke/2013/mr20130312a.html (accessed March 25, 2014).

Carr, M.H., Neigel, J.E., Estes, J.A., Andelman, S., Warner, R.R., Largier, J.L., 2003. Comparing marine and terrestrial ecosystems: implications for the design of coastal marine reserves. Ecol. Appl. 13 (Suppl.), S90–S107.

Carson, R., 1962. Silent Spring. Houghton Mifflin, Boston.

CBD, 1992. Convention on Biological Diversity. Secretariat, Convention on Biological Diversity, Montreal. Available: http://www.cbd.int/doc/legal/cbd-en.pdf (accessed August 29, 2012).

CBD, 2005. Towards Effective Protected Area Systems: An Action Guide to Implement the Convention on Biological Diversity Programme of Work on Protected Areas. Secretariat, Convention on Biological Diversity, Montreal. Available: https://www.cbd.int/doc/publications/cbd-ts-18.pdf (accessed April 1, 2014).

CBD, 2006. Protected Area Provisions in the Convention on Biological Diversity. Secretariat, Convention on Biological Diversity, Montreal. Available: http://www.cbd.int/protected/pacbd/ (accessed August 19, 2011).

Commonwealth of Australia, 1992a. Intergovernmental Agreement on the Environment. Department of the Arts, Environment, Sport and Territories, Canberra.

Commonwealth of Australia, 1992b. National Forest Policy Statement: A New Focus for Australia's Forests. Australian Government Publishing Service, Canberra. Available: http://www.daff.gov.au/__data/assets/pdf_file/0019/37612/nat_nfps.pdf (accessed April 24, 2014).

De'ath, G., Fabricius, K.E., Sweatman, H., Puotinen, M., 2012. The 27-year decline of coral cover on the Great Barrier Reef and its causes. PNAS 109, 17995–17999.

Deloitte, 2013. Economic Contribution of the Great Barrier Reef. GBRMPA, Townsville. Available: http://www.gbrmpa.gov.au/__data/assets/pdf_file/0006/66417/Economic-contribution-of-the-Great-Barrier-Reef-2013.pdf (accessed April 1, 2014).

Ehrlich, P.R., 1968. The Population Bomb. Ballantine Books, New York.

Elliott, M., 2011. Marine science and management means tackling exogenic unmanaged pressures and endogenic managed pressures—a numbered guide. Mar. Pollut. Bull. 62, 651–655.

Flannery, T., 2012. After the future: Australia's new extinction crisis. Q. Essay 48, 1–80.

Fletcher, W.J., Wise, B.S., Kearney, R.E., Nash, W.J., in press. Large-scale expansion of no-take closures within the Great Barrier Reef, Australia did not enhance fishery production.

GBRMPA, 1978a. Great Barrier Reef Marine Park Authority Annual Report 1976–77. Australian Government Publishing Service, Canberra. Available: http://www.gbrmpa.gov.au/resources-and-publications/publications/annual-reports (accessed July 15, 2013).

GBRMPA, 1978b. Great Barrier Reef Marine Park Authority Annual Report 1977–78. Australian Government Publishing Service, Canberra. Available: http://www.gbrmpa.gov.au/resources-and-publications/publications/annual-reports (accessed April 30, 2014).

GBRMPA, 1979. Great Barrier Reef Marine Park Authority Annual Report 1978–79. Australian Government Publishing Service, Canberra. Available: http://www.gbrmpa.gov.au/resources-and-publications/publications/annual-reports (accessed March 26, 2014).

GBRMPA, 1980. Great Barrier Reef Marine Park Authority Annual Report 1979–80. Australian Government Publishing Service, Canberra. Available: http://www.gbrmpa.gov.au/resources-and-publications/publications/annual-reports (accessed March 26, 2014).

GBRMPA, 1984. Annual Report 1983–84. Commonwealth of Australia, Townsville. Available: http://www.gbrmpa.gov.au/resources-and-publications/publications/annual-reports (accessed March 26, 2014).

GBRMPA, 1986. Annual Report 1985–86. Commonwealth of Australia, Townsville. Available: http://www.gbrmpa.gov.au/resources-and-publications/publications/annual-reports (accessed March 26, 2014).

GBRMPA, 1988. Annual Report 1987–88. Commonwealth of Australia, Townsville. Available: http://www.gbrmpa.gov.au/resources-and-publications/publications/annual-reports (accessed May 2, 2014).

GBRMPA, 1989. Annual Report 1988–89. Commonwealth of Australia, Townsville. Available: http://www.gbrmpa.gov.au/resources-and-publications/publications/annual-reports (accessed May 2, 2014).

GBRMPA, 1990. Annual Report 1989–1990. Great Barrier Reef Marine Park Authority, Townsville. Available: http://www.gbrmpa.gov.au/resources-and-publications/publications/annual-reports (accessed April 22, 2013).

GBRMPA, 1992. 1991/92 Annual Report. Great Barrier Reef Marine Park Authority, Townsville. Available: http://www.gbrmpa.gov.au/resources-and-publications/publications/annual-reports (accessed April 22, 2014).

GBRMPA, 1993. 1992/93 Annual Report. Great Barrier Reef Marine Park Authority, Townsville. Available: http://www.gbrmpa.gov.au/resources-and-publications/publications/annual-reports (accessed April 22, 2013).

GBRMPA, 1994. 1993/94 Annual Report. Great Barrier Reef Marine Park Authority, Townsville. Available: http://www.gbrmpa.gov.au/resources-and-publications/publications/annual-reports (accessed April 22, 2014).

GBRMPA, 1995. 1994–1995 Annual Report. Great Barrier Reef Marine Park Authority, Townsville. Available: http://www.gbrmpa.gov.au/resources-and-publications/publications/annual-reports (accessed March 26, 2014).

GBRMPA, 1997. Annual Report 1996–1997. Great Barrier Reef Marine Park Authority, Townsville. Available: http://www.gbrmpa.gov.au/resources-and-publications/publications/annual-reports (accessed April 8, 2014).

GBRMPA, 2000. Annual Report 1999–2000 [Online]. Great Barrier Reef Marine Park Authority, Townsville. Available: http://www.gbrmpa.gov.au/resources-and-publications/publications/annual-reports (accessed July 8, 2014).

GBRMPA, 2002. Technical Information Sheet # 3: Do No-Take Areas Work? Great Barrier Reef Marine Park Authority, Townsville. Available: http://www.gbrmpa.gov.au/__data/assets/pdf_file/0017/6209/tech_sheet_03.pdf (accessed April 29, 2014).

GBRMPA, 2009. Great Barrier Reef Outlook Report 2009. Great Barrier Reef Marine Park Authority, Townsville. Available: http://www.gbrmpa.gov.au/outlook-for-the-reef/great-barrier-reef-outlook-report (accessed April 1, 2014).

GBRMPA, 2010. Annual Report 2009–10. Great Barrier Reef Marine Park Authority, Townsville. Available: http://www.gbrmpa.gov.au/resources-and-publications/publications/annual-reports (accessed March 26, 2014).

GBRMPA, 2011a. About the Reef: Commercial Tourism. Great Barrier Reef Marine Park Authority, Townsville. Available: http://www.gbrmpa.gov.au/about-the-reef/Managing-multiple-uses/commercial-tourism (accessed April 28, 2014).

GBRMPA, 2011b. Annual Report 2010–11. Great Barrier Reef Marine Park Authority, Townsville. Available: http://www.gbrmpa.gov.au/resources-and-publications/publications/annual-reports (accessed March 26, 2014).

GBRMPA, 2011c. Monitoring the Ecological Effects of the 2004 Rezoning of the Great Barrier Reef Marine Park. Great Barrier Reef Marine Park Authority, Townsville. Available: http://www.gbrmpa.gov.au/zoning-permits-and-plans/rap/monitoring-the-ecological-effects-of-the-2004-rezoning-of-the-great-barrier-reef-marine-park (accessed April 29, 2014).

GBRMPA, 2011d. Outlook for the Reef: Fishing. Great Barrier Reef Marine Park Authority, Townsville. Available: http://www.gbrmpa.gov.au/outlook-for-the-reef/Managing-multiple-uses/fishing (accessed April 28, 2014).

GBRMPA, 2011e. Overview of the Representative Areas Program. Townsville. Available: http://www.gbrmpa.gov.au/zoning-permits-and-plans/rap (accessed April 30, 2014).

GBRMPA, 2011f. Visit the Reef: Fishing—When Fishing. Great Barrier Reef Marine Park Authority, Townsville. Available: http://www.gbrmpa.gov.au/visit-the-reef/responsible-reef-practices/fishing (accessed July 24, 2014).

GBRMPA, 2012. Annual Report 2011–12. Great Barrier Reef Marine Park Authority, Townsville. Available: http://www.gbrmpa.gov.au/resources-and-publications/publications/annual-reports (accessed March 26, 2014).

GBRMPA, 2013. Annual Report 2012–13. Great Barrier Reef Marine Park Authority, Townsville. Available: http://www.gbrmpa.gov.au/resources-and-publications/publications/annual-reports (accessed March 26, 2014).

Goñi, R., Hilborn, R., Diaz, D., Mallol, S., Alderstein, S., 2010. Net contribution of spill-over from a marine reserve to fishery catches. Mar. Ecol. Prog. Ser. 400, 233–243.

Government of South Australia, 2007. Marine Parks Act 2007. Adelaide. Available: http://www.legislation.sa.gov.au/LZ/C/A/MARINE%20PARKS%20ACT%202007/CURRENT/2007.60.UN.PDF (accessed April 30, 2014).

Government of South Australia, 2011. Letter from the Minister for Environment and Conservation (reference 11MEC0703). Government of South Australia, Adelaide.

Hardin, G., 1968. The Tragedy of the Commons. Science 162, 1243–1248.

Hatton, T., Cork, S., Harper, P., Joy, R., Kanowski, P., Mackay, R., Mckenzie, N., Ward, T., 2011. State of Environment 2011. Canberra. Available: http://www.environment.gov.au/soe/2011/report/index.html (accessed September 4, 2012).

Hewitt, C., Campbell, M., 2010. The Relative Contribution of Vectors to the Introduction and Translocation of Invasive Marine Species: Keeping Marine Pests Out of Australian Waters. Commonwealth of Australia, The Department of Agriculture, Fisheries and Forestry, Canberra.

Hilborn, R., 2006. Faith-based fisheries. Fisheries 31, 554–555.

Holling, C.S., 1978. Adaptive Environmental Assessment and Management. John Wiley and Sons, London.

IUCN, 2013. World Nearing 3% of Ocean Protection. IUCN, Gland. Available: http://www.iucn.org/media/news_releases/?13912/World-nearing-3-of-ocean-protection (accessed April 30, 2014).

IUCN WCPA, 2005. WCPA Strategic Plan 2005–2012. IUCN World Commission on Protected Areas, Gland.

IUCN WCPA, 2010. 50 Years of Working for Protected Areas: A Brief History of IUCN World Commission on Protected Areas. IUCN, Gland.

Jameson, S.C., Tupper, M.H., Ridley, J.M., 2002. The three screen doors: can marine "protected" areas be effective? Mar. Pollut. Bull. 44, 1177–1183.

Kareiva, P., Lalasz, R., Marvier, M., 2011. Conservation in the Anthropocene: beyond solitude and fragility. In: Schellenberger, M., Nordhaus, T. (Eds.), Love Your Monsters: Postenvironmentalism and the Anthropocene. Breakthrough Institute, Oakland.

Kearney, R., 2012. Faith, conservation and science. In: Banks, P., Lunney, D., Dickman, C. (Eds.), Science Under Siege: Zoology Under Threat. Royal Zoological Society of New South Wales, Mosman.

Kearney, R., 2013a. Governments Are Not Protecting the Great Barrier Reef. The Conversation, Sydney. Available: https://theconversation.com/governments-are-not-protecting-the-great-barrier-reef-16107 (accessed July 23, 2014).

Kearney, R., 2013b. Australia's out-dated concern over fishing threatens wise marine conservation and ecologically sustainable seafood supply. Open J. Mar. Sci. 3, 55–61.

Kearney, R., Buxton, C.D., Farebrother, G., 2012a. Australia's no-take marine protected areas: appropriate conservation or inappropriate management of fishing? Mar. Policy 36, 1064–1071.

Kearney, R., Buxton, C.D., Goodsell, P., Farebrother, G., 2012b. Questionable interpretation of the precautionary principle in Australia's implementation of 'no-take' marine protected areas. Mar. Policy 36, 592–597.

Kearney, R., Farebrother, G., Buxton, C.D., Goodsell, P., 2013. How terrestrial management concepts have led to unrealistic expectations of marine protected areas. Mar. Policy 38, 204–211.

Lambert, J.A., Elix, J.K., Chenoweth, A., Cole, S., 1995. Bioregional Planning for Biodiversity Conservation. Australian Government Department of the Environment, Melbourne. Available: http://www.environment.gov.au/archive/biodiversity/publications/series/paper10/pubs/elix.pdf (accessed March 31, 2014).

Margules, C.R., Pressey, R.L., 2000. Systematic conservation planning. Nature 405, 243–253.

Mcclanahan, T.R., Kaunda-Arar, B., 1996. Fishery recovery in a coral-reef marine park and its effect on the adjacent fishery. Conserv. Biol. 10, 1187–1199.

Mccook, L.J., Ayling, T., Cappo, M., Choat, J.H., Evans, R.D., Freitas, D.M.D., Heupel, M., Hughes, T.P., Jones, G.P., Mapstone, B., Marsh, H., Mills, M., Molloy, F.J., Pitcher, C.R., Pressey, R.L., Russ, G.R., Sutton, S., Sweatman, H., Tobin, R., Wachenfeld, D.R., Williamson, D.H., 2010. Adaptive management of the Great Barrier Reef: a globally significant demonstration of the benefits of networks of marine reserves. PNAS 107, 18278–18285.

Newton, K., Côté, I.M., Pilling, G.M., Jennings, S., Dulvy, N.K., 2007. Current and future sustainability of island coral reef fisheries. Curr. Biol. 17, 655–658.

NSW Government, 2013. A New Approach to Managing the NSW Marine Estate FAQs 4—Marine Estate Management Strategy. New South Wales Government, Sydney. Available: http://www.marine.nsw.gov.au/__data/assets/pdf_file/0018/501903/FAQs-4-Marine-Estate-Management-Strategy.pdf (accessed April 29, 2014).

NSW Government, n.d. NSW Marine Parks [Online]. New South Wales Government, Marine Parks, Sydney. Available: http://www.mpa.nsw.gov.au/index.html (accessed April 22, 2014).

Pauly, D., Christensen, V., Dalsgaard, J., Froese, R., Torres, F., 1998. Fishing down marine food webs. Science 279, 860–863.

Pauly, D., Christensen, V., Guénette, S., Pritcher, T.J., Sumaila, U.R., Walters, C.J., Watson, R., Zeller, D., 2002. Towards sustainability in world fisheries. Nature 418, 689–695.

Pears, R.J., Morison, A.K., Jebreen, E.J., Dunning, M.C., Pitcher, C.R., Courtney, A.J., Houlden, B., Jacobsen, I.P., 2012. Ecological Risk Assessment of the East Coast Otter Trawl Fishery in the Great Barrier Reef Marine Park: Summary Report. Great Barrier Reef Marine Park Authority, Townsville. Available: http://elibrary.gbrmpa.gov.au/jspui/bitstream/11017/1147/1/ECOTF_ERA_Summary_web.pdf (accessed April 1, 2012).

Penn, J.W., Fletcher, W.J., 2010. The Efficacy of Sanctuary Areas for the Management of Fish Stocks and Biodiversity in WA Waters: Fisheries Research Report No. 169. Department of Fisheries, Western Australia.

Pitcher, C.R., Burridge, C.Y., Wassenberg, T.J., Hill, B.J., Poiner, I.R., 2009a. A large scale BACI experiment to test the effects of prawn trawling on seabed biota in a closed area of the Great Barrier Reef Marine Park, Australia. Fish. Res. 99, 168–183.

Pitcher, T., Kalikoski, D., Pramod, G., Short, K., 2009b. Not honouring the code. Nature 457, 658–659.

Possingham, H., 2011. Does fishing kill fish. Australas. Sci. (2000) 32, 46.

Queensland State Archives, 2013. Royal Commission into Exploratory and Production Drilling for Petroleum in the Area of the Great Barrier Reef (ID1748). Queensland Government, Brisbane. Available: http://www.archivesearch.qld.gov.au/Search/AgencyDetails.aspx?AgencyId=1748 (accessed April 30, 2014).

RIS, 2003. Zoning Plan for the Great Barrier Reef Marine Park Regulatory Impact Statement. Great Barrier Reef Marine Park Authority, Canberra. Available: http://www.gbrmpa.gov.au/__data/assets/pdf_file/0017/6173/RIS_25-11-03.pdf (accessed March 27, 2014).

RIS, 2012. Completing the Commonwealth Marine Reserves Network: Regulatory Impact Statement. Australian Government: Department of the Prime Minister and Cabinet, Canberra. Available: http://ris.dpmc.gov.au/2012/06/22/completing-the-commonwealth-marine-reserves-network-regulation-impact-statement-department-of-sustainability-environment-water-population-and-communities/ (accessed March 27, 2014).

Russ, G.R., 1991. Coral reef fisheries: effects and yields. In: Sale, P.F. (Ed.), The Ecology of Fishes on Coral Reefs. Academic Press, San Diego.

Russ, G.R., Alcala, A.C., Maypa, A.P., 2003. Spillover from marine reserves: the case of Naso vlamingii at Apo Island, the Philippines. Mar. Ecol. Prog. Ser. 264, 15–20.

Russ, G.R., Cheal, A.J., Dolman, A.M., Emslie, M.J., Evans, R.D., Miller, I., Sweatman, H., Williamson, D.H., 2008. Rapid increase in fish numbers follows creation of world's largest marine reserve network. Curr. Biol. 18, R514–R515.

Starck, W., 2014. Are Academics Above The Law? Quadrant Online. Available: http://quadrant.org.au/opinion/doomed-planet/2014/03/academics-law/ (accessed April 30, 2014).

Steele, J.H., 1985. A comparison of terrestrial and marine ecosystems. Nature 313, 355–358.

Steele, J.H., 1991. Can ecological theory cross the land-sea boundary? J. Theor. Biol. 153, 425–436.

Taylor, G.R., 1970. The Doomsday Book. Thames and Hudson Ltd., London.

The White House, 2010. Deepwater BP Oil Spill. USA Government, Washington, DC. Available: http://www.whitehouse.gov/deepwater-bp-oil-spill/ (accessed April 1, 2014).

UNESCO, 2012. Mission Report: Great Barrier Reef Australia (N 154). UNESCO, Saint Petersburg. Available: http://whc.unesco.org/en/list/154/documents/ (accessed March 26, 2014).

UNESCO, 2013. World Heritage Committee Thirty-Seventh Session Phnom Penh, Cambodia. UNESCO, Paris. Available: http://whc.unesco.org/archive/2013/whc13-37com-7B-en.pdf (accessed July 11, 2013).

UNESCO, 2014. World Heritage Committee Thirty-Eighth Session Doha, Qatar. UNESCO, Paris. Available: http://whc.unesco.org/archive/2014/whc14-38com-7B-en.pdf (accessed May 2, 2014).

Woinarski, J.C.Z., Armstrong, M., Brennan, K., Fisher, A., Griffiths, A.D., Hill, B., Milne, D.J., Palmer, C., Ward, S., Watson, M., Winderlich, S., Young, S., 2010. Monitoring indicates rapid and severe decline of native small mammals in Kakadu National Park, northern Australia. Wildl. Res. 37, 116–126.

Worm, B., Barbier, E.B., Beaumont, N., Duffy, J.E., Folke, C., Halpern, B.S., Jackson, J.B. C., Lotze, H.K., Micheli, F., Palumbi, S.R., Sala, E., Selkoe, K.A., Stachowicz, J.J., Watson, R., 2006. Impacts of biodiversity loss on Ocean ecosystem services. Science 314, 787–790.

CHAPTER EIGHT

Establishment, Management, and Maintenance of the Phoenix Islands Protected Area

Randi Rotjan*,[1], Regen Jamieson*, Ben Carr[†], Les Kaufman*,[†],[‡],
Sangeeta Mangubhai*, David Obura*,[§], Ray Pierce[¶], Betarim Rimon[‖],
Bud Ris*,[#], Stuart Sandin**, Peter Shelley[††], U. Rashid Sumaila[‡‡],
Sue Taei[§§], Heather Tausig*, Tukabu Teroroko[‖],[#], Simon Thorrold[¶¶],
Brooke Wikgren*, Teuea Toatu[#], Greg Stone[‡],[#]

*New England Aquarium, 1 Central Wharf, Boston, MA, 02110, USA
[†]Boston University, Department of Biology, Boston, Massachusetts, USA
[‡]Conservation International, Arlington, Virginia, USA
[§]CORDIO East Africa, P.O. Box 1013, Mombasa, Kenya
[¶]EcoOceania, Speewah, Queensland, Australia
[‖]Phoenix Island Protected Area Office, Ministry of Environment, Lands and Agriculture Development,
P.O. Box 234, Tarawa, Kiribati
[#]Phoenix Islands Protected Area Conservation Trust, P.O. Box 366, Tarawa, Kiribati
**Scripps Institution of Oceanography, UC San Diego, La Jolla, California, USA
[††]Conservation Law Foundation, Boston, Massachusetts, USA
[‡‡]The University of British Columbia Fisheries Centre, Vancouver, British Columbia, Canada
[§§]Conservation International Pacific Islands and Oceans Programme, P.O. Box 2035, Apia, Samoa
[¶¶]Woods Hole Oceanographic Institution, Woods Hole, Massachusetts, USA
[1]Corresponding author: e-mail address: rrotjan@neaq.org; randi.rotjan@gmail.com

Contents

Advances in Marine Biology, Volume 69
ISSN 0065-2881
http://dx.doi.org/10.1016/B978-0-12-800214-8.00008-6

Abstract

The Republic of Kiribati's Phoenix Islands Protected Area (PIPA), located in the equatorial central Pacific, is the largest and deepest UNESCO World Heritage site on earth. Created in 2008, it was the first Marine Protected Area (MPA) of its kind (at the time of inception, the largest in the world) and includes eight low-lying islands, shallow coral reefs, submerged shallow and deep seamounts and extensive open-ocean and ocean floor habitat. Due to their isolation, the shallow reef habitats have been protected *de facto* from severe exploitation, though the surrounding waters have been continually fished for large pelagics and whales over many decades. PIPA was created under a partnership between the Government of Kiribati and the international non-governmental organizations—Conservation International and the New England Aquarium. PIPA has a unique conservation strategy as the first marine MPA to use a conservation contract mechanism with a corresponding Conservation Trust established to be both a sustainable financing mechanism and a check-and-balance to the oversight and maintenance of the MPA. As PIPA moves forward with its management objectives, it is well positioned to be a global model for large MPA design and implementation in similar contexts. The islands and shallow reefs have already shown benefits from protection, though the pending full closure of PIPA (and assessments thereof) will be critical for determining success of the MPA as a refuge for open-ocean pelagic and deep-sea marine life. As global ocean resources are continually being extracted to support a growing global population, PIPA's closure is both timely and of global significance.

Keywords: Kiribati, Open ocean, Phoenix, Reef, Zone, Protection, Fisheries

1. INTRODUCTION TO THE PHOENIX ISLANDS PROTECTED AREA

The Phoenix Islands Protected Area (PIPA) is a unique Marine Protected Area (MPA) in many regards, but is most known for its size and innovative conservation strategy. First created in 2008, PIPA has accomplished much in a short period. At the time of creation, it was the world's largest MPA, and as of 2014 still remains the largest and deepest UNESCO World Heritage site. It was the first MPA to contain substantial deep-water, pelagic and seamount habitat in addition to shallow reefs and critical terrestrial habitat for nesting seabirds. PIPA was established by the Republic of Kiribati, an UN-designated Least Developed Country, through a collaboration between the government and two partner institutions—the New England Aquarium (NEAq) and Conservation International (CI). In less than a decade since its inception, PIPA has achieved a number of key milestones. These include the creation of the PIPA Conservation Trust to support a sustainable financing mechanism for the MPA, and the signing of a conservation contract between Kiribati and the PIPA Trust, which has accelerated the timeline of PIPA's planned phased closure from an original 12,714 km^2 (3.1%) no-take to 405,755 km^2 (99.4%) no-take effective 1 January 2015. The remaining ~0.6% will remain a restricted use zone around Kanton Island to accommodate subsistence fishing for a small care-taker population. Below, the details of PIPA geography, protection, extraction and remaining challenges are reviewed.

2. GEOGRAPHY AND ECOSYSTEMS

Kiribati is an ocean nation covering 3,500,000 km^2 in the central Pacific on both sides of the equator, approximately midway between Australia and Hawaii (Figure 8.1). Kiribati's marine area encompasses three island archipelagos, the Gilbert, Phoenix and Line Islands. These total 33 islands with a combined land area of only 811 km^2. With its land area well less than 1% of its sovereign domain, Kiribati is truly an oceanic nation.

The Phoenix Island group straddles the central region of Kiribati (where the equator crosses the International Dateline), positioned distantly between the Gilbert Islands to the west and the Line Islands to the east. Two of the Phoenix group members, Howland and Baker, are low-reef islands in adjacent territory of the United States to the north of Kiribati. Located directly north of the Tonga-Kermadec ocean trench, the Phoenix Islands sit atop the Tokelau Ridge, incorporating some of the many Tokelau volcanoes aligned along the ridge. Beyond the Tokelau seamounts there is a well-defined cluster of volcanoes to the east. Beyond these volcanoes lies a typical flat terrain mid-ocean sea floor. Rising from an average depth of 4500 m and a maximum depth of 6147 m, a dozen of these massive volcanoes host shallow reefs and the highest peaks reach the surface, capped by low-elevation islands and atolls. Four are

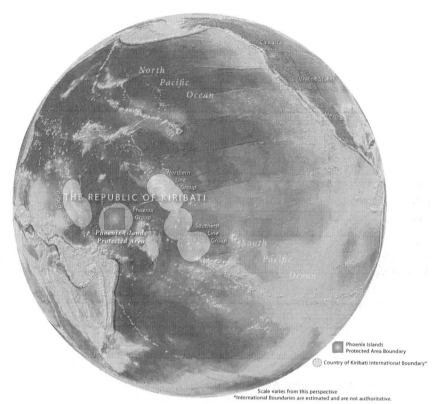

Figure 8.1 The Phoenix Islands protected area (teal box) (dark grey in the print version) is located in the central of Kiribati's three archipelagos.

topped by true coral atolls (Orona, Manra, Kanton and Nikumaroro), six present as low-reef islands (Birnie, Rawaki, McKean and Enderbury in Kiribati and Baker and Howland in U.S. territory) and two are shallow submerged reefs (Carondelet and Winslow). The eight atolls and low-reef islands and the two submerged reefs of PIPA (Figure 8.2) represent only the highest of numerous large- and long-extinct volcanoes. Most of the large volcanoes are greater than 200 m below the surface and are therefore technically classified as seamounts. The catalogued seamounts include Fautasi, Siapo, Polo, Tai, Tanoa, Tau Tau, Gardner and four unnamed seamounts.

The terrestrial habitats range from simple plant herb–shrub communities on the low-reef islands to forest communities on the atolls, especially on the three southern islands which experience the highest rainfall. Atolls and low-reef islands in PIPA are small, ranging from 1.03 to 17.5 km across, but are surrounded by some of the most untouched coral reefs in the world (Obura

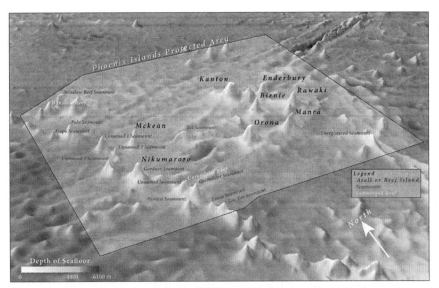

Figure 8.2 Three-dimensional view of PIPA (box) showing surface islands, seamounts and submerged reefs.

et al., 2011a,b). These support a remarkable abundance of marine life (Obura, 2011). Over 500 fish species have been recorded, including abundant populations of large parrotfishes and wrasses that have elsewhere been severely overfished (Allen and Bailey, 2011). The coral species assemblage is typical of the Central Pacific region and expeditions to the Phoenix Islands early in this century found the reefs to be in excellent condition with coral cover averaging 45.1% and 58.1%, respectively, with a maximum cover of 100% live coral (Obura and Stone, 2002; Obura et al., 2011). This is twice the recent average for southern and western Pacific coral reefs, and broadly overlapping with what can be considered near pristine conditions (Bruno and Selig, 2007; Sandin et al., 2008). The same can be said of the fish biomass, which averaged in the vicinity of 250 gm/m^2 during 2009 surveys (values for historically fished coral reefs range from 60 to as low as 20 gm/m^2, (Ssandin Pers.comm)).

3. ESTABLISHMENT OF THE MPA

3.1. Placement and zonation

The Phoenix Islands are located within the Kiribati 200–mile Exclusive Economic Zone (EEZ). PIPA comprises 54.9% of the EEZ surrounding the Phoenix Islands, and 11.3% of the total Kiribati EEZ. The first phase of MPA management was established in 2008, with multiple use zones. In addition to full protection of all the terrestrial resources, no-take zones were initially established around seven of the eight islands, totalling 3.1% of the MPA (12,714 km^2). There was also a 12-nautical mile restricted use zone around Kanton Island totalling 0.6% of the MPA (2495 km^2) that allowed for subsistence fishing by Kanton inhabitants. Finally, there was a Kanton Purse Seine Exclusion Zone totalling 9.1% of the MPA (37,197 km^2). At the time of inception, the rest of the MPA was open to commercial tuna fishing, with plans for phased increases in no-take zone areas (planned to increase from 3.1% to 28.1% in the second phase). In February 2014, the Kiribati cabinet re-affirmed a decision to accelerate the timeline of no-take protection and to bypass the originally planned phased approach. The decision of the Kiribati government to enact full closure of PIPA to all commercial fishing, including distant-water-fishing-nations (DWFNs), will be effective from 1 January 2015 (Figure 8.3).

3.2. Partnerships and legislation

The collaboration between the Government of Kiribati, NEAq and CI first developed in the early 2000s after exploratory expeditions by the NEAq to the Phoenix Islands under its 'Primal Oceans' project—a project led by Dr. Gregory S. Stone to find some of the last virtually untouched areas in the global ocean. The results of these expeditions, including film and under-water photography, were presented to Kiribati officials and a dialogue on the protection of the Phoenix Islands was initiated. Later, CI was brought in for technical expertise and funding support and a Memorandum of Understanding was signed between the three entities in 2005 and is still in use today.

PIPA is managed by a Management Committee (PIPA MC), a government multi-ministerial committee with representatives from the government stakeholders with jurisdiction over the various aspects of the protected area. The PIPA Office, established within the Ministry of Environment, Lands, and Agriculture Development (MELAD) and based in the capital Tarawa, provides logistical support.

PIPA was the first and is still the only formal MPA in Kiribati. Because there was no precedent for creation or management of MPAs, Kiribati legislation could not accommodate or legally recognize special status for PIPA. Consequently, MELAD, in coordination with the Office of the Attorney General and the Kiribati Cabinet, undertook a major revision of Kiribati

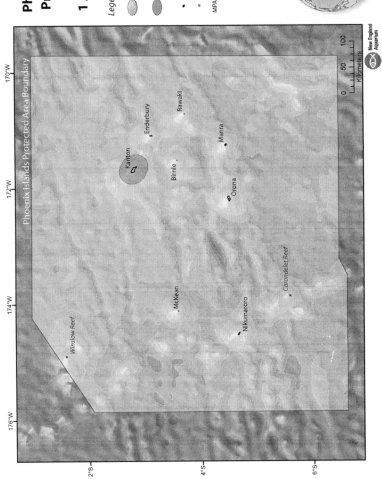

Phoenix Islands Protected Area Phase 2 1 January 2015

Legend

No Take Zone (405,755 km² or 99.4%)

Kanton Restricted Use Zone (2,495 km² or 0.6%)

• **Low-lying Island/Atoll**

× **Submerged Reef**

MPA total area: 408,250 km²

Figure 8.3 PIPA map showing the pending Phase 2 zonation, effective 1 January 2015.

law, beginning in 2005 and culminating in the passage of the Environmental
(Amendment) Act 2007, which took effect on 4 September 2007. Acting
under this new authority, then MELAD Minister Tetabo Nakara, acting
on the advice of the Kiribati Cabinet formally declared the boundaries and
enacted the new legal framework for PIPA with the formal adoption of
the PIPA Regulations 2008. These regulations defined the boundaries of
PIPA, approved the nomination for the area to be inscribed on the list of
World Heritage sites, identified PIPA as a Category 1b Wilderness Area under
IUCN's 1994 Guidelines for Protected Area Management Categories and
required the development of a PIPA Management Plan which was completed
in 2009. These 2008 regulations also created the PIPA MC, which oversees
and enforces that all persons and public authorities undertaking any activities
within the protected area boundary acts in accordance with the Environment
(Amendment) Act, the PIPA Regulations and the PIPA Management Plan.

3.3. Sustainable financing

To embrace the idea of creating a large MPA in the Phoenix Islands, the
Kiribati Government stipulated that the MPA should not have any negative
impacts to the national economy, which was highly dependent on fishing
revenues. In 2011, issuing of fishing licenses to foreign fleets accounted
for over 40% of Kiribati's national revenue (International Monetary Fund
Kiribati, 2011). Creating a sustainable financing mechanism to replace
any lost revenue Kiribati might experience from closing the Phoenix Islands
to commercial fishing, while simultaneously compensating Kiribati for any
increased costs for PIPA management activities, was central to the strong
political support from Kiribati and its people. The model of the forest conservation concession was proposed: instead of leasing land to logging companies, landowners are paid not to log their forests (e.g. de Koning et al.,
2011). Under this model, compensation and management costs could be
covered by the revenues earned from a conservation endowment, securing
dedicated funding for the long-term management of the MPA that was not
subject to volatility of government budgets. This idea led to the enactment
of the PIPA Trust Act by Parliament in 2009, which legalized the establishment of the PIPA Conservation Trust as the 'financing mechanism' of the
MPA as a charitable non-government organization.

   The PIPA Conservation Trust has multiple activities, including provision of financial support for the management of PIPA and paying any fee
that might be required to compensate the government for demonstrated

declines in national fishing revenues as a result of the PIPA closures. These activities are managed and accomplished through the mechanism of a conservation contract, which was signed by the government of Kiribati and the PIPA Conservation Trust in 2014. Performance under the contract is measured by the government's compliance with the provisions of the Management Plan and the success of the government's actions taken to prohibit pelagic or other fishing in the designated PIPA no-take zones. This is the first marine example of such a conservation contract. The PIPA Conservation Trust is governed by a board of directors, with the three founding PIPA partners having permanent seats. In 2011, the Trust hired an executive director and established an office in Tarawa. In October 2013, the PIPA Trust was capitalized with USD $5,000,000, half of which was gifted by the Kiribati government. This contribution was matched by CI through its Global Conservation Fund.

The initial USD $5,000,000 capitalization of the PIPA Trust was a major step towards the goal of reaching a 2014 target of USD $13,500,000, which was originally intended to catalyze the second phase of protection, allowing an increase to the pelagic the no-take zone by an additional 25%. Kiribati President Anote Tong decided to support measures to accelerate the PIPA closure timeline. In January 2014, the Kiribati cabinet re-affirmed its decision to fully close the whole of PIPA to commercial fishing and approved 31 December 2014, as the effective date for full closure of the protected area to all forms of fishing including tuna. As part of the first conservation contract with Kiribati, a Tuna Working Group is being established to analyse the potential revenue impacts of the PIPA closure over the next 5 years.

3.4. International and regional context

The creation of PIPA was a remarkable achievement for a Small Island Developing State (SIDS) and sets an example as the world's largest MPA (at the time of creation) with a novel partnership structure and the first marine model of using a conservation contract with a sustainable financing mechanism. The establishment of PIPA moves Kiribati closer to its commitments under the Convention on Biological Diversity, and Aichi Target 11, which states that 'by 2020, at least 17 percent of all terrestrial and inland water areas, and 10 percent of coastal and marine areas ... are conserved through effectively and equitably managed, ecologically representative and well-connected systems of protected areas and other effective area-based conservation measures'.

Following the inception of PIPA, a number of other large MPAs have been created, including several in the Pacific. This 'race to protect the oceans' is launching a new era of ocean protection and management (Leenhardt et al., 2013). To date, 28 countries have now exceeded 10% coverage of their waters as designated MPAs, but this area still represents less than 2.3% of the total global ocean area as protected (Spalding et al., 2013). Even within this 2.3%, not all MPAs are governed effectively or with similar conservation goals. A recent meta analysis of 87 MPAs found that conservation benefits increased exponentially with the accumulation of five attributes: no take, well enforced, old (>10 years), large (<100 km^2) and isolated by deep water or sand (Edgar et al., 2014). PIPA will soon meet three of these criteria (no-take, large and isolated). As of 2018, PIPA will formally be a decade old. However, enforcement remains a challenge due to the remoteness of the area (detailed below).

4. WHY PROTECT THE PHOENIX ISLANDS?

Viewed against the background of profound human impacts on oceanic islands and marine ecosystems globally, the relative intactness of PIPA is remarkable (Obura et al., 2011). Notwithstanding evidence of prehistoric and recent failed attempts at human colonization on some of the atolls, the Phoenix group is sufficiently remote and inhospitable to human colonization as to be exceptional in terms of the minimal evidence of the impacts of direct anthropogenic stressors both on the atolls and in the adjacent seas. PIPA is more than 600 km across and covers an area of 408,250 km^2, which is a very large managed area even by marine standards. PIPA's remoteness and the protected status of its reefs make it one of a very few remaining places on earth where it is possible to examine the ecological dynamics of coral reef habitats that remain largely unimpacted by intense local human impacts. Due to their remoteness, PIPA waters have been largely *de facto* protected from typical anthropogenic stressors such as coastal development and sedimentation, pollution, high-human traffic and invasive alien species. Fishing has largely been focused on large pelagic such as tuna, and reef-fish extraction and exploitation have been minimal compared to nearly all other Pacific Islands (NEAq, unpublished data).

Biodiversity on PIPA reefs is high despite the small amount of reef area, and the abundance of commonly overfished organisms is notable (e.g. giant clams, bumphead parrotfish and Napoleon wrasse). The total known shallow reef-fish fauna of the Phoenix Islands now stands at 518 species, consisting of the following: 192 species originally recorded by Shultz (1943), 100 additional species recorded by the year 2000 expedition (Stone et al., 2001), 9 additional species reported in various generic revisions and 217 new records from the 2002 expedition (Allen and Bailey, 2002a,b). A formula for predicting the total reef-fish fauna based on the number of species in six key indicator families (G. Allen, unpublished data) indicates that at least 576 species, over 50 more than currently listed, can be expected to eventually be found in the Phoenix Islands. A new species of damselfish, *Chrysiptera albata*, was collected in 42–50 m depth at Nikumaroro Island (Allen and Bailey, 2002a). Other previously undescribed species were found in the genera *Eviota* (Gobiidae), *Trimma* (Gobiidae) and *Myripristis* (Holocentridae; Randall et al., 2003).

Given their relative health prior to legal protection, it is reasonable to ask whether and why conservation measures were necessary for the Phoenix Islands. Despite their remoteness, PIPA waters are threatened by climate change, by increasing fishing pressure on declining stocks of large pelagics like tuna and by potential increase in human usage given the acute over-crowding on several islands in Kiribati, plus the general increases in global human population, affluence and mobility (Lambin and Meyfroidt, 2011), all of which have the potential to enable increased traffic and use. Further-more, because of its large size and relative isolation, PIPA has the ability to act as a critical benchmark for identifying, understanding and evaluating ecological change in tropical oceans, heightening the urgent need to study and protect this remote oceanic baseline. The ambitious creation of PIPA by a SIDS was both timely and of global significance, serving to protect these coral ecosystems and to set a global conservation example for extending MPAs into the pelagic realm to protect deep-water habitats and associated diversity.

4.1. PIPA as a natural laboratory

With ecosystems around the world in a rapid decline due to both local and global anthropogenic impacts (Burke et al., 2011; Doney et al., 2012), the very remote, reasonably intact and legally protected oceanic environments of PIPA are of considerable importance as a baseline for healthy intact eco-systems. PIPA offers opportunities to identify and to monitor the effects of sea level change, acidification and warming on coral reef systems without the confounding influences of pollution, watershed disruption or resource extraction. In this regard, it is notable that within Kiribati there are otherwise similar reef systems that are subject to varying levels of human occupation (Sandin et al., 2008) that also vary in their historical exposure to severe ocean warming anomalies and coral bleaching. For coral reefs, the nation of Kiribati is a planetary Rosetta Stone, offering the opportunity to translate and to interpret the possibilities and limitations for coral reef resilience in an era of anthropogenic climate change.

4.2. Protection of marine source populations

PIPA is biogeographically situated in the centre of the equatorial Pacific, thus likely playing a pivotal role in the movements and dispersal of marine animals and larvae across the Pacific. Few studies of gene flow and connectivity have been carried out across the whole Pacific that includes samples from the Phoenix Islands. However, the limited data available to suggest that there is some gene exchange between the Central Pacific equatorial regions, although some other Pacific regions, like Hawaii, are strongly isolated (Polato et al., 2010). In *Porites lobata* corals, which have a relatively robust (symbiotic) larval stage, no significant genetic differentiation was detected between the Phoenix Islands (Enderbury) and other equatorial archipelagos in the Central Pacific (Baums et al., 2012). However, evidence of an Eastern Pacific genotype was found in the PIPA samples, suggesting that occasional gene flow is possible even across regions that are thought to be isolated from the Central Pacific (Baums et al., 2012). Thus, PIPA may serve as both a source and a sink for *P. lobata* larvae from various regions.

Little is known about the effect of these islands on the surrounding pelagic marine species and systems. The islands of PIPA support internationally important seabird colonies and numerous migratory birds. Pierce et al. (2006) suggested that the Phoenix Islands are affecting and supporting the pelagic marine life and seabird ecology by increasing nutrient loads, thereby potentially impacting the food chain in the local marine environment.

4.3. Protection of novel habitat

PIPA's marine wilderness contains terrestrial low-lying island habitats, shallow, mesophotic and deep reef habitats, large tracts of open-ocean and ocean seafloor and a large suite of seamounts. Among designated MPAs, this is unusual as most are focused on coastal habitats. At the time of inception, PIPA was the only MPA to contain considerable amounts of blue water, seamounts and abyssal plain.

4.4. Protection from climate change

The Phoenix Islands are subject to climatic variability associated with the El Niño–Southern Oscillation (ENSO) index and positive ENSO events (El Niños) that occur every 2–7 years and last for 18–24 months. During these events, the westward trade winds are reduced, water currents experience variations and even reversal and the eastern Pacific thermocline deepens. In particular, the Phoenix Islands are located within the region of the Central Pacific in which a warm pool of surface waters develops at the onset of El Niño phases. Consequently, PIPA can experience persistent hotspots, as occurred in 2002–2003. In that case, a sea surface temperature (SST) hotspot developed (Figure 8.4) and remained over the central Pacific from June 2002 to March 2003 (Alling et al., 2007). Data from *in situ* temperature loggers showed the highest maximum SST was recorded in November 2002, being about 0.5–1 °C warmer than the following 2 years (Obura and Mangubhai,

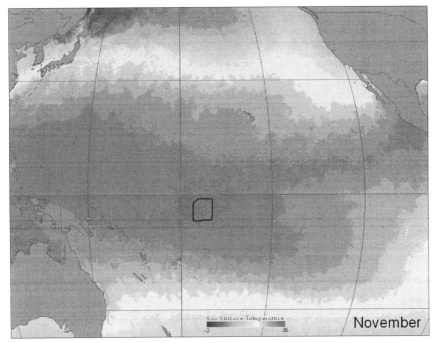

Figure 8.4 Sea surface temperatures (SSTs) in the Phoenix Islands protected area (PIPA; centre box) in 2002–2003, where a prolonged (7 months) high-temperature event occurred in 2002–2003. Red dots (light grey in the print version) denote temperatures at or in excess of Figure 8.1. The PIPA (teal box) (dark grey in the print version) is located in the central of Kiribati's three archipelagos.

2011). The hotspot remained in place for an unprecedented period of 21 Degree Heating Weeks (DHW), where each DHW is equivalent to 1 week of SST 1 °C warmer than the expected summer-time maximum.

Post-bleaching surveys in 2005 documented a reduction in live coral cover to 12.1% overall in the Phoenix Islands, down from 58% in 2000 (Obura and Mangubhai, 2011). This level of coral loss is consistent with those observed in other bleaching events (e.g. Vargas-Ángel et al., 2011). Most likely as a result of this intensive bleaching event, subsequent surveys showed a persistent (consistent across all sites and all fish grouped) decline in reef-fish abundance between 2002 and 2005, with significant declines for the Carangidae, Chaetodontidae and the serranid subfamily Epinephelinae (Mangubhai et al., 2014). Interestingly, changes in fish abundance varied spatially within and across islands, suggesting that there are critical lessons to be learned about differential resilience and recovery in natural (relatively untouched) environments (Mangubhai et al., 2014). The exposure to warm water pools may also exert unusual selective pressures on marine organisms relevant to climate change adaptation.

The establishment of the protected area so soon after significant bleaching will ensure that the future trajectory of PIPA reefs should be unimpaired by fishing-induced loss of reef-fish diversity, abundance or biomass. Post-bleaching re-growth of corals and crustose coralline algae (a necessary precursor for settlement of many coral species) was observed in 2009, suggesting that PIPA reefs have both resilience and rapid recovery potential (Stone et al., 2009). As noted above, PIPA is also a valuable natural laboratory. The successional dynamics of fast-growing and brooding coral species as the dominant re-colonists post-bleaching is a prime example of what can be learned regarding the natural dynamics of reef recovery in the absence of confounding disturbance. However, ongoing thermal stress and associated habitat loss will continue to be important players in determining the future fate PIPA reefs. A 2010 bleaching event in the U.S. Phoenix Islands (Vargas-Ángel et al., 2011) and corresponding high-temperature event in the Kiribati Phoenix Islands (Mangubhai et al., 2012) will likely impact the recovery trajectory of the Phoenix Islands post-2002–2003, but also heightens the importance of this region for the opportunity to examine long-term reef response to repeated disturbance in the absence of local human stressors.

4.5. Protection of indigenous populations/traditional rights

The Kiribati Phoenix Islands have no permanent inhabitants, although most islands have a recent cultural history extending over the past 150 years.

Kanton is the only atoll that has a non-permanent population of usually less than 50 people comprising government employees and their families engaged in protection and management of Kiribati interests in the region. To recognize the importance of marine life as a critical protein source for the highly remote Kanton population, allowance of subsistence fishing was an important initial consideration in the creation of the phased MPA closure for PIPA (detailed in later sections).

4.6. Protection from extraction and exploitation

Protection of remote areas is increasingly important as the global human population grows (Lee, 2011). While potential settlement of uninhabited lands everywhere becomes increasingly possible as technology enables remote living, the biggest threat from rising population comes from increased need for ocean resources, for example, to ensure food security (e.g. Brewer et al., 2013). Increased global wealth can also give rise to new and niche markets, for example, tourism, that have the potential to induce habitat loss and degradation. In the past decade alone, several foreign interests have expressed the desire to develop tourism in the Phoenix Islands, for example, as a site for recreational fishing of bonefish or amateur radio destination. Thus far, the Phoenix Islands have remained relatively intact in the region due to their remoteness and isolation, but the global reach of commercial fishing, settlement and development did not wholly bypass the Phoenix Islands (detailed below). The creation of PIPA was thus both timely and necessary to prevent future exploitation and extraction. Under the management plan, a sustainable use plan will be developed for Kanton to ensure that any increase in population (i.e. for planned research stations, increased management staffing and ecotourism) will be well considered.

5. EXTRACTION, EXPLOITATION AND USE OF PHOENIX ISLANDS RESOURCES

Archaeological evidence indicates that there have been a few settlements in the Phoenix Islands and that because of their isolation from larger population centres, these early settlers never stayed for very long. For the canoe explorers who originally mapped the South Pacific some 3000–5000 years ago, the Phoenix Islands must have been diminutive in comparison to the lushness and largesse offered by other Pacific Island groups such as Fiji, Samoa and Hawaii. The Phoenix Islands exhibit a legacy of temporary and sporadic human use and settlement over several hundred years, with species introductions both of plants (e.g. coconuts), animals (e.g. rats, rabbits and cats) and remnants of guano mining. None of the islands have hosted long-term human settlements due to the scarcity of fresh water and terrestrial resources.

The early Polynesian settlers who came and went in varying periods between AD 950 and 1500 left stone building foundations that resembled the *marae* of eastern Polynesia. Ancient stone weirs and fish traps were also discovered on some of the Phoenix Islands. There is also evidence that the Phoenix Islands were visited by Caroline Islanders (Micronesians). Most archaeological structures have been found on Orona and Manra. However, though the Phoenix Islands were repeatedly deemed uninhabitable and therefore abandoned, remoteness and Spartan conditions failed to dissuade later attempts to use the Phoenix Islands as a locale for whaling, shark fin-ning, guano extraction, copra harvesting and tuna fishing, all prior to the establishment of PIPA.

5.1. Whaling

Great whales, and in particular sperm whales (*Physeter macrocephalus*), once abounded in the Central Pacific including in the Phoenix Islands. The islands may have been visited as early as 1794 by European and North American whaling vessels and developed in earnest with the expansion of the American whale fleet into the Pacific in the early 1800s, when American whalers from Nantucket and New Bedford, Massachusetts dis-covered the rich concentration of sperm whales in the Pacific. Much of the historic Pacific whaling grounds were located along the 'on-the-line' grounds, which cut through what were to become Kiribati territorial waters. At times, more than 600 whale ships plied these waters and whalers were so effective that even today the Phoenix Islands appear to be largely devoid of sperm whales. Hutchinson (1950) compared sperm whale pop-ulation with zooplankton density across the equatorial Pacific. Zooplank-ton densities were the highest at about 2°S (the latitude passing through the Phoenix Islands) and that corresponded with the peak in sperm whale abundance.

In recent years, opportunistic sightings of sperm whales near the Phoenix Islands have been rare. For example, Stone et al. (2001) noted that during the 2000 NEAq expedition to the Phoenix Islands, few cetaceans of any sort were seen. Odontocetes were the most common sub-order observed and bottlenose dolphins (*Tursiops truncatus*) the most common species. Pierce et al. (2006) recorded no whales during a period of 27 days, which is a concern considering the extensive time spent observing pelagic seabirds in Phoenix Island waters. Bottlenose and spinner dolphins were the most commonly observed cetaceans during five trips by Pierce et al. in 2006–2013 with only one whale, a probable beaked whale, being seen within PIPA waters (R. Pierce, pers. comm.). It should be noted, however, that no dedicated surveys for cetaceans have been carried out, a gap in survey effort that needs to be remedied. To gain more insight into the spatial distribution of sperm whales in PIPA, researchers at NEAq are currently studying historic vessel logbooks, which feature detailed records of the New Englanders' whaling voyages in the 1800s. Building on the previous studies that compiled and digitized such records (Smith et al., 2012), these researchers are mapping data of whale sightings and catches in the central Pacific to determine potential sperm whale hotspots throughout PIPA waters where special whale conservation strategies may be needed. PIPA, by virtue of its large size, could offer significant potential benefit for cetacean conservation. However, as for any other migratory species, the potential contribution of PIPA to cetacean conservation remains to be assessed.

5.2. Shark finning and exploitation

Early accounts of shark populations within the Phoenix Islands speak of incredible abundance, with little or no exploitation. A 1929 eyewitness account from a stranded shipwreck passenger aboard the *Norwich City* (Nikumaroro) writes '... the lagoon and the whole of this inland water was absolutely infested by sharks ...'. A 1937 report (Maude) notes that on Kanton Island, 'sharks were plentiful everywhere'. The Waikiki Aquarium of Hawaii found it easy to collect baby blacktip sharks from Kanton Island in the 1980s, likely facilitated by their abundance in the lagoon and shallow reef areas, and by the presence of a then-functioning air strip.

Early expeditions (2000) to the Phoenix Islands by NEAq noted substantial populations of grey reef (*Carcharhinus amblyrhynchos*), whitetip reef (*Triaenodon obseus*) and blacktip reef (*Carcharhinus melanopterus*) sharks on all of the shallow reefs, and especially on Nikumaroro Island, which had the highest overall abundances (Stone et al., 2000a,b). Deep-water sharks including six-gill (*Hexanchus griseus*) and Pacific sleeper (*Somniosus pacificus*) sharks were also observed via baited camera deployments—the first confirmed records for these species in the Central Pacific (Stone et al., 2000a,b). However, in 2002, the NEAq noted severe depletions of shark populations due to a single foreign fishing vessel (the *Maddee*, based in American Samoa) that was permitted to harvest shark fins by longlining around four islands for ~9 months (Obura and Stone, 2002). As a result, the frequency of blacktip reef shark sightings decreased from 64% (across all sites) in 2000 to 16% in 2002. Orona may have suffered additional exploitation from a short-lived government resettlement scheme begun in early 2001 (the Kakai settlement scheme, which ended in 2004), with 200 people harvesting island resources, including shark fins, for market sale. However, no quantitative data on extraction from that time are available. Yet, the 2002 NEAq Expedition noted that not a single shark was recorded in their circular transects on Orona (Obura and Stone, 2002). The differences in abundance between 2000 and 2002 were staggering. Overall shark densities at fished islands were much lower than at those that were not fished (Figure 8.5; Obura and Stone, 2002). This demonstrates the extreme vulnerability of small-island populations—and PIPA in particular—to shark extraction lasting only a few months and may be regarded as a proxy for the likely impacts of increased resource exploitation there. This level of exploitation is typical of the far-reaching impacts of the shark fin market, which has expanded the elasmobranch hunt to the most remote regions of the globe (Dulvy et al., 2008). Now, over a third of elasmobranch species in the southwest Pacific region are at risk of extinction (Jupiter et al., 2014).

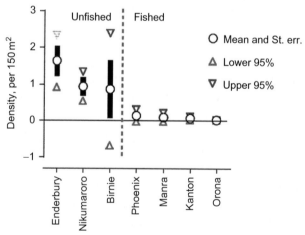

Figure 8.5 Densities of reef sharks (whitetip, blacktip and grey reef sharks) sampled in 150 m² circular transects, Phoenix Islands, 2002 (mean, standard error and upper/lower 95% confidence intervals). *Reproduced with permission from Obura and Stone (2002).*

5.3. Seabird colonies and guano extraction

As is typical for remote bits of land in a vast ocean, the Phoenix Islands are important for nesting seabirds. PIPA hosts forested and arid seabird islands of high-global significance (>40 breeding colonies with several of the world's largest seabird breeding populations), including at least 19 breeding species, 51 species reported and a total population exceeding 1 million birds (Pierce and Teroroko, 2011). In particular, the islands support globally important breeding colonies or concentrations of the following species: Audubon's Shearwater (*Puffinus l'herminieri*), Christmas Shearwater (*Puffinus nativitatis*), Phoenix Petrel (*Pterodroma alba*), White-throated Storm petrel (*Nesofregetta fuliginosa*), Great Frigatebird (*Fregata minor*), Lesser Frigatebird (*Fregata ariel*), Brown Booby (*Sula leucogaster*), Masked Booby (*Sula dactylatra*), Sooty Tern (*Sterna fuscata*), Greybacked Tern (*Sterna lunata*) and Blue Noddy (*Procelsterna cerulea*). Most of these species forage primarily within PIPA waters reflecting the local diversity and abundance of marine resources. Over centuries, these birds deposited guano that was later extracted from the Phoenix Islands in the late 1800s for use in the United States and Europe mainly as fertilizer. Guano was first discovered and used as an effective fertilizer in Peru. However, Peruvian supplies were tightly regulated and highly priced and thus could not meet international demand. This spawned a global search for new and rich guano deposits. Guano exploitation in the Phoenix Islands began in 1870 but lasted for only a decade, by which time over 70,000 tonnes of guano were extracted, representing more than 20% of all imported sources to the United States (Skaggs, 1995).

5.4. Commercial tuna fishing

Reef-associated tuna species (dogtooth tuna) were likely spared from severe exploitation until the early 2000s, when legal fishing was permitted for sharks (detailed above) and reef tuna in 2001. Post-extraction, the decline in tuna was highly noticeable with, for example, dogtooth tuna decreasing from being present in 75% of surveys in 2000 down to 0% in 2002 (Obura and Stone, 2002).

Longline and purse seine fishery licenses constitute an important part of the Kiribati economy. Starting when under the United States and British control, what are today Kiribati waters were extensively fished for tuna and other large pelagics. Kiribati grants the majority of its licenses to foreign fishing vessels flying the flags of the United States, EU countries, South Korea, China, Japan and other nations. Purse seine tuna fishing is the major extraction method, with a purse seiner catching on average 32 tonnes of tuna per day, mostly skipjack (Fonteneau et al., 2000). Tuna purse seine licensing is currently regulated under the Nauru Agreement, a subregional agreement spanning eight countries including Kiribati that make up the Parties to the Nauru Agreement (PNA) established in February 1982. The Nauru Agreement was created and signed in part to address fisheries stock exploitation through the cooperation and coordination of licensing and EEZ regulations by the eight nations who signed the international agreement; these countries control 25–30% of the global tuna supply (Jupiter et al., 2014). Since 2007, most foreign fishing vessels are able to fish Kiribati waters subject to the PNA under the so-called vessel day scheme (VDS) whereby vessel owners can purchase and trade days fishing at sea as set and allocated between PNA nations for 1-year periods, up to 3 years in advance. United States vessels are controlled under a multilateral treaty that is still being finalized at the present time; EU vessels are controlled under an agreement between Kiribati and the EU that came into effect in 2012. Both the United States fishery and the EU fishery are based on the VDS approach as well.

The most recent stock assessment information on the target tuna species, skipjack (*Katsuwonus pelamis*), yellowfin (*Thunnus albacares*) and bigeye (*Thunnus obesus*) and economic information is used to determine the annual tuna harvest levels and to assess the allocation of fishing days. The VDS is apportioned based on vessel length to account for differential fishing power and/or efficiency (catch) of the various vessels in the fleet, and ranges from 0.5 to 1.5 fishing days depending on vessel length. The VDS is implemented as part of the Western and Central Pacific Fisheries Commission's (WCPFC) *Conservation and Management Measure for Bigeye and Yellowfin Tuna in the Western and Central Pacific Ocean* (CMM2005-01). The VDS replaces the previous purse seine management scheme that was based on vessel number limits of 205 vessels under Annex 1 of the *Palau Arrangement for the Management of the Western Pacific Purse Seine Fishery* (Palau Arrangement).

The catch of tuna and other large pelagic species in the PIPA area is accomplished using both longlines and purse seines. Catches of highly migratory species are reported by the WCPFC, which monitors fishing efforts that occur in 337 of its $5° \times 5°$ management units. These have been reported on monthly from the 1950s to the present. Eight species are included in the WCPFC Aggregated Catch/Effort data: yellowfin, bigeye, albacore (*Thunnus alalunga*), skipjack tuna, striped marlin (*Kajikia audax*), black marlin (*Istiompax indica*), blue marlin (*Makaira mazara*) and swordfish (*Xiphias gladius*). All other landings by the fleets are reported in the

catch–all category of 'other'. Sumaila et al. (2014) determined that the
WCPFC management area tuna fishery was worth USD $4.1 billion in
2009 while also examining the subsidies and incentives that help drive
the fishing in this highly productive portion of the ocean.

The closest management block does not perfectly align with the boundaries of the PIPA, but it does give insight into the catch in this small region of
the Pacific (Tables 8.1 and 8.2). The 5° × 5° management block used to estimate catch in PIPA spans 175°W 05°S to 180°E 00°N.

While the exact landed value of fish caught in the Pacific is hard to deter-
mine due to the various home ports of vessels, transshipping and the multiple
markets receiving/selling catch, landed value can be estimated using the
U.S. National Marine Fisheries Office of Science and Technology, Annual
Commercial Landing Statistics Valuations, for fish brought to market on the
U.S. Pacific coast (Table 8.3).

All eight reported species are caught within the management unit that
encompasses PIPA. These have produced a landed value ranging from
USD $3,960,000 in 2001 to USD $221,360,000 in 2009, the last reported
year. This range represents a 40-fold change in effort by the longline fleet as
measured by the number of hooks fished and a nearly 17-fold increase by the

Table 8.1 Reported vessel day scheme days and landings in metric tonnes of the three principal tuna species within the PIPA management block
PIPA purse seine catch in metric tonnes

Year	Fishing days	Skipjack tuna	Yellowfin tuna	Bigeye tuna
2000	166	1763	917	215
2001	68	750	119	41
2002	363	11,440	182	55
2003	156	2731	502	183
2004	139	1669	332	148
2005	305	4685	8129	762
2006	358	16,868	700	372
2007	92	4294	355	88
2008	Not reported	Not reported	Not reported	Not reported
2009	1153	56,704	1710	561

Source: WCPFC (2011).

Table 8.2 Reported hundreds of longline hooks fished and landings in metric tonnes of the seven principal tuna and billfish species within the PIPA management block

PIPA longline catch in metric tonnes

Year	Hundreds of hooks	Albacore tuna	Yellowfin tuna	Bigeye tuna	Striped marlin	Black marlin	Blue marlin	Swordfish
2000	18,151	22	581	264	3	–	32	17
2001	1825	8	28	33	0.2	–	9	2
2002	26,959	69	294	703	8	3	116	38
2003	21,934	18	217	367	4	2	88	29
2004	58,559	54	1063	1167	10	3	186	48
2005	40,442	26	582	829	12	2	66	32
2006	24,035	17	339	636	18	2	127	31
2007	38,018	107	783	800	3	3	56	35
2008	40,281	45	549	890	5	2	63	66
2009	71,971	90	958	1778	20	6	280	124

Source: WCPFC (2011).

Table 8.3 US Pacific port landings valuations for the eight species reported by the WCPFC adjusted for inflation to USD $2010
Pacific landing prices in 2010 US$ per KG

Year	Albacore tuna	Yellowfin tuna	Bigeye tuna	Skipjack tuna	Striped marlin	Black marlin	Blue marlin	Swordfish
2000	2.66	5.45	9.96	2.17	–	–	–	5.72
2001	2.53	5.49	9.72	3.20	–	–	–	5.10
2002	1.88	4.71	7.39	2.50	4.42	3.19	3.21	4.75
2003	1.84	5.50	8.42	1.91	2.27	2.97	2.33	4.58
2004	2.29	4.70	7.69	2.12	3.79	3.85	3.26	5.07
2005	2.71	5.09	8.38	1.71	3.20	2.87	2.55	5.99
2006	2.13	6.23	8.30	3.13	2.64	2.62	2.44	5.15
2007	2.07	5.08	7.80	2.60	4.25	3.04	2.92	5.35
2008	2.75	6.04	8.55	3.00	2.36	1.74	2.55	4.39
2009	2.39	5.75	8.46	3.25	3.40	6.10	2.67	4.48

Source: NOAA Fisheries (2014).

purse seine fishery based on the number of fishing days (WCPFC, 2011). While the value of catch per unit effort (CPUE) by longline vessels in the area was fairly constant throughout the 2000s, with an average of USD $280 for each 100 longline hooks fished, the purse seine fishery saw a dramatic increase in the value of the CPUE with a low of USD $44,738/fishing day in 2004 and a high of USD $182,459/day in 2005 based on landings and effort records from the WCPFC (2011) and valuations from NOAA Fisheries (2014).

The landed value of the PIPA catch is highly skewed, with yellowfin and bigeye tuna accounting for over 93% of the landed value of the longline catch in the region between 2000 and 2009. The value of the purse seine catch, which is dominated by skipjack, is between 4 and 10 times that of the total longline catch in the region (NOAA Fisheries, 2014; WCPFC, 2011).

The direct effect of the pending closure of the PIPA region to the commercial fishing industry is difficult to estimate precisely, but the mobility of both fishes and fishers, and the access that both enjoy to displace any prior PIPA effort to other management units (including around the perimeter of PIPA), likely mean that the PIPA closure will have minimal economic impact on the blue water fleet. The question as to the direct economic impact to the I–Kiribati people is an unknown, as the area to be closed is just 11.3% of the island nation of Kiribati's 3.5 million km^2 EEZ and the loss of funds for licenses to fish in PIPA may be compensated for by increases in the market price of licenses to fish in other areas of the EEZ. Kiribati currently licenses more than 400 foreign flagged vessels (Uwate and Onorio, 2011) to fish within the EEZ, though no data are provided as to the spatial makeup of that fleet other than the number of hooks fished or vessel days at sea reported in each management area cell by the WCPFC.

As far as the longline fishery is concerned, a PIPA closure will likely have a very small effect, if any, on the overall effort in the WCPFC management area as longline fishing within the management unit containing PIPA accounts for on average less than 0.5% of total regional effort as measured by number of hooks fished, and on average less than 0.8% of longline-landed value for the WCPFC region. The proportionate purse seine effort is similar to that of the longline fishery with PIPA accounting for less than 0.25% of the average effort between 2000 and 2007, though a drastic increase in reported vessel days in 2009 resulted in 1.17% of all purse seine fishing effort being directed in PIPA. This is also shown in the landed value of purse seine catch: PIPA accounts for less than 0.6% of average value between 2000 and 2007, but jumps to 3.1% of WCPFC purse seine value in 2009 where 1153 vessel days were reported in just this management unit. As mentioned above, under the Conservation Contract signed in 2014 between Kiribati and the PIPA Conservation Trust, the economic impact of full closure requires more attention and a careful analysis that can only be truly assessed post-closure; these impacts will be studied by the newly formed Tuna Working Group over a 5-year post-closure period.

6. IMPACTS OF THE MPA
6.1. Social impacts

The Phoenix Islands have no permanent inhabitants. The one currently inhabited atoll, Kanton, has a non-permanent population of approximately 40 people comprising government employees including a policeman, a nurse, two primary teachers, a meteorological officer and their families engaged in protection and management of Kiribati interests in the region. As such, the social impacts of PIPA establishment for Kanton local community residents are functionally negligible. On the other hand, the declaration of PIPA has been a source of national pride for all-Kiribati citizens. Great strides have been made to ensure knowledge of PIPA by Kiribati citizens in the 'neighbouring' archipelagos in the Gilbert and Line Islands, where the entire ~110,000-person nation resides (mostly on Tarawa in the Gilbert Islands group).

News of PIPA is now regularly broadcast on the government-owned all-Kiribati radio station, Kiribati Broadcasting and Publications Authority (AM, 1440), and national pride is apparent. Signs celebrating PIPA are displayed in Kiribati International Airports on both Tarawa and Kiritimati Island, and songs, including 'PIPA You Are My Gift To Humanity' have been written and are regularly sung to celebrate major PIPA milestones and events. Outreach initiatives also include messaging in schools that extend beyond PIPA issues alone, with a focus on encouraging a conservation mindset, building in-country capacity and expertise in areas relevant to ocean conservation and research and promoting knowledge of all-of-Kiribati geography, since very few Kiribati citizens have ever been to the Phoenix Islands. The I-Kiribati also are invoking the local word 'okai' with regards to PIPA, which means a traditional storehouse where reserved foods and treasures are kept for future use—especially in times of prolonged draughts and bad times. Considering PIPA as an okai for potential food security as well as a bank of Central Pacific biodiversity has been an important part of the outreach programme to enable Kiribati residents to think about the multiple local and global benefits of ocean stewardship. Coupled with growing social media outreach and press coverage, PIPA is having an increasing impact on both domestic and international populations.

6.2. Ecological impacts of protection

Scientific research in the Phoenix Islands has historically been limited, but all indications pointed towards the island group's diversity and abundance of marine life. The NEAq's 2000 expedition launched a decade of systematic and repeated exploration, but prior to this there are only a few known examples of science within the Phoenix Islands. These included a visit by the research vessel Bushnell in 1939 that resulted in a taxonomic collection of fishes (Shultz, 1943), studies on seabirds (Clapp, 1964), turtles (Balazs, 1982) and the corals from McKean island in the early 1970s (Dana, 1979). It was not until 1972–1973 that detailed marine surveys were conducted. A comprehensive study of Kanton atoll was undertaken (Smith and Henderson, 1978), including work on lagoon circulation and biogeochemistry (Smith and Jokiel, 1978), coral taxonomy and biogeography (Maragos and Jokiel, 1978) and lagoon and leeward reef coral distributions and assemblages (Jokiel and Maragos, 1978). Since 2000, regular terrestrial and marine expeditions have documented the biodiversity of algae, invertebrates, fishes and corals, as well as their changes in abundance over time (including over two known episodes of high-temperature-induced coral bleaching). Since the implementation of no-take reef areas in 2008, an increase in shark populations have been noted, the most direct impact of protection seen to date (Mangubhai et al., 2012; Stone et al., 2009).

Other major impacts of protection include the ability of reefs to rapidly recover post-bleaching in the absence of local anthropogenic impacts, a phenomenon attributed in part to the persistent abundance of herbivorous reef fishes (chiefly Scaridae and Acanthuridae). Following the 2002–2003 bleaching event, PIPA reefs experienced dramatic coral mortality down to 12.1% overall (Obura and Mangubhai, 2011) from 58.1% reported in 2002 (Obura and Stone, 2002). Yet, expeditions in 2009 and 2012 reported resilience and recovery following bleaching events in 2002–2003 and 2010 (Mangubhai et al., 2012; Stone et al., 2009). Though coral population recovery followed the same timeline as legal protection of PIPA reefs, it is unlikely that reef recovery was a direct result of PIPA protection, since coral habitat degradation was not under threat from local populations during this time period. Instead, the synchrony suggests that the protection of PIPA was timely and necessary to ensure a natural reef successional trajectory and recovery, as it is well documented on other reefs that local threats can impair recovery processes (e.g. Burke et al., 2011). Fragile reef ecosystems are in no way guaranteed to recover from major disturbances, and recovery potential likely goes down with repeated disturbance. In addition, the removal of terrestrial invasive alien species from three islands has been followed by the recovery of biota there. This has included vegetation recovery particularly on Rawaki post rabbit (*Oryctolagus cuniculus*). Thus, the creation of PIPA offers the best risk-management potential for an already-recovering ecosystem, which cannot be protected from the boundary-free threats of global change.

6.3. Economic impacts of protection

The single-most significant economic issue with respect to PIPA is any potential loss in fishing revenues arising from full closure to commercial pelagic fishing, which, as already discussed above, will be effective from 1 January 2015. The current debate on PIPA has been somewhat polarized by fears of the economic losses that may occur as a result of the closure of PIPA from commercial fishing and all forms of extractive activities. Some Kiribati observers are arguing the case for compensation. Less is understood or being discussed about the potential 'win–win' outcomes and the economic benefits of PIPA for Kiribati, and the global community that will invariably more than offset the losses in fishing revenues. While there is as yet no established or generally agreed figure on the exact amount of fishing revenues derived from PIPA per annum, crude estimates indicate that this could range from USD $1 million to USD $4 million in any given year, depending on seasonality and climatic conditions during the period in which the harvest takes place.

The potential socio-economic benefits that PIPA can bring to Kiribati are an often-overlooked prospective counterbalance to the arguments about potential revenue losses. There are many potential economic benefits of PIPA, including provision of ecosystem services through intact marine food webs and habitats, potential replenishment of fisheries stocks and potential revenue generated by future tourism. By closing off PIPA from all forms of extractive activities including commercial fishing, these protected grounds and related habitats may indirectly benefit neighbouring fishable areas of the Kiribati EEZ. Once fully no-take, the PIPA region may facilitate recruitment and replenishment of tuna stocks, with possible spillover benefits to adjacent regions. However, the potential costs and benefits of closing PIPA to all commercial fishing have yet to be determined; they will be examined over a 5-year period following the 1 January 2015 full closure, and in any event, will likely emerge over a period of decades.

PIPA also has ecotourism potential that could benefit the Kiribati economy, though reservations have been expressed about the long-term economic viability and sustainability of tourism development in the Phoenix Islands given its isolation and remoteness. A growing niche tourism market might generate enough revenue to support the necessary infrastructure, as there are several potential niche markets. For example, PIPA is believed to be the final resting place of the famous American aviators Amelia Earhart and Fred Noonan lost in the 1930s, and PIPA has historical interests related to WWII, the NASA (Apollo) satellite tracking station, shipping and guano days and as historical whaling grounds. The 2010 UNESCO inscription of PIPA as the largest and deepest World Heritage site may also help to foster a niche tourism market. The presence of a World War II era airstrip on Kanton that remains in fundamentally sound condition makes such an opportunity a realistic future goal for PIPA. Sadly, atolls in such an untouched state are likely to become increasingly scarce elsewhere notwithstanding the growing efforts to protect them, simply because of the unrelenting challenges of population growth and coastal development. While such tourism would have some inherent negative environmental impacts in PIPA, there would also be important benefits from having both additional interests in Kiribati that economically benefit and that would help with the challenge of monitoring and enforcement against illegal fishing and other activities in this remote zone.

It is important to note that, without the PIPA project, most of the islands in the Phoenix group would likely have served as only idle assets. The creation of PIPA therefore represents the government of Kiribati's interest in exploring new development models for Kiribati's natural resources that are less dependent on resource exploitation and consumption.

7. POST-MPA ESTABLISHMENT: REMAINING CHALLENGES

7.1. Fisheries

PIPA was established in 2008 and immediately the shallow reefs were protected as 'no take' around 7 of the 8 islands, equalling about 3.1% of the whole MPA. At this time, fisheries extraction on the shallow reef environments ceased, though some illegal activity still occurred. For example, a bunker vessel from Singapore, *Hai Soon* 28, was caught and fined $4.73 million USD for conducting bunkering 20 miles south of Nikumaroro without a license. Some illegal shark finning has also likely occurred, as indirectly suggested by smaller-than-expected shark abundance and biomass post-reef closure to fishing. Overall, however, the benefits of reef closure to fishing has already been observed on monitoring trips conducted by the NEAq after the 2008 closure; a seeming resurgence of shark populations (especially small juveniles) was noted on some of the islands, especially at Kanton, which had previously been heavily fished for shark fins (Stone et al., 2009). Continued evidence for resurgence of juvenile sharks continued in 2012, but the marked overabundance of juveniles compared to adults could be an indication of an illegal fishing event between 2009 and 2012 (Mangubhai et al., 2012).

In the surrounding blue water environment, commercial fishing of tuna and large pelagics has remained legal since MPA inception in 2008 and will remain so until 31 December 2014 (when the whole MPA becomes no take). From 2008 to 2014, PIPA waters were fished under the Nauru Agreement (detailed above) and in recent times, license days were sold for Kiribati waters (without exclusion of PIPA). Hundreds of vessels have fished legally inside PIPA under the Nauru Agreement, and no impact of protection for pelagic stocks will be relevant until full closure begins in 2015.

7.2. Enforcement

Effective enforcement is one of the key features of successful MPAs (Edgar et al., 2014). Unfortunately, it remains one of the hardest issues to tackle, especially in remote MPA regions. Isolation notwithstanding, there are several enforcement measures already in place, and more are planned. Kiribati requires 100% observer coverage on all purse seine vessels legally within Kiribati waters, and observers have responsibility of ensuring legality and enforcement of in-country rules, including the PIPA closures and the reporting of illegal activities onboard. Mandatory 100% observer coverage on DWFN vessels is required for all eight signatory nations to the Third Arrangement to the Nauru Agreement, together with seven other Pacific Island states. For PIPA enforcement, this means that all legal vessels will have observer coverage and be able to be tracked via video monitoring systems (VMS) in real time; these legal fishing boats can also act as an additional surveillance tool to detect illegal/non-licensed vessels. In order to assist in the detection of illegal activities, visitors (tourists and researchers) and residents are also required (under their permit for visitation) to assist with monitoring, control and surveillance and to report any suspicious activities.

The government of Kiribati already undertakes annual patrol visits within PIPA boundaries, and currently Australia, New Zealand and France provide aerial surveillance of the Phoenix Islands area. Kiribati also has a 'ship-riders' agreement with the United States whereby Kiribati enforcement personnel are present on U.S. Navy and Coast Guard vessels operating in the area and the United States fleet participates with Kiribati in investigating, reporting and arresting ships for illegal activities. The aforementioned bunker ship from Singapore was escorted to the Port of Betio in Tarawa by the U.S. Coast Guard under this agreement. Additionally, Kiribati has signed a Declaration on Deep-Sea Bottom Trawling to Protect Biodiversity in the High Seas (Nadi Communiqué, Pacific Islands Forum, October 2006), which commits the members of the Pacific Islands Forum to urgently take actions consistent with international law to prevent destructive fishing practices on seamounts in the Western Tropical Pacific Islands Area.

Additional plans for enforcement are also underway. For example, the recently signed conservation contract between Kiribati and the PIPA Conservation Trust (April 2014) states that the Ministry of Fisheries and Marine Resources Development will submit a request to the Pacific Islands Forum Fisheries Agency to create a virtual electronic perimeter using geofence technology that conforms to the boundaries of the PIPA set forth in the PIPA Regulations 2008. Vessel monitoring and reporting of any vessel movements within PIPA are expected to continue and will be enhanced by this virtual fencing technology. Furthermore, a new Kiribati Fisheries and Wildlife Conservation Unit is planned to be stationed on Kanton Island, which is hoped to significantly contribute to surveillance and enforcement of management rules. In May 2013, a PIPA Community Agreement was signed with Kanton Island caretaker residents whereby suspicious or illegal activities must now be reported. In this agreement, all members of the Kanton community indicated their commitment to safeguarding and protecting the universal values of Kanton Island. Finally, new technologies are being explored to further aid in surveillance and enforcement of PIPA boundaries. These include satellite surveillance and unmanned drone technology, and possibilities for use are currently being scoped to determine costs, feasibility and effectiveness.

7.3. Invasive alien species

Prior to first discovery, the Phoenix Islands had been isolated for millennia, which enabled seabirds and other fauna to live and nest safely in the absence of invasive pests and exploit the food rich seas around the islands. Biological invasions likely began soon after the discovery of the Phoenix Islands by early Pacific seafarers and continued through the guano mining years and subsequent settlement and development of the islands (detailed above). The deliberate and accidental introductions of invasive species during all of these periods have been significant in the Phoenix Islands, and the faunal invaders include Pacific and Asian rats, rabbits, cats, ants, pigs and dogs; floral invaders include coconuts, lantana and other weeds. Some of the negative impacts of these biological invasions include the elimination of native sea-birds through the destruction of eggs and young, and the elimination of native plants via competition for limited resources. Mammalian pests have been the most damaging invasives in PIPA, threatening the bird populations and modifying the entire natural island ecosystem. Rabbits, cats and different rat species have had notable impacts over time, with some birds (e.g. Phoe-nix petrels and White-Throated storm-petrels) and other biota being reduced to critically low numbers (Pierce, 2013). The most sensitive fauna species survived on only the one island (Rawaki) that escaped rat infestation (Pierce, 2013).

Since 2006 an eradication programme has been implemented to remove invasive mammals from the atolls in PIPA, which has resulted in the gradual restoration of the islands, their habitats and fauna populations (Pierce, 2013). Most recently, in May 2013, a terrestrial team visited five PIPA islands to assess invasive species status, biosecurity issues and biota responses to previous pest eradications in 2008 and 2011 and plan for further work (Pierce, 2013; Pierce and Kerr, 2013). Three islands (Rawaki, Birnie and McKean) are now free of invasive species (Pierce and Teroroko 2011), but the rat eradication efforts on Enderbury conducted in 2011 were unsuccessful. In March 2013, a biosecurity training team visited Kiritimati to work with MELAD staff and community in improving biosecurity and quarantine measures for Kiritimati and Kiribati generally with a big emphasis on protecting the PIPA from further invasions (Nagle et al., 2013). It is hoped that through these eradication and biosecurity programmes, threats to the terrestrial ecology of the islands will be reversed, minimized, mitigated and, most importantly, prevented.

7.4. Impacts of vessel groundings

The Phoenix Islands have had numerous vessel groundings over the years. One of the earliest recorded groundings was the whale ship Canton on Kanton, its namesake, in 1854. The most famous is of the SS Norwich City, which ran around on Nikumaroro in 1929 and is still prominently visible today. In addition to coral damage during grounding and break-up, it is now also becoming clear that rusting shipwrecks add iron to the surrounding seawater environment. In iron-limited regions such as the Central Pacific, iron addition can result in a phase shift from coral-dominated reefs to reefs dominated by iron-enriched microbial mats and turf algae (Kelly et al., 2012). Oil and other chemical spills remain a persistent threat with ship traffic, though thus far the Phoenix Islands have been spared from any catastrophic oil spills. However, even small and non-commercial vessel groundings have potential negative impacts via nutrient pollution and other forms of plastic pollutants. Shipwrecks also introduce terrestrial dangers, as evidenced by a recent fishing vessel wreck at McKean Island in c. 2001 that resulted in the introduction of the Asian rat (*Rattus tanezumi*), which in turn exterminated many bird species from the atoll. Elimination of commercial fishing may eliminate some risk of vessel grounding due to decreased vessel traffic, though increased tourism traffic may pose even higher threats as they will likely spend more time near islands (instead of in blue water habitats).

8. MEASURES OF SUCCESS/FAILURE

The PIPA has already celebrated many conservation and ecological milestones in the past decade. Most laudably, PIPA is the largest and deepest UNESCO World Heritage site on earth, recognized for its outstanding natural value. As ocean resources are increasingly being called upon to support a growing global population, PIPA's upcoming closure sets an important international precedent. Scheduled for 1 January 2015, full closure to commercial fishing will bring the opportunity to assess the efficacy of this type of management intervention in pelagic fisheries stock improvement, one of the first such experiments in conservation history. One other potential measure of success is that PIPA is no longer the largest marine MPA in the Pacific (as it was at the time of its creation): since then, other MPAs have been announced that have eclipsed PIPA in size, which in essence denotes a 'trend' of ocean protection.

Recognizing and reconciling the many challenges of creating and maintaining large MPAs has become the agenda for a newly formed entity called 'Big Ocean', which is a network of managers and partners of existing and proposed large-scale marine managed areas. The success or failures of PIPA, and of large MPAs in general, will be shared through Big Ocean as part of their objective to use lessons learned to advance and improve the effectiveness of large MPAs (Wagner, 2013). Globally, however, success will likely be measurable only with time: do stocks of commercially important fishes within PIPA improve? Will PIPA be effectively enforced and maintained? Will the sustainable financing model be successful in the long term? Will PIPA be heralded as a model of marine conservation? And finally, even with all of the protections afforded to PIPA to mitigate habitat loss and overfishing, will the ecosystems succumb to the inevitable impacts of climate change, and even if so, will PIPA ecosystems have at least fared better than their unprotected neighbours as a result of its MPA status? These and many other questions remain to be answered in the coming decades.

9. OVERVIEW AND LOOKING AHEAD

Though there are still objectives to be met, PIPA is well positioned to be a global model for large MPA development and maintenance. Whether attributable to legal protection or simply a product of their extreme remoteness, the shallow reef and terrestrial ecosystems of PIPA have already shown benefits associated with protection. PIPA offers a novel conservation mechanism for protection in the waters of economically challenged but ecologically rich nations like Kiribati. This can serve as a model for island states globally, and indeed, it already has. The pending closure of PIPA, and the rigour of the subsequent assessments of that closure, will be critical for determining success of the MPA as a refuge for open-ocean pelagic and deep-sea marine life. Of the remaining challenges, some are capable of being addressed (e.g. enforcement) and others cannot (e.g. climate change). Though these challenges still remain, the act of creating very large MPAs is still justified under the precautionary principle, and these MPAs (PIPA and others) will together serve as a beacon of hope in marine conservation initiatives now, and in the years to come.

REFERENCES

Allen, G.R., Bailey, S., 2002a. *Chrysiptera albata*, a new species of damselfish (Pomacentridae) from the Phoenix Islands, central Pacific Ocean. Aqua J. Ichthyol. Aquat. Biol. 6 (1), 39–43.

Allen, G., Bailey, S., 2002b. Reef fishes of the Phoenix Islands, Central Pacific Ocean, Results of survey in the Phoenix Islands June/July 2002.

Allen, G., Bailey, S., 2011. Reef fishes of the Phoenix Islands, Central Pacific Ocean. Atoll Res. Bull. 589, 83–97.

Alling, A., Doherty, O., Heather, L., Feldman, L., Dustan, P., 2007. Catastrophic coral mortality in the remote central Pacific Ocean: Kiribati, Phoenix islands. Atoll Res. Bull. 551, 1–19.

Balazs, G.H., 1982. Status of Sea Turtles in the Central Pacific Ocean. Pp 243–251. In: Bjorndal, K.A. (Ed.), Biology and Conservation of Sea Turtles. Smithsonian Press, Washington, DC, p. 583.

Baums, I.B., Boulay, J.N., Polato, N.R., Hellberg, M.E., 2012. No gene flow across the Eastern Pacific Barrier in the reef-building coral *Porites lobata*. Mol. Ecol. 21, 5418–5433. http://dx.doi.org/10.1111/j.1365-294X.2012.05733.x.

Brewer, T.D., Cinner, J.E., Green, A., Pressey, R.L., 2013. Effects of human population density and proximity to markets on coral reef fishes vulnerable to extinction by fishing. Conserv. Biol. 27, 443–452.

Bruno, J.F., Selig, E.R., 2007. Regional decline of coral cover in the Indo-Pacific: timing, extent, and subregional comparisons. PLoS One 2 (8), e711.

Burke, L., Reytar, K., Spalding, M., Perry, A., 2011. Reefs at Risk Revisited. World Resources Institute, Washington, DC.http://www.wri.org/publication/reefs-risk-revisited.

Clapp, R., 1964. Smithsonian Biological Survey. 01-18-14-90-81.

Dana, T.F., 1979. Species-numbers relationships in an assemblage of reef-building corals: McKean Island, Phoenix Islands. Atoll Res. Bull. 228, 1–42.

De Koning, F., Aguiñaga, M., Bravo, M., Chiu, M., Lascano, M., Lozada, T., Suarez, L., 2011. Bridging the gap between forest conservation and poverty alleviation: the Ecuadorian Socio Bosque program. Environ. Sci. Pol. 14 (5), 531–542.

Doney, S.C., Ruckelshaus, M., Duffy, J.E., Barry, J.P., Chan, F., English, C.A., Galindo, H.M., Grebmeier, J.M., Hallowed, A.B., Knowlton, N., Polovina, J., Rabalais, N.N., Sydeman, W.J., Talley, L.D., 2012. Climate change impacts on marine ecosystems. Annu. Rev. Mar. Sci. 4, 11–37.

Dulvy, N.K., Baum, J.K., Clarke, S., Compagno, L.J.V., Cortes, E., Domingo, A., Fordham, S., Fowler, S., Francis, M.P., Gibson, C., Martinez, J., Musick, J.A., Soldo, A., Stevens, J.D., Valenti, S., 2008. You can swim but you can't hide: the global status and conservation of oceanic pelagic sharks and rays. Aquat. Conserv. Mar. Freshwat. Ecosyst. 10, 1009–1033.

Edgar, G.J., et al., 2014. Global conservation outcomes depend on marine protected areas with five key features. Nature 506, 216–220.

Fonteneau, A., Pallares, P., Pianet, R., 2000. A worldwide review of purse seine fisheries on FADs. In: Pêche thonière et dispositifs de concentration de poissons, Caribbean-Martinique, pp. 15–19.

Hutchinson, G.E., 1950. Survey of existing knowledge of biogeochemistry. 3. The biogeochemistry of vertebrate excretion. Bull. Am. Mus. nat. Hist. 96, 554.

International Monetary Fund Kiribati, 2011. Article Iv Consultation-Staff Report, Informational Annexes, Debt Sustainability Analysis, Public Information Notice on the Executive Board Discussion, and Statement by the Executive Director for Kiribati. International Monetary Fund, Washington, DC.

IUCN, 2008. Standards and Petitions Working Group. Guidelines for Using the IUCN Red List Categories and Criteria Version 7.0. Downloaded 18 November 2013 from, http://intranet.iucn.org/webfiles/doc/SSC/RedList/RedListGuidelines.pdf.

Jokiel, P., Maragos, J., 1978. Reef corals of Canton atoll: II. Local distribution. Atoll Res. Bull. 221, 71–97.

Jupiter, S.D., Mangubhai, S., Kingsford, R.T., 2014. Conservation of biodiversity in the Pacific Islands of Oceania: challenges and opportunities. Pacific Conservation Biology 20 (2), 206.

Kelly, L.W., Barott, K.L., Dinsdale, E., Friedlander, A.M., Nosrat, B., Obura, D., et al., 2012. Black reefs: iron-induced phase shifts on coral reefs. ISME J. 6, 638–649.

Lambin, E.F., Meyfroidt, P., 2011. Global land use change, economic globalization, and the looming land scarcity. Proc. Natl. Acad. Sci. 108 (9), 3465–3472.

Lee, R., 2011. The outlook for population growth. Science 333 (6042), 569–573.

Leenhardt, P., Cazalet, B., Salvat, B., Claudet, J., Feral, F., 2013. The rise of large-scale marine protected areas: conservation or geopolitics? Ocean Coast. Manage. 85, 112–118.

Mangubhai, S., Rotjan, R.D., Obura, D.O., 2012. Phoenix Islands Protected Area Expedition 2012 – Assessment Report. New England Aquarium, 44 pgs.

Mangubhai, S., Strauch, A.M., Obura, D.O., Stone, G., Rotjan, R.D., 2014. Short-term changes of fish assemblages observed in the near-pristine reefs of the Phoenix Islands. Rev. Fish Biol. Fish. 24 (2), 505–518.

Maragos, J., Jokiel, P., 1978. Reef corals of Canton atoll: I. Zoogeography. Atoll Research Bulletin 221, 55–69.

Nagle, Gruber, Pierce, 2013. Draft report on biosecurity training at Kiritimati March 2013. Eco Oceania Report.

NOAA Fisheries, 2014. Annual Landings. Available from: http://www.st.nmfs.noaa.gov/st1/commercial/landings/annual_landings.html.

Obura, D.O., 2011. Coral reef structure and zonation of the Phoenix Islands. Atoll Res. Bull. 589, 63–82.http://hdl.handle.net/10088/16742.

Obura, D., Mangubhai, S., 2011. Coral mortality associated with thermal fluctuations in the Phoenix Islands, 2002–2005. Coral Reefs 30 (3), 607–619.

Obura, D., Stone, G.S., 2002. Phoenix Islands. Summary of Marine and Terrestrial Assessments, Conducted in the Republic of Kiribati, June 5–10, Primal Ocean Project Technical Report: NEAq-03-02.

Obura, D.O., Stone, G., Mangubhai, S., Bailey, S., Yoshinaga, A., Holloway, C., Barrel, R., 2011a. Baseline marine biological surveys of the Phoenix Islands, July 2000. Atoll Res. Bull. 589, 1–62. http://hdl.handle.net/10088/16741.

Obura, D.O., Mangubhai, S., Yoshinaga, A., 2011b. Sea turtles of the Phoenix Islands, 2000–2002. Atoll Res. Bull. 589, 119–124. http://hdl.handle.net/10088/16744.

Pierce, R., 2013. Opportunities to restore the terrestrial ecosystems in the Phoenix Islands, Kiribati, by removing invasive alien species. EcoOceania Pty Ltd report for PIPA December 2013.

Pierce, R.J., Kerr, V., 2013. Report on a biota survey of the Phoenix Islands atolls, Kiribati, May 2013. Eco Oceania Pty Ltd. Report for Government of Kiribati and CI-CEPF.

Pierce, R.J., Etei, T., Kerr, V., Saul, E., Teatata, A., Thorsen, M., Wragg, G., 2006. Phoenix Islands conservation survey and assessment of restoration feasibility: Kiribati. Report prepared for: Pacific Invasives Initiative, CEPF and Conservation International, Samoa.

Pierce, R., Teroroko, T., 2011. Enhancing biosecurity at the Phoenix Islands Protected Area (PIPA), Kiribati. In: Veitch, C.R., Clout, M.N., Towns, D.R. (Eds.), 2011. Island invasives: eradication and management. IUCN, Gland, Switzerland, pp. 481–486.

Polato, N.R., Concepcion, G.T., Toonen, R.J., Baums, I.B., 2010. Isolation by distance across the Hawaiian Archipelago in the reefbuilding coral Porites lobata. Mol. Ecol. 19, 4661–4677.

Randall, J.E., Allen, G.R., Robertson, D.R., 2003. Myripristis earlei, a new soldierfish (Beryciformes: Holocentridae) from the Marquesas and Phoenix islands. Zool. Stud. 43 (3), 405–410.

Sandin, S.A., Smith, J.E., DeMartini, E.E., Dinsdale, E.A., Donner, S.D., et al., 2008. Baselines and degradation of coral reefs in the Northern Line Islands. PLoS One 3 (2), e1548. http://dx.doi.org/10.1371/journal.pone.0001548.

Shultz, L.P., 1943. Fishes of the Phoenix and Somoan Islands collected in 1939 during the Expedition of the U.S.S. "Bushnell" Smithsonian Institution, United States National Museum, Bulletin vol. 180, 315 pages.

Skaggs, J., 1995. The Great Guano Rush: Entrepreneurs and American Overseas Expansion. St. Martin's Press, New York, NY.

Smith, S.V., Henderson, R.S., 1978. An environmental survey of Canton Atoll Lagoon, 1973. Atoll Res. Bull. 221, 192.

Smith, S.V., Jokiel, P.L., 1978. Water composition and biogeochemical gradients in the Canton Atoll lagoon. Atoll Res. Bull. 221, 17–53.

Smith, T.D., Reeves, R.R., Josephson, E.A., Lund, J.N., 2012. Spatial and seasonal distribution of American whaling and whales in the age of sail. PLoS One 7 (4), e34905. http://dx.doi.org/10.1371/journal.pone.0034905.

Spalding, M., Meliane, I., Milam, A., Fitzgerald, C., Hale, L., 2013. Protecting marine spaces: global targets and changing approaches. Ocean Yearb. 27, 213–248.

Stone, G., Clarke, A., Earle, S., 2000a. Deep ocean frontiers. Mar. Technol. Soc. J. 33 (4), 93.

Stone, G., Obura, D., Bailey, S., Yoshinaga, A., Holloway, C., Barrel, R., Mangubhai, S., 2000b. Marine Biological Journeys of the Phoenix Islands, Summary of Expedition Conducted from June 24–July 15.

Stone, G., Obura, D., Bailey, S., Yoshinaga, A,., Holloway, C., Barrel, R., Mangubhai, S., 2001. Marine Biological Surveys of the Phoenix islands. New England Aquarium report, 107 pp.

Stone, G.S., Obura, D.O., Rotjan, R.D., 2009. Phoenix Islands Protected Area Expedition 2009 – Assessment Report. New England Aquarium 35 pp.

Sumaila, U.R., Dyck, A., Baske, A., 2014. Subsidies to tuna fisheries in the Western Central Pacific Ocean. Mar. Policy 43, 288–294.

Uwate, K.R., Onorio, B., 2011. The Relationship Between the Phoenix Islands and the US Pacific Remote Islands Marine National Monuments (PRIMNM), and Possibilities of Kiribati Vessels Fishing Within the US EEZ and the PRIMNM. NOAA, Honolulu, HI.

Vargas-Ángel, B., Looney, E.E., Vetter, O.J., Coccagna, E.F., 2011. Severe, Widespread El Niño–Associated Coral Bleaching in the US Phoenix Islands. Bulletin of Marine Science 87 (3), 623–638.

Wagner, D. (Ed.), 2013. Big Ocean: A shared research agenda for large-scale marine protected areas. Big Ocean Network Report, 21 pp.

WCPFC, 2011. Aggregated Catch/Effort Data. Western and Central Pacific Fisheries Commission, Federated States of Micronesia.

CHAPTER NINE

Diverging Strategies to Planning an Ecologically Coherent Network of MPAs in the North Sea: The Roles of Advocacy, Evidence and Pragmatism in the Face of Uncertainty

Alex J. Caveen[*,†,1], **Clare Fitzsimmons**[†], **Margherita Pieraccini**[‡], **Euan Dunn**[§], **Christopher J. Sweeting**[†], **Magnus L. Johnson**[¶], **Helen Bloomfield**[‖], **Estelle V. Jones**[†], **Paula Lightfoot**[†], **Tim S. Gray**[#], **Selina M. Stead**[†], **Nicholas V.C. Polunin**[†]

[*]Seafish Industry Authority, Grimsby, United Kingdom
[†]School of Marine Science and Technology, Newcastle University, Newcastle upon Tyne, United Kingdom
[‡]Law School, University of Bristol, Bristol, United Kingdom
[§]Royal Society for the Protection of Birds, Bedfordshire, United Kingdom
[¶]Centre for Environmental and Marine Sciences, University of Hull, Scarborough, United Kingdom
[‖]School of Environmental Sciences, University of Liverpool, Liverpool, United Kingdom
[#]School of Geography Politics and Sociology, Newcastle University, Newcastle upon Tyne, United Kingdom
[1]Corresponding author: e-mail address: alex.caveen@seafish.co.uk

Contents

Advances in Marine Biology, Volume 69
ISSN 0065-2881
http://dx.doi.org/10.1016/B978-0-12-800214-8.00009-8

325

Abstract

The North Sea is one of the most economically important seas in the world due to productive fisheries, extensive oil and gas fields, busy shipping routes, marine renewable energy development and recreational activity. Unsurprisingly, therefore, the use of marine protected areas (here defined widely to include fisheries closed areas and no-take marine reserves) in its management has generated considerable controversy—particularly with regards to the design of a regional *ecologically coherent* MPA network to meet international obligations.

Drawing on three MPA processes currently occurring in the UK North Sea, we examine the real-world problems that make the designation of MPA networks challenging. The political problems include: disagreement among (and within) sectors over policy objectives and priorities, common access to fisheries resources at the EU level increasing the scale at which decisions have to be made and lack of an integrated strategy for implementing protected areas in the North Sea. The scientific problems include the patchy knowledge of benthic assemblages, limited knowledge of fishing gear–habitat interactions, and the increased risk of unforeseen externalities if human activity (predominantly fishing) is displaced from newly protected sites. Diverging stakeholder attitudes to these problems means that there is no consensus on what ecological coherence actually means.

Ultimately, we caution against 'quick-fix' solutions that are based on advocacy and targets, as they create confusion and undermine trust in the planning process. We argue for a more pragmatic approach to marine protection that embraces the complexity of the social and political arena in which decisions are made.

Keywords: North Sea, Planning, Network, Natura 2000, MCZ, Consultation, Habitat

1. INTRODUCTION

The North Sea is used by a number of different countries and sectors often with widely divergent views on how human activities should be managed. Additionally, the North Sea ecosystem is highly variable both spatially and temporally, and despite being comparatively well studied there remain many gaps in our understanding of how it functions (Galparsoro et al., 2014). These characteristics make marine planning and management challenging. Despite these challenges, an ecologically coherent MPA network is currently being developed (OSPAR, 2012).

Broader policy issues make the planning of MPAs highly contested. A major issue is that MPAs are often seen as part of an ecological worldview that has generally promoted them as a 'technical fix' for solving the problem of overfishing (see Degnbol et al., 2006). It has long been recognised, however, that political disputes underpin many technical arguments (Mazur, 1981). In reality, debates on natural resource management are often a mixture of extreme scientific uncertainty and conflicting value priorities (see Salomon et al., 2011) making politics inevitable. Such problems have often been characterised as being 'wicked' (see Balint et al., 2011 for overview) in that they *are complex, persistent or reoccurring and hard to fix because they are linked to broader social, economic and policy issues*' (Khan and Neis, 2010). Particularly important to MPA planning are wider value-laden debates on food security and protection of biodiversity (Fischer et al., 2014; Godfray et al., 2010), as well as environmental justice conflicts between the rights of local resource users and wider society (Jones, 2009).

In this technocratic age, some scholars have argued that the environmental movement has persistently sought to define the underlying causes of environmental problems as technical deficiencies rather than value conflicts (Shellenberger and Nordhaus, 2004). This is also generally true for debates on MPAs (Caveen et al., 2013). Whilst the use of MPAs (or closed areas) in marine management has occurred since the early twentieth century, the debate on MPAs has been reframed by marine ecologists since the early 2000s; it is no longer ecologically desirable to have MPAs but ecologically necessary (see *Ecological Applications* 2003; 13[1]). This high-level belief has been reflected in a focus during the 2000s on the scientific basis of designing *ecologically coherent* MPA networks to protect marine biodiversity (see Olsen et al., 2013 for latest thinking). This worldview promoting MPAs as a technical fix has influenced high-level policy, as is evident from planning documents (e.g. Natural England's Ecological Network Guidance), and legal mandates (e.g. Marine Strategy Framework Directive (MSFD) and OSPAR) (see Section 3).

It is often assumed that there will be net benefits to society from the designation of MPA networks through enhanced ecosystem goods and services (Hussain et al., 2010; see Chapter 7), though there remains considerable uncertainty in the evidence base, and how benefits and costs are distributed among stakeholders (Potts et al., 2014). Calls for MPA networks have however been largely challenged by the fishing industry (Devillers et al., 2014), particularly the mobile gear sector. Drawing on three MPA planning processes predominantly occurring in the UK North Sea (Section 4), we look at the real-world challenges facing marine planners in developing an ecologically coherent MPA network (Section 5).

1.1. Historical insights from the North Sea

The North Sea (Figure 9.1) has a long history of exploitation dating back 13,000 years to Mesolithic coastal societies that established themselves on the northern edges of Doggerland and tracked the slowly flooding coastline (now submerged under the shallow southern North Sea) (Coles, 2000). These societies largely undertook half-tide fisheries, harvested seal, shellfish and coastal fish (Fleming, 2008), activity that persisted in coastal regions despite the advent of farming during the Neolithic.

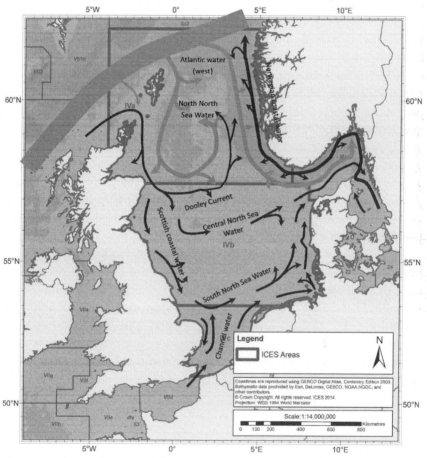

Figure 9.1 The North Sea as defined by ICES rectangles IVa, b, c (bold outline) also showing direction of main currents. Arrow width represent the magnitude of volume transport. Red arrows (grey in the print version) indicate the flow of Atlantic water and black arrows water of other types. *Adapted from Turrell (1992) and Paramor et al. (2009).*

The settlement of the Vikings in Northern Scotland (ca. 800 AD) coincided with the appearance of cod, ling and haddock bones in coastal middens and the ability to prosecute sea fisheries probably with hook and line (Barrett et al., 2000). Viking maritime knowledge coupled with the decline of freshwater fish stocks across eleventh century Europe, led to a shift from consumption of freshwater to marine fish (Barrett et al., 2004; Roberts, 2007) and the development of markets between coastal and landlocked settlements for dried cod and herring (Barrett et al., 1999, 2004).

Towed fishing gears appeared ca. 1370. The wondyrechaun, a predecessor to the beam trawl, caused controversy for damaging 'the spat of oysters, mussels and other fish upon which the great fish feed', and was prohibited in 1377 (Davis, 1923) making it among the first of the North Sea's many subsequent fishing gear exclusions. The gear remained banned until the early seventeenth century (Johnston, 1987). Trawling was separately in operation on continental coastlines by the fifteenth century and was banned in Flanders (northwest Belgian, northern France and the Netherlands) in 1499, a ban that was renewed repeatedly over the following century (Roberts, 2007).

Sail powered sea fishing with a light beam trawl emerged in the late seventeenth century from ports on the English channel, ultimately reaching the southern North Sea by the early nineteenth century and from there to the rest of the North Sea, reaching peak fleet size in the 1880s (Engelhard, 2009). From the 1820s, the development of a railway network opened up large markets in inland cities for cheap supplies of fish, further stimulating the rapid expansion of trawling (Thurstan et al., 2013). The first commercially viable steam trawler is thought to originate from a steam tug conversion in North Shields, North East England (Anon, 1908). The experiment was such a success that by 1878, 53 registered steam trawlers operated out of North Shields and, from 1881 onward, purpose built steam trawlers appeared in Hull and Grimsby.

Rapid fishing power expansion reopened complaints that beam trawling was causing the destruction of seabed habitats and unnecessary destruction of young fish (Thurstan et al., 2013). Recognising the rapid changes occurring in the fishing industry, in 1885, the Scottish Fishery Board suspended fishing in coastal waters *for the purpose of conducting scientific experiments and collecting fisheries statistics'* (Fulton, 1895). This was perhaps the first example of using an MPA as a scientific tool, a concept which is only now re-emerging with Marine Conservation Zone (MCZ) References Areas. Ultimately, no effect of closure was detected (Fulton, 1895), although the associated discussion would be familiar to current practitioners and included suitability of controls (inshore vs. adjacent offshore design), high noise-to-signal ratios and the necessity of extended sampling campaigns, external larval supply and adult movement, and wider fishing mortality particularly on spawning grounds (Beare et al., 2013; Bloomfield et al., 2012).

At the start of the twentieth century, the number of steam trawlers at European North Sea ports was estimated to be around 1700 vessels (Board of Agriculture and Fisheries, 1908). Steam trawlers capable of operating over a larger range, in deeper water and at greater speed led to the introduction of the otter trawl, and the emergence of distant water trawl fisheries before the end of the nineteenth century. Technical innovation also saw conflicts develop among towed (steam) and static gear (largely sail/oar) sectors. Relics of this period of conflict still remain such as the Yorkshire Prohibited Trawling Areas (PTAs) that were established at the turn of the century to spatially segregate sectors (Bloomfield et al., 2012), although these PTAs have been subsequently re-designated as MCZs to meet conservation objectives.

Among the more novel MPA assessments, the *de facto* closure of the North Sea during World Wars I and II constituted two 'great fishing experiments' and were unparalleled in showing what happens when fishing effort is substantially reduced. Over the closure period of WWI plaice stock abundance doubled, and modal length increase by approximately 50%, from 25 cm in 1914 to 37 cm in 1919 (Borley et al., 1923). In WWII, plaice abundance tripled (Margetts and Holt, 1948) and increases were seen in other species such as haddock (24.8–28.7 cm mean length, 11–59 kg h^{-1} of fishing between 1939 and 1945; Beare et al., 2010). At 575,000 km^2, these closures exceeded the UK's largest no-take MPA established around the Chagos Islands (see Chapter 3). This recovery was, however, quickly eroded with the resumption of fishing.

Fish landings from the North Sea peaked during the mid-1970s at around 3.5 million tonnes, although these catch levels could not be sustained (Figure 9.2). Generally, fleet overcapacity during the latter half of the twentieth century has led to a loss of stock productivity and profitability of North Sea fisheries (Thurstan et al., 2010). The problem of overfishing was recognised as far back as the 1930s (Russell, 1942), with these concerns increasing with the rapid expansion of the post-war North Sea fleet powered by rapid technological advancement and fleet modernisation. A brief opportunity to integrate area-based management proposed by some delegations to the 1946 International Overfishing Conference[1] met strong resistance and,

---

[1] This was the forerunner of the North East Atlantic Fisheries Commission (NEAFC) (Holden, 1994).

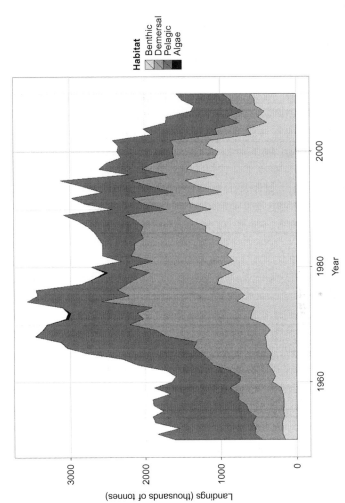

Figure 9.2 Cumulative landings of all species recorded by ICES for the North Sea region between 1950 and 2012. Benthic species include those that spend the majority of their time of the sea bed such as lobsters, crabs, gastropods and flatfish; demersal species include those that live in association with the sea bed but rarely on it such as cod, haddock and whiting; pelagic species include those that are generally found in mid-water and often shoal such as mackerel, sprat and squid.

instead, throughout the 1950s management favoured efficiency restrictions, particularly mesh size regulation. When, by 1960, mesh restrictions alone had proved ineffective, catch restrictions were introduced in the form of total allowable catch (TAC) and dominated management discourse into the 1980s, only substantially challenged as TAC-managed fisheries in their turn entered crises in the early 1990s with fishing effort being reduced through limiting days-at-sea and decommissioning programmes (Gezelius, 2008; Holden, 1994) (see Section 2 for overview of current fisheries management).

In addition to declines in fisheries yields, the ecology of the North Sea has changed considerably since the dawn of industrial-scale fishing (see Roberts, 2007 for a general overview). Bottom-trawling has reduced the diversity of seafloor habitats (FSBI, 2004) and associated fauna (Callaway et al., 2007), and the distribution and abundance of species vulnerable to overfishing such as elasmobranches (McHugh et al., 2011; Molfese et al., 2014; Rogers and Ellis, 2000). Pelagic fisheries have made big predatory fish such as bluefin tuna regionally extinct (MacKenzie and Myers, 2007). Generally, the mean body size of fish has decreased (Daan et al., 2005), as too have the diversity and evenness of demersal fish species (Rijnsdorp et al., 1996).

Change in the North Sea ecosystem is not unique. With growing concerns globally about the loss of marine biodiversity and fisheries sustainability (Godfray et al., 2010; Halpern et al., 2008; Worm et al., 2006), concepts such as the ecosystem approach to environmental and fisheries management and marine spatial planning (Crowder and Norse, 2008; Douvere, 2008) have influenced the development of marine policies that encourage greater conservation of marine resources, of which discourse on MPAs currently dominates (Caveen, 2013; Halpern et al., 2010).

1.2. Present fisheries characteristics and impacts

Pelagic species comprise the majority of fish caught from the North Sea (Figure 9.2). UK pelagic vessels over 40 m in length landed 248,036 tonnes of seafood worth £159.1 million in 2010 (Curtis and Anderson, 2010). Bottom trawlers, however, dominate the UK fleet in terms of vessel numbers and days spent at sea, and the target roundfish and flatfish species are caught as part of a mixed fishery. The food web of the North Sea is complex, and there is a trade-off in the yields of different species. For example, low fishing mortality on cod is predicted to lead to an increase in its spawning stock biomass (SSB) but a decrease in the SSB and yield for whiting and haddock through their increased direct predation by cod. Effects may cascade, with the potential to also increase the SSB and yield for herring, sandeel, Norway pout and sprat due to their reduced predation by whiting and haddock (ICES, 2013). Figures 9.3 and 9.4 present the ICES data for the top 25 finfish species caught in the North Sea and the top 19 recorded invertebrate species. The data are a fascinating window on the social, ecological and economic history of the North Sea. Some major trends evident, such as the disappearance of herring and the gadoid outburst are examples of nature determining landings (Heath et al., 2012). Others, such as the catches of blue whiting, mackerel and wolf fish show how technological advance or culinary fashions can encourage fishers to target new species. Latterly declines in landings of some species are clearly, at least in part, a result of over-exploitation (e.g. cod and whiting). Other declines may be the result of climate change, managing down exploitation or mis-management (Beare et al., 2013; Hilborn, 2007; Hilborn and Ovando, 2014; Mackinson, 2014). Whilst it is thought that there is still excess fishing power in the North Sea fleet (Engelhard, 2009), natural mortality is now once again becoming the dominant source of mortality for many stocks in the North Sea (ICES, 2013).

In much of Europe, and especially the United Kingdom, there was a move away from targeting finfish and species where there was a requirement to own quota using mobile fishing gear in the 1970s and 1980s (Pawson, 1989; Pawson and Benford, 1983; Walmsley and Pawson, 2005), to fishing for invertebrate species using static gear, often from boats of less than 10 m. This is evident in the rise in landings for some shellfish species in the late 1970s and mid-1980s (Figure 9.4). Other increases in landings for species such as velvet crabs and whelks are a result of the increasing globalisation, improved food transport infrastructure and procedures incentivising European fishermen to fish and find markets for species that have little in the way of a local market (Paust and Rice, 1999; Valentinsson et al., 1999). Decreases in landings of bivalves such as mussels and oysters reflect a shift from fisheries to mariculture for these species (David Jarrad, Shellfish Association of Great Britain, pers. com.).

Fishing activities that interact with seafloor habitats are widespread in the North Sea (Nielsen et al., 2013). However, fishing effort is not evenly distributed across the region (Jennings and Lee, 2012; Jennings et al., 1999; Rijnsdorp et al., 1998; STEFC, 2012), and there are considerable differences in total footprints depending upon fishing gear (Bloomfield et al., n.d.; Vanstaen and Silva, 2010) and the country to which vessels are registered

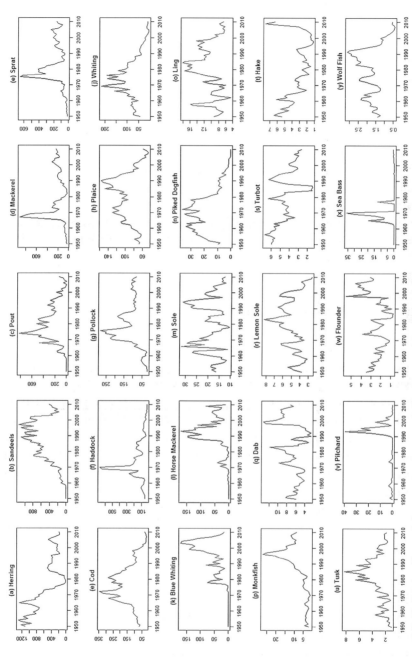

Figure 9.3 Finfish landings of the top 25 species recorded by ICES for the North Sea region between 1950 and 2012. Landing figures are in thousands of tonnes.

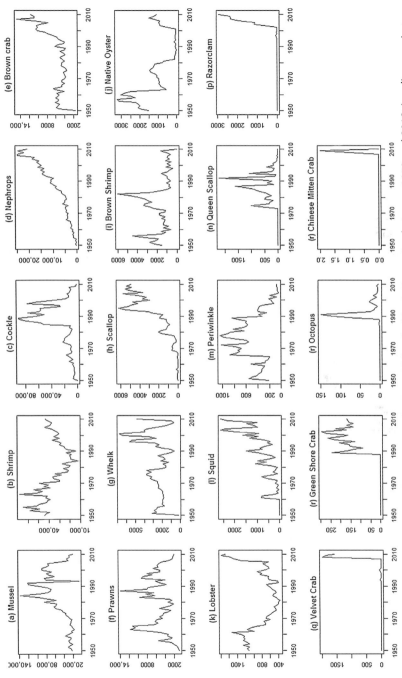

Figure 9.4 Invertebrate landings for 19 species recorded by ICES for the North Sea region between 1950 and 2012. Landings are in tonnes.

(Jennings et al., 2012), the latter presumably influenced by the distance to the main port from which a vessel operates (Stelzenmuller et al., 2008). By their nature, the physical footprints of mobile fishing gears, such as otter and beam trawling, are generally larger than those for static gears such as shellfish pots, trammels and gill nets (Nielsen et al., 2013; Figure 9.5), which tend to be operated from smaller vessels whose activities are more patchily distributed and closer inshore (Nielsen et al., 2013). Recent studies demonstrate inter-annual variation in the extent and location of fishing grounds, and the distribution of fishing effort therein (Jennings et al., 2012; Johnson et al., 2014). Within year (seasonal), variations in the distribution of fishing effort have also been noted for some gears (Greenstreet et al., 1999). These fishing footprints are also affected by the local seabed substrate type. Mixed sediments are the predominant bottom type in the North Sea. Coarser sands are dominant in the shallower more tidally active south. Mud is generally more prevalent in the deeper northern parts, with substantial areas of exposed bedrock found mainly around the coasts of Scotland and Norway.

At the finer scale, there remain significant gaps in our empirical understanding about both the spatial and temporal effects of fishing gear–habitat interactions. This uncertainty is manifest when attempts are made to value the economic benefits and costs of restricting fishing over large spatial scales (Hussain et al., 2010), with assumptions made on the impact of different fishing gears reflecting pre-conceived biases. Indeed, this knowledge deficit has become increasingly relevant to the debates over management measures for MPAs, particularly the ongoing revision of fishing activities to be restricted in European Marine Sites (EMSs),[2] and current debates on the management restrictions to be placed in MCZs. Importantly, regional scale modelling exercises reveal that there are trade-offs in designing MPA networks to achieve different sets of conservation objectives (Greenstreet et al., 2009).

2. FISHERIES MANAGEMENT IN THE NORTH SEA

Much of the North Sea is under the fisheries jurisdiction of the European Union (excluding the EEZ of Norway, and UK territorial waters stretching up to 12 nm from the coast). The Common Fisheries Policy (CFP) lays out the regulatory framework for EU fisheries management. The United Kingdom has exclusive fishing rights within 6 nm of their coastline. Between 6 and 12 nm, vessels from other EU countries may have historic rights of access to fish. Beyond 12 nm, all EU Member States have fishing rights on the principle of relative stability (EC, 2013). Non-EU

[2] http://www.marinemanagement.org.uk/protecting/conservation/ems_fisheries.htm.

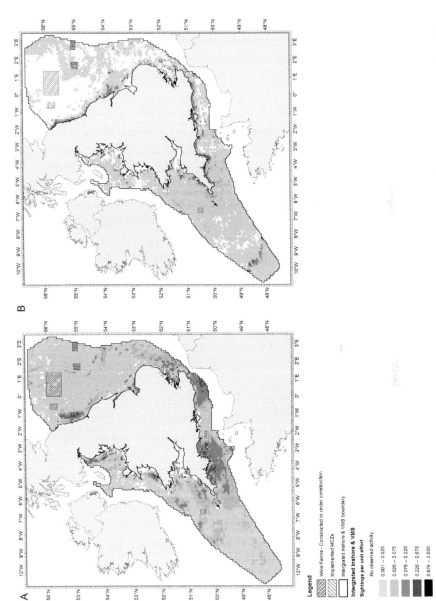

Figure 9.5 Relative fishing effort for English inshore and offshore fishing activities for 2007–2009; (A) mobile gears and (B) static gears (from Vanstaen and Silva, 2010). Implemented MCZs are also shown. *Produced from data provided by the Inshore Fisheries and Conservation Authorities, Welsh government and Marine Management Organisation.*

countries may fish EU waters through third party agreements and vice versa. In the North Sea, this primarily relates to the 'northern agreements' held bi- and tri-laterally with Norway and the Faroe Islands (EC, 2014). ICES provides the underpinning scientific advice for the North East Atlantic region, including the North Sea (ICES management area IV). Harvesting strategy for the majority of major commercial fish stocks (e.g. the negotiation of TACs) is decided at the Council of Fisheries Ministers from each EU Member State who meet in December each year.

The CFP has been criticised for failing to deal adequately with the socio-political drivers leading to overfishing (Symes and Hoefnagel, 2010; Symes and Phillipson, 2009; Urquhart et al., 2011), and also the wider ecosystem impacts of fishing (Gascuel et al., 2012; Jennings and Kaiser, 1998; Mesnil, 2012; Raakjaer et al., 2014) (see Section 1.1). In addition to reduced fisheries productivity and economically inefficient fisheries, unwanted impacts on the marine environment have also resulted. Historically, fleet overcapacity, lack of political will, continuous disregard of scientific advice and early warnings, absence of clear objectives and a decision-making process geared towards short-term gains have all been cited as barriers to delivering sustainable fisheries for Europe (Daw and Gray, 2005; Frost and Andersen, 2006; Khalilian et al., 2010). Nevertheless, recent evidence suggests that several fish stocks are beginning to rebuild in the North Sea (Fernandes and Cook, 2013; Hilborn and Ovando, 2014) and that implies management measures administered through the CFP are beginning to have some positive effect.

The CFP has been reviewed and reformed approximately every 10 years since its inception in 1983. The latest reforms made radical changes and reached political agreement in May 2013. Major changes include: commitments to eliminate discards on a defined timeline; technical and procedural measures to protect marine biodiversity; decentralisation of certain elements of decision-making allowing Member States to agree appropriate measures for shared fisheries under multiannual plans, including adjustments in fishing capacity; and requirements to set fishing rates at sustainable levels (recognising the need to rebuild depleted stocks). Greater coherence with other EU marine environmental policies is also sought; for example, ecosystem-based fisheries management is listed as one of the objectives under Article 2 of the CFP (Raakjaer et al., 2014) in accordance with the EU Integrated Maritime Policy and MSFD.

The multi-national nature of the EU demands complex governance arrangements. Administered centrally, the CFP is no exception. Before 2002, it was frequently criticised for a top-down approach to management (Suarez de Vivero et al., 2008); although Member States are responsible for implementation and enforcement, the EU creates proposals and makes decisions, setting TACs and allocating national quotas (Burns and Stahr, 2011; Khalilian et al., 2010). Seeking to address criticisms of this top-down structure, in the 2002 reform, legal provision for Regional Advisory Councils (RACs) was made. These provided a mechanism for the incorporation of knowledge from fishers and other stakeholders. The North Sea RAC (NSRAC) was the first to be established in 2004. This initiated a shift towards participation by regional stakeholders, but while it promoted discussion, decision-making power was retained centrally (Griffin, 2010). In 2013, RACs were reformed to Advisory Councils, designed to sustain a newly strengthened consultative relationship with the European Commission and Member States, continuing the move towards participation and with the intention of increasing the influence of regional stakeholders (Hatchard and Gray, 2014). One of seven management units based on biological criteria, the North Sea Advisory Council (NSAC[3]) provides advice on behalf of its members: fisheries organisations and other stakeholders including environmental organisations (EC, 2004).

Intergovernmental agreements support trans-boundary management. In the North Sea, this primarily relates to the 'northern agreements' held bi- and tri-laterally with Norway and the Faroe Islands (EC, 2014). These are implemented in the form of annual fisheries arrangements, which allows for the setting of TACs for joint stocks, transfers of fishing possibilities, joint technical measures and issues related to control and enforcement. A bilateral arrangement with Norway covers the North Sea and the Atlantic; a trilateral agreement covers Skagerrak and Kattegat (Denmark, Sweden and Norway) and the neighbourhood arrangement covers the Swedish fishery in Norwegian waters of the North Sea.

The principal tool currently used for management of the North Sea is a quota based TAC restriction. TACs are agreed annually for the main commercial stocks as described above with the International Council for the Exploration of the Sea (ICES) providing fish stock assessments that form part of the basis of EU fisheries negotiations (Da Rocha et al., 2012). Member States then allocate their shares to their home fleets. The European Commission proposed management based on maximum sustainable yield (MSY), designed to ensure economic, environmental and socially sustainable exploitation of fish stocks over the longer term.[4] By 2015, fishing limits

---

[3] http://www.nsrac.org/.
[4] Communication from the European Commission to the Council and the European Parliament: implementing sustainability in EU fisheries through maximum sustainable yield [COM(2006) 360].

must not exceed the exploitation rate consistent with FMSY where possible, and at the latest by 2020 for all stocks. Plans define the fishing rate appropriate for many stocks, and TACs are usually negotiated on this basis (Froese and Quaas, 2013; Khalilian et al., 2010). EU quota is proportionally shared among Member States based on historical catch, known as the principle of relative stability. The main implication of the MSY commitment is felt during TAC negotiations; arguments to raise TACs purely on socio-economic grounds will no longer carry weight. It has been commented that current MSY-based targets play to a tendency to seek simple and straightforward solutions to fishery management issues (Fulton et al., 2014); managers still like a single number. The scientific limitations of such approaches have received sustained criticism; the common refrain being that conventional single-species fisheries management has failed and new approaches are needed (Shepherd, 2003). It has also been suggested that ecosystem-based approaches to fisheries management lack credibility unless used in tandem with empirical approaches (Mackinson, 2014). It is also argued, however, that managers and politicians have had access to the appropriate information to manage stocks and avoid collapses, and that failures were caused by a lack of political will to withstand pressure from the fishing industry (Cardinale and Svedang, 2008).

Most multiannual plans also include fishing effort and/or technical restrictions in addition to annual TACs, and specific control rules; e.g. mesh size, days-at-sea and closed areas (Da Rocha et al., 2012; Salomon et al., 2014). Within the 12 nm limit, countries may add additional national legislation where it is in line with the CFP and not discriminatory against other Member States (EC, 2014; Ratz et al., 2010). They may also take unilateral measures, affecting their own fleet. Management is regulated by the European Union Council of Ministers and European Parliament within the framework of controls in the CFP, and the EC operates a surveillance system to ensure compliance (Da Rocha et al., 2012). Member States are responsible for proper enforcement of the CFP within their sectors and can be fined for failure to do so (Khalilian et al., 2010). Such fisheries management structures differ for each nation fishing the North Sea. Here, relevant UK practices in England and Scotland are outlined as illustrative of different national level management practices (Figure 9.6 from Sweeting et al., 2006).

Laws and regulations are set by Member States' competent fisheries departments including the Department for Environment, Food and Rural Affairs (DEFRA) in the United Kingdom. In English waters, and for English vessels operating outside those waters, the Marine Management Organisation (MMO) is the delegated authority for regulation and licensing of

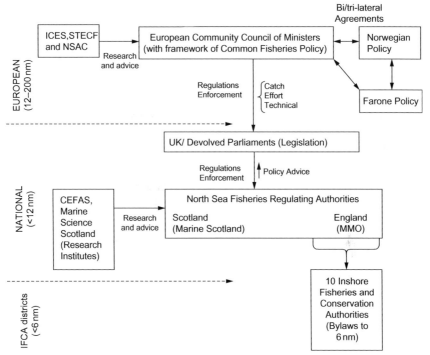

Figure 9.6 Overview of fisheries decision-making in the United Kingdom.

fishing, and coordinates the enforcement programme. The EU 'control' regulation includes the use of vessel monitoring systems and electronic recording systems, as well as a range of other requirements which the MMO must deliver for England. The Centre for Environment, Fisheries and Aquaculture Science (CEFAS) researches and monitors fish and shellfish stocks. In England, there is an additional tier, where 10 Inshore Fisheries and Conservation Authorities (IFCAs, which were the Sea Fisheries Committees) can make bylaws to regulate local fisheries, and are responsible for the sustainable management of fisheries in their districts (0–6 nm). The devolved administrations, Scotland in the case of the North Sea, manage fisheries in their own territorial waters (to 12 nm) working with the UK Government in negotiating fishing opportunities through the European Union and in other international negotiations. Marine Scotland manages quota for fish stocks and all inshore fisheries within the 12 nm limit for the Scottish Government. It is also responsible for controlling the activities of fishing vessels and fishing effort (days spent at sea) in the North Sea, West of Scotland and Faroese waters. Marine Scotland Science provides research and technical advice.

UK enforcement is carried out by agencies and inspectorates within the competent fisheries departments and is supported by the Royal Navy and aerial surveillance through a commercial company. Fisheries measures that are enforced include the following: access rules for foreign fishing vessels in British fishery limits; TACs and quotas for fish stocks; technical conservation measures for juvenile and spawning fish, such as minimum landing sizes, mesh sizes for nets and closed areas; special stock recovery measures for at risk fish stocks; control measures for monitoring and inspecting fishing vessels and their catches, such as fishing logbooks and landing declarations and effort limitation measures limiting the days vessels may fish. Satellite monitoring is applied to all vessels over 15 m and all vessels over 10 m must maintain a logbook of capture quantities, time and location. These are inspected remotely or at port, through boardings and cross checking with other paperwork such as landing declarations and sales notes. Generally, the UK policy leans towards funding selective catching gear to help stop discards, and research to improve the economic and environmental sustainability of the fishing industry.

3. LEGAL FRAMEWORK FOR NORTH SEA MPAs AND POLICY DRIVERS

Historically, there is a bifurcation between nature conservation and fisheries management in the EU (De Santo, 2010). The responsibility for fisheries management was transferred to the European Commission in 1970s, whilst Member States have retained jurisdiction over the conservation of nature within their territories. Fundamentally, MPAs are spatial protection measures, and because the geographic distribution of most fish species found in the North Sea varies seasonally the use of MPAs in fisheries management has, with the notable exception of the 'sandeel box'[5] been largely restricted to protecting spawning and nursery areas (e.g. the Plaice Box, EU Council Regulation 4193/88), and congregations of juvenile fish in temporary real-time closures (i.e. CFP Basic Regulation 1380/2013, Article 7(2)(d)).

3.1. Nature conservation

Both legally binding and soft law instruments have called for the establishment of MPAs networks to protect marine biodiversity. At the international

[5] A 20,000 km[2] box off east Scotland/NE England, created in 2000 to conserve sandeel but in response to severe decline of a sandeel-dependent predator, the kittiwake. As such, it's the first case of a seabird used as a proxy for a fish stock conservation method.

level, a requirement to establish a representative network of MPAs by 2012 in accordance with international law and based on best scientific information can be found in the Plan of Implementation of the World Summit on Sustainable Development 2002 (para 32 (c)).[6]

Regional level obligations to establish a network of MPAs occur in the Convention for the Protection of the Marine Environment of the North East Atlantic (OSPAR, 1992). The North Sea is covered in its entirety under the OSPAR Convention. Article 3(1)(b)(ii) requires the OSPAR Commission *'to develop means, consistent with international law, for instituting protective, conservation, restorative or precautionary measures related to specific areas or sites or related to particular species or habitats'* and OSPAR Recommendation 2003/3 specifically referred to the establishment of a coherent ecological network of MPAs by 2010. Two sets of guidelines in 2003 and 2006, respectively, have been published to support the establishment of the OSPAR network. First, guidelines for identification and selection of MPAs in the OSPAR maritime area (Reference number 2003-17) set out detailed ecological criteria/considerations to be applied in the first stage of identification of an ecologically coherent network, whose aim is to: *'protect, conserve and restore species, habitats and ecological processes which are adversely affected as a result of human activities; b. prevent degradation of and damage to species, habitats and ecological processes, following the precautionary principle; c. protect and conserve areas that best represent the range of species, habitats and ecological processes in the OSPAR area'*. Second, the Guidance on Developing an Ecologically Coherent Network of OSPAR marine protected areas (Reference number 2006-3) set out principles to assist contracting parties in interpreting the concept of an ecologically coherent network. These principles were further elaborated in a self-assessment checklist (Reference number 2007-6) (see Table 9.1 for summary). As will be discussed in Section 4, these principles play an important role in designing the ecologically coherent network of MPAs at the domestic level. However, noting that the target to establish an OSPAR ecologically coherent network by 2010 was not reached, Recommendation 2010/2 amended and updated Recommendation 2003/3.

In Europe, the regional requirements of OSPAR have also been reinforced through the MSFD 2008/56/EC. With the exception of

---

[6] Noted in para 19 of Decision VII/5 of the COP 7 to the Convention on Biological Diversity and reiterated in the form of an encouragement in para 21 of Decision X/31 of the COP 10 to the Convention on Biological Diversity.

Table 9.1 Summary of OSPAR assessment criteria for designing ecologically coherent MPA networks (Reference number 2007-6)

Assessment criteria	Description
Adequacy/ viability	Shape and size of MPAs in the network maximise the effectiveness of the network to achieve its ecological objectives, and the MPA network consists of several no-take areas
Representativity	The MPA network represents all (or almost all) of the range of species and/or habitats known in the study area
Replication	Known features within the area are protected in spatially separated replicate MPAs
Connectivity	Ecological linkages between MPAs in the network are enhanced

Norway, all North Sea coastal states are EU Member States, hence bound to implement the MSFD. The MSFD adopts an ecosystem-based approach and protected areas are recognised as one spatial mechanism to achieve Good Environmental Status (GES) of EU waters by 2020. GES is the key concept of the MSFD and it is defined in Article 3(5) of the MSFD to mean *'the environmental status of marine waters where these provide ecologically diverse and dynamic oceans and seas which are clean, healthy and productive within their intrinsic conditions, and the use of the marine environment is at a level that is sustainable, thus safeguarding the potential for uses and activities by current and future generations···'*. Article 3 also states that adaptive management on the basis of the ecosystem approach shall be applied with the aim to achieve GES. The definition of GES is therefore very inclusive, bridging the divide between purely ecological aims and socio-economic uses of the environment and employing a dynamic understanding of the marine governance. GES is to be determined at the regional or sub-regional level as referred in Article 4 on the basis of 11 qualitative descriptors in Annex I, the first of which requires the maintenance of all biological diversity in EU waters.

Key components of the MPA network in the EU are marine Natura 2000 sites, as recognised by Article 13(4) of the MSFD.[7] Natura 2000

---

[7] Other protected areas are Ramsar sites under the Convention of wetlands of international importance designated under the Convention on Wetlands of International Importance especially as waterfowl Habitat 1971. The wetlands covered in the Convention include marine water the depth of which at low tide does not exceed 6 m (Art. 1(1)).

(see Article 3(1)) is a network of sites comprising both terrestrial and marine sites hosting the natural habitat types listed in Annex I and habitats of the species listed in Annex II of the Habitats Directive 92/43/EEC, termed Special Areas of Conservation (SACs) and Special Protected Areas (SPAs) classified by the Member States pursuant to the Wild Birds Directive 2009/147/EC. However, the number of marine habitats and species covered by the Habitats Directive is much lower than the terrestrial counterparts as recognised by the European Commission[8] and the majority of EMSs are within territorial waters, leaving an offshore gap. The Natura 2000 network can therefore only in part contribute to the achievement of the GES under the MSFD because its remit, especially in relation to the marine environment, is very modest (focusing on a small number of particular sites and species). Therefore to fulfil the broad aims of the MSFD, the Natura 2000 network is not sufficient.[9] Article 13(4) of the MSFD also refers to '*marine protected areas as agreed by the Community or Member States concerned in the framework of international or regional agreements to which they are parties*'. This includes MPAs established under OSPAR. As for other marine protected areas, Article 13(4) is silent but this silence can be interpreted positively, in the sense that the MSFD does not preclude the establishment of other MPAs in the North Sea (beyond Natura 2000 sites and OSPAR sites).

In Sections 4 and 5, we look at the implications of the above for scientifically informed decision-making. Two issues arise: first, some authors have noted that there is an insufficient evidence base to support OSPAR's definition of ecological coherence and what it actually means in practice (Jones and Carpenter, 2009); second, there is often insufficient local knowledge of seabed habitats, species distributions and fisher behaviour to design such MPA networks to be sure that they will achieve their intended objectives (see Sections 4 and 5).

3.2. Fisheries conservation

Spatial closures play a role in the context of fisheries management too. At the European level, the CFP contains provisions for environmental protection. The legal basis for this is Article 11 of the Treaty on the Functioning of the European Union explicitly requesting the integration of environmental

---

[8] European Commission, Guidelines for the Establishment of the Natura 2000 Network in the Marine Environment: Application of the Habitats and Birds Directives, 2007, p. 14.

[9] This is why the introduction of new MPAs in the United Kingdom under the Marine and Coastal Access Act 2009 and the Marine (Scotland) Act 2010 are welcome to broaden the range of sites to form an EU network.

protection measures in the definition and implementation of the EU policies. The Basic Regulation (1380/2013) contains a series of provisions relevant in this respect. The objectives of the CFP stated under Article 2 of the
basic Regulation are: (1) ensuring environmentally sustainable fisheries and
aquaculture activities in the long term, (2) requiring the CFP to apply a precautionary approach to fisheries management and (3) implementing the
ecosystem-based approach. Article 7 lists the types of conservation measures
that can be adopted by the EU for the purpose of achieving the objectives of
the CFP. For the present discussion on protected areas and spatial closures,
technical measures under Article 7(2)(c) and (d) are of interest. Article 7(2)
(c) describes limiting or prohibiting the use of certain fishing gears or activities in certain areas, and Article 7(2)(d) relates to requiring fishing vessels to
stop operating in a defined area for a specific minimum period for the protection of temporary aggregations of vulnerable marine resources. There are
also emergency measures that Member States can adopt to alleviate threat to
the conservation of marine biological resources (Article 13).

The CFP should not be seen as separate from the nature conservation
legislation examined previously. The links between the two are explicitly
taken into account under Article 11, that empowers Member States to adopt
conservation measures not affecting fishing vessels of other Member States
and that are applicable to waters under their jurisdiction to comply with the
Habitats, Wild Birds and MSFD provided they are compatible with objectives in Article 2. In cases in which other Member States have a direct management interest in the fisheries to be affected, the European Commission
has the power to implement the measures required by the Member State
under Article 11(1) by means of delegated act.[10]

4. CURRENT PLANNING OF MPAs IN THE NORTH SEA

At the time of writing, there are several MPA planning processes
occurring simultaneously in the North Sea (Figure 9.7), all for the purpose
of nature conservation. Our focus here is largely what is happening in the
UK EEZ, with the exception of the Dogger Bank SAC[11] which provides
an interesting case study of the challenges of managing international relations
when designating trans-boundary MPAs. In discussing these case studies, we

[10] For more information about the links between fisheries management measures and Natura 2000 sites,
see a 2008 European Commission guidance entitled Fisheries Measures for Marine Natura 2000 Sites
at: http://ec.europa.eu/environment/nature/natura2000/marine/docs/fish_measures.pdf.

[11] Designated through the Habitats Directive 92/43/EEC.

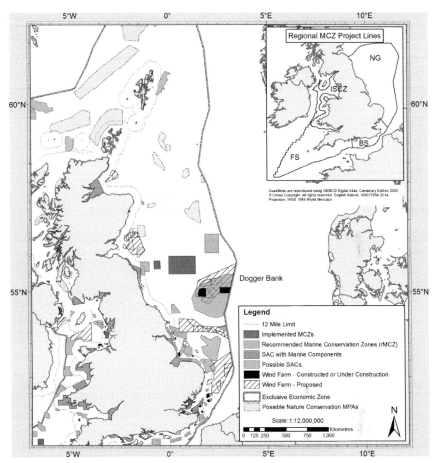

Figure 9.7 MPAs in the UK North Sea, along with constructed and proposed wind farm areas. The English half of the Dogger Bank SAC is also shown. The insert map (top right) shows the four English regional MCZ planning areas; Net Gain (NG), Balanced Seas (BS), Finding Sanctuary (FS), Irish Sea Conservation Zones (ISCZ).

examine the roles that central government, conservation NGOs and the fishing industry have taken in the designation of sites, focusing particularly on the use of information *sensu lato* (i.e. natural science and socio-economics) in site planning.

The Marine and Coastal Access Act 2009[12] was intended to contribute to the fulfilment of several regional obligations outlined in Section 3.1, in

---

[12] The Marine and Coastal Access Act 2009, including parallel acts for Scotland and Northern Ireland, includes detailed provisions to designate and protect nationally important marine areas as MCZs (Part 5, Chapter 1, Articles 116–148).

particular to contribute to an ecologically coherent and representative network of MPAs by 2016 under the MSFD, and also OSPAR commitments.
First, we discuss the relatively well documented English MCZ process (see
Jones, 2012a,b; Pieraccini, 2013) that followed the principle that users
should be fully involved with the design of the MCZ network in order
to balance conservation with socio-economic impacts and promote user
support. The MCZ process involved four regional planning workshops in
England, and here we focus on the regional project for the North Sea,
Net Gain, drawing on observations made by the first author at the meetings.

4.1. Designation of MCZs in the English North Sea

The MCZ process combined top-down policy steer administered through
Natural England and the Joint Nature Conservation Committee's (JNCC)
Ecological Network Guidance (ENG) (Natural England and JNCC, 2010),
with bottom-up regional stakeholder planning groups who ultimately chose
the location of MCZs within the constraints set by the ENG. The ENG provided a range for the percentage of each habitat type that should be protected
in each region, as well as a description of the principles guiding the design of
the network. Twenty four stakeholder workshops were held from March
2010 to June 2011 to plan the North Sea MCZ network. These workshops
involved more than 6000 h of stakeholder planning, and recommended
18 MCZs and 13 no-take reference areas covering an area of 13,000 km^2,
with targets for 19 of the 23 broad-scale habitat types being met. The workshops relied heavily on two types of data to inform the choice of MCZs:
broad-scale habitat data and fishing effort data.

Habitat data used as the basis for planning varied greatly in age, quality
and extent. Recent collaborative projects collated and interpreted a range of
environmental data to produce predictive maps of broad-scale habitats in the
North Sea using the EUNIS habitat classification system[13] (Cameron and
Askew, 2011; McBreen et al., 2011), but levels of confidence in predictive
maps vary. The quality and resolution of the underlying data and the number
of habitat classes used in interpretation are major drivers of the quality of the
outputs; reducing the number of classes to broad categories (e.g. EUNIS
levels 3 and 4) increases the accuracy of predictions, but resultant maps
are unlikely to contain sufficient detail to support robust MPA planning,
since references to specific taxa only begin to be introduced at level 4 to

[13] European Nature Information System: http://eunis.eea.europa.eu/habitats.jsp.

discriminate rocky habitats, and discrimination of soft substrata is still based on physical and zonal attributes (as in level 3).

More detailed ground truth data, such as from diver surveys, video tows and benthic grabs, tended to be spatially biased as a result of surveys undertaken for environmental impact assessments for oil and gas platforms and windfarms or recreational diver site preferences, although a limited number of seabed surveys have also been undertaken specifically to inform the designation of MCZs. An example is the dive survey of the North Norfolk chalk reef conducted by Seasearch[14] volunteers in 2010. A combination of side scan sonar and diver surveys were used to map and measure the extent of the chalk, producing detailed records of geological features, biotopes and associated species and identifying it as potentially the largest chalk reef in Europe (Spray and Watson, 2011).

Commercial fishing data were initially confined to those derived from VMS, producing an incomplete and biased picture of activity as it only mapped activity of vessels of 15 m and over. This was later augmented by FisherMap interviews which collected local data from those engaged in the commercial fishing sector. Interviews from other recreational sea users (sea anglers, charter boat operators among others) were also undertaken by the Net Gain during the period March to October 2010. In total, 53% of mobile gear vessel owners, and 75% of static gear vessel owners were interviewed (Net Gain, 2011). These data were subsequently mapped in GIS and made available to the regional hub meetings in January 2011.

International fishing representatives initially convened at a meeting of the NSRAC in March 2010. Once appointed, JNCC Fisheries Liaison Officers worked with Net Gain to develop a strategy to engage with those non-UK commercial fishers who had shown an interest in providing data to the project. Representatives from the Dutch and Danish fleets subsequently attended a number of the regional hub meetings. Data on Dutch and Danish fishing activity were fairly comprehensively mapped and passed through to Net Gain during October 2011 (Net Gain, 2011). Information relating to the Belgium and German fleets was limited to VMS only. Only very limited information was obtained relating to the activities of Scottish fishing vessels.

At the start of the planning process, the meetings were fractious, with protagonists prejudging what each other stood for, and there was a general

---

[14] Seasearch is a UK and Ireland-based project that trains volunteer SCUBA divers to record marine species and habitats, providing a cost-effective source of verified data to inform the conservation and management of the marine environment. http://www.seasearch.org.uk/.

air of mistrust between them. Particularly, tense was the relationship between the mobile gear sector and conservation organisations, the former believing their activities to be unfairly discriminated against in the preliminary advice from the statutory conservation agencies. There were also complaints from the fishing industry that they were struggling to understand the overarching rationale for having a MCZ network. Ambiguity over the management objectives of individual sites also created further uncertainty. As the meetings progressed, emphasis on minimising socio-economic conflict took precedence over ensuring that MCZs were protecting key habitat features.

There was some criticism from the Science Advisory Panel (SAP), as independent assessors, when they convened to scrutinise the outcomes of the MCZ regional groups. They noted that the UK Government had committed to making decisions on MCZ site selection and regulation based on the '*best available evidence*', and stated that '*lack of full scientific certainty should not be a reason for postponing proportionate decisions on site selection*'. Regional groups therefore used the '*best available evidence*' as the basis for selecting sites. As a result the SAP noted that some sites '*relied greatly on socio-economic considerations with biodiversity often of secondary consideration*' (SAP, 2011) and that some sites with little ecological data were recommended. Criticisms were reinforced by concerns over the accuracy of data employed.[15] That the choice of MCZs was predominantly made on broad-scale habitat data did not sit well with some NGOs who argued that some data categories (e.g. seabirds and marine mammals) were overlooked.

Though the majority of targets set to achieve an ecologically coherent network of MPAs were considered by the SAP to have been met overall, the North Sea received specific criticism such as '*Net Gain stakeholders appear to have minimised the areas designated*' (SAP, 2011); minimum proportions of habitat in the region (specifically sub-tidal mud) were not protected. Deficiencies in robustness of data and lack (number and size) of recommended reference areas (rRAs) were also highlighted. Provisional management objectives of 'maintain' versus 'recover' were viewed as simplistic; in the absence of monitoring data, the former was criticised for maintaining the status-quo, whilst the latter was criticised for being based on the tenuous assumption that removal of anthropogenic pressures will return a feature to some pre-industrial 'good' status; the 'erroneous equilibrium paradigm' (see Section 1). It was felt that gaps might require re-engagement with

---

[15] https://www.gov.uk/government/uploads/system/uploads/attachment_data/file/69451/sap-mcz-final-report.pdf.

stakeholders. As socio-economic considerations were generally prioritised over biodiversity, re-adjustment of this balance could be seen to devalue stakeholder input, undermining confidence in the process. This is particularly problematic as Net Gain stakeholders were critical of the quality of evidence, challenging habitat data on the basis of their own knowledge. When delegates were asked at the end of the process whether the best available data had been used, they returned an average score of only 5.7 out of 10 (Net Gain, 2011).

If the proposals of Net Gain were fully adopted, 30% of the lower and middle English North Sea region would be protected in some form (two thirds of this via existing MPAs etc.) (SAP, 2011). Despite receiving over 40,000 responses to the public consultation on the four English proposals, 73% of which called for the designation in English seas of all 127 recommended MCZs in 2013, the government rejected 96 of the recommended MCZs claiming that gaps in the evidence base and potential misidentification of features made it impossible to define the management measures and take effective conservation action. It was felt that more evidence was required to support the designation of more than 31 sites to prevent legal challenges. Economic impact assessments were used to discriminate between the only two viable policy options perceived by the government: designating all recommended MCZs, or adopting a staged approach initially designating the 31 sites for which a good cost-benefit case and robust evidence existed. The latter prevailed, and at the time of writing, DEFRA are using a staged approach to designate sites in tranches and fund additional survey work to support MCZ designations. The first tranche of 27 MCZs was designated on 21 November 2013 (three in the Net Gain area of the North Sea, three in the outer Thames and one at Thanet[16]), with a further 37 proposed for consideration in tranche two (with consultation in 2015) with seven in the North Sea and one in the outer Thames. Given the minimum recommendations of the SAP, this still falls far short of that required for an ecologically coherent network.

North Sea specific concerns expressed during the SAP review process are of greater concern in this reduced network. Specifically, lack of representativity across northern and southern parts of the area due to the influence of the Flamborough-Helgoland[17] Front (e.g. north–south larval transport),

---

[16] These four MCZs were planned by the Balanced Seas workshop.

[17] The Flamborough-Helgoland Front stretches offshore from Yorkshire's Flamborough Head. This site is known as an area of food-rich waters, important for a whole range of marine wildlife, including seabirds—including kittiwakes, puffins, gannets, razorbills and guillemots.

poor links with adjacent Scottish and the Balanced Seas[18] areas and connectivity problems, given the predominance of sedimentary habitats in the North Sea and the importance of widespread small occurrences of other habitats with the species that they sustain, are significant outstanding issues. Given the limited implementation to date, and the fact that even the 127 recommended sites were considered to be the minimum required by the SAP, it may be debated whether current designations and proposals will form an ecologically coherent network.

4.2. Scottish designation of marine nature conservation MPAs

Scotland's approach to planning their MPA network initially appears to have been more top-down. Selection of Nature Conservation MPAs has been clearly stated to be 'science-led'.[19] One crucial difference is that the Scottish MPA project has not been driven by percentage targets of seabed habitat to protect.

The Scottish MPA Project was set up as a joint initiative between Marine Scotland, JNCC, Historic Scotland, the Scottish Environment Protection Agency (SEPA) and Scottish Natural Heritage (SNH). A series of five MPA workshops during 2011–2012 focused on engaging national and international level organisations. Attendees included representatives from local authorities, community councils, universities, recreational and conservation organisations, and a range of industries including renewable energy, fishing and tourism. The Marine Strategy Forum, established in 2009 with a membership comprising formal organisations from government and industry bodies, used quarterly liaison meetings to engage marine users and interest groups at a more regional and local level. This was the main mechanism that allowed input by informal stakeholders before a 16-week public consultation from July to November 2013. Currently, civil servants in Marine Scotland are going through the consultation responses to prioritise sites where there is strong evidence that protection is needed, and then to find out what activities take place there.

It is envisaged that users would be more involved during discussion on management of individual MPAs with engagement increasingly involving local communities and direct interest groups to identify appropriate local management measures on the principle that 'management of activities in MPAs

---

[18] The Eastern English Channel MCZ planning group.
[19] http://jncc.defra.gov.uk/page-5510.

will be determined on a site-by-site basis and there will be opportunities for individuals to get involved in developing site specific management plans'.[20]

Seventeen proposed MPAs (pMPAs) in Scotland fall, at least partially, within the North Sea Region. Eight of these are offshore, eight within territorial waters, and one spans both. Of these, one remains a 'search location' pending full assessment against the Scottish MPA Selection Guidelines to enable further work before formal advice to Scottish Ministers is provided to allow designation by July 2014. This primarily relates to mobile species features, and, in the Southern Trench[21] in the North Sea, also burrowed mud.[22] Marine Scotland, SNH and JNCC held 56 coastal events and drop-in sessions for people to discuss MPAs and over 14,000 responses to the MPA consultation were received, a large proportion from NGO-led electronic campaigns.[23]

4.3. European designation of SACs and SPAs

The development of MCZs has sought to capitalise upon existing designations to create the ecologically coherent network of MPAs required. European legislation under the Habitats and Species Directive (Council Directive 92/43/EEC of 21 May 1992) and the Birds Directive (Council Directive 79/409/EEC of 2 April 1979) already required the establishment of marine SACs for listed habitats and species, and SPAs for listed birds within the jurisdiction (≤ 200 nm offshore) of Member States. Each EU Member State compiled a list of its most important wildlife areas, following criteria set out in the Habitats Directive. A UK Marine SACs project ran from May 1996 to October 2001, focusing on 12 marine SACs around the UK coastline to build knowledge and establish the best practice needed for the management and monitoring of EMSs.

The first SACs were established during the 1990s and 2000s when initial marine SACs and SPAs were identified and designated to create a network of marine protected areas across European waters known as the Natura 2000 network and also referred to as EMSs. Once the list of candidate SACs in England was approved by the UK Government, it was submitted to the European Commission. Adopted in 2004, they became Sites of Community

---

[20] http://jncc.defra.gov.uk/pdf/Partnership_Project_Web_Ready.pdf.
[21] The Southern Trench is off NE Scotland and one of the most biologically productive areas in the North Sea, and it's features also include fronts and shelf deeps.
[22] http://www.scotland.gov.uk/Publications/2013/08/2591/3.
[23] www.scotland.gov.uk/Resource/0044/00445967.docx.

Importance, before being formally designated as SACs on 1 April 2005. Further SACs have been created in subsequent iterations by NE and JNCC. The Habitats Directive requires that any activities that may significantly affect a SAC must be subject to special scrutiny and require an initial detailed 'appropriate assessment' to determine that it will have no adverse effect on site integrity. The UK currently has 108 SACs and 108 SPAs with marine components (although only three of the SPAs are entirely marine) in addition to the 27 MCZs. Combined, these cover 9.5% of the UK marine area. Sites of Special Scientific Interest (SSSIs) with marine components and Ramsar sites also contribute to the existing UK MPA network.[24]

This interface between European and Member State specific legislation offers specific challenges to management due to the competing interests of national fishing fleets. The case of the Dogger Bank is now discussed.

4.3.1 Challenges facing the management of the Dogger Bank SAC
The Dogger Bank is the largest sandbank (17,600 km^2) in the North Sea, straddling the EEZs of the United Kingdom, Netherlands, Germany and Denmark. It is a major commercial fishing ground and hosts six benthic communities (Moorsel, 2011). The Dutch FIMPAS project (Fisheries Measures in Protected Areas) initiated in 2009 agreed on an inter-governmental approach to managing the Dogger Bank because, apart from the Netherlands SAC, Germany and the United Kingdom were also seeking to designate SACs in the area, creating a jigsaw of adjoining Natura 2000 sites. The aim was to achieve a coherent, integrated approach to managing fisheries across the SAC complex. The Dogger Bank Steering Group (DBSG) of Member States was formed and invited Denmark to join because of its sandeel fishing interests in the area.

This international cooperation was greatly welcomed by the NSRAC as a means of avoiding vessels encountering a different SAC management regime every time they crossed a national border.

The DBSG members each had different conservation objectives for their respective SACs and therefore focussed on developing a joint recommendation to the European Commission on fisheries management measures for protecting Habitat H1110, a marine interest feature in Annex 1 of the Habitats Directive, namely 'sandbanks which are slightly covered by sea water all the time'. With the conservation status of H1110 accepted as 'unfavourable' in exhibiting an excess of short-lived, opportunistic species, and a deficit of

---

[24] http://jncc.defra.gov.uk/page-1445.

long-lived, slow-growing species, it was agreed to restrict bottom-contacting gears (especially beam- and otter trawling) with the aim of restoring and conserving the habitat.

The DBSG first invited the NSRAC to devise a stakeholder-led fisheries zoning proposal in 2011. It was a challenge for the NSRAC to resource this task but, fortunately, it was facilitated and supported by the Wageningen University led EU project 'MASPNOSE' (Marine Spatial Planning in the North Sea). The RAC made significant progress towards potential zoning scenarios but failed to reach consensus, with the fishing industry seeking much smaller area closures to bottom trawling which were unacceptable to the environmental NGOs.

The DBSG gave the NSRAC more time to derive a proposal but now based on strict terms of reference, including:

(i) applying a concept of two zones—a 'free zone' in which all legal gears within the CFP are allowed, and a 'management zone' in which bottom-contacting gears likely to result in habitat deterioration are prohibited.

(ii) establishing a management zone to cover 25–55% of the total SAC area, representing all five benthic communities.

(iii) adopting the perspective of the entire Dogger Bank rather than individual national EEZs, while acknowledging Germany's aim of protecting 50% of its SAC.

The NSRAC stakeholders found the 25–55% range an impossibly wide gulf to bridge. The percentage was vague because there is little understanding of how much of a sandbank should be closed in order to meet the conservation objectives. The breadth of the target, however, gave comfort to both industry and NGOs for their respective positions. With the benefit of hindsight, it is clear that such a critical Term of Reference should have been addressed right at the start of the process, not half way through.

By April 2012, the NSRAC's final report had still not reached consensus, the industry seeking around 22% closure and the NGOs around 45%. Two new challenges now emerged. Firstly, the fishermen sought to trade-off closures in the management zone against potential exclusions from the windfarm being developed inside the UK SAC. However, the United Kingdom was not willing to enter into any such discussion lest it prove legally prejudicial to the windfarm consenting process. Secondly, it was disputed whether Scottish seines (flyshoots) should be considered potentially damaging bottom gears; the fishing sector (along with United Kingdom and Denmark) claimed they would have negligible effect on achieving the conservation objectives, whereas the NGOs (along with Germany and the Netherlands) opposed such claims.

The NSRAC process having run its course, the DBSG (with NSRAC observers) now took responsibility for developing a joint recommendation during 2012–2013, building on NSRAC's groundwork. The DBSG sought to close about one-third of the Dogger Bank area, based on a literature survey of other closures. However, these precedents were mainly for protection of fishing resources rather than, as in Dogger Bank, MPAs for nature conservation. The DBSG (as did NSRAC earlier) took, as its baseline, fisheries landing data for 2007–2009 by grid square. Where squares in the management zone had equal nature conservation value, losses to the industry were minimised by selecting those producing lowest landings.

After incorporating ICES advice in 2012, the DBSG was on the verge of submitting its joint recommendation to the Commission in July 2013 when the Netherlands Parliament sought to test the proposed measures, and suggested significant changes. As of early-2014, the Dutch ministry was still trying to find a compromise with the other Member States. With this delay, the proposal now fell under the 2014 reform of the CFP which changed the rules of engagement. Firstly, the proposal must now be approved by all Member States with 'a direct management interest', i.e. any Member State with an entitlement to fish on the Dogger Bank. In effect, the DBSG's joint recommendation must also now be approved by at least Sweden, Belgium and France. Secondly, as the new CFP created regionalised decision-making bodies, in the case of the Dogger Bank this wider sign-off must be sought in the coalition of North Sea Member States known as the 'Scheveningen Group'. A new challenge is for this group to establish a modus operandi for engaging, as it must, with the North Sea Advisory Council (NSAC). Thirdly under the new CFP, a 'joint recommendation' means that not only must the Dogger Bank proposal be agreed by all Member States with a direct management interest but it can only be accepted by the Commission in its entirety. Therefore as an example the Scottish seine issue must be resolved by the Member States before submission to the Commission.

4.4. Uncertainties of future changes in climate and exploitation

Predicted changes in climate are significantly affecting outcomes of MPA management measures in the North Sea. Albeit in a spatially complex pattern, overall surface and bottom waters of the North Sea have shown a net warming over 30 years or more (Simpson et al., 2011). Attributing changes in fish abundances, diversity and community structure to specific variables such as water temperature is difficult because other large-scale changes have also occurred, such as in the North Atlantic Oscillation, Gulf Stream Index and North Sea inflow (Dulvy et al., 2008). There is evidence of substantial spatial shifts, in particular northwards in the North Sea (Perry et al., 2005) and to greater mean depths within it (Dulvy et al., 2008). Particularly, strong are increases in fish abundances (Simpson et al., 2011) and species richness (Hiddink and ter Hofstede, 2008). These increases trends are most strongly linked to warming, although it is likely that decline in overall fishing mortality over the last 25 years has contributed to the increases in at least some fishery-target species (Simpson et al., 2011). These changes, both climate and exploitation related suggest many implications, not least that the assemblages of organisms present at the time a MPA is set up are likely to shift over time; this will be especially true of small MPAs. Another implication is that the relative abundance of target species are likely to shift over time, as indicated by the developing boarfish fishery (Simpson et al., 2011). These shifts in abundance of major predators must be having wider implications for food webs and marine biodiversity as a whole but understanding of these, for example through trophic cascades remains poor.

The extent to which existing legal regimes on MPAs are capable of responding to such climate challenges and climate unknowns is debatable. Many conservation laws have been adopted prior to climate change becoming a prominent issue in the international environmental policy agenda. For example, at the European level, the Habitats and Wild Birds Directives do not mention climate change but there are certain provisions, especially those related to the establishment of Natura 2000 network that are relevant to issues of climate change adaptation. For example, connectivity is addressed specifically in Articles 3(3) and 10 of the Habitats Directive. Links between the Natura 2000 network and climate change adaptation have also been made in environmental policy such as the 2009 White Paper on Climate Change Adaptation (Communication COM(2009) 147, n. 89). There is a growing legal literature that has provided a critical appraisal of conservation law in relation to climate change (Cliquet et al., 2009; Troubworst, 2009). Although a detailed appraisal is outside the scope of this paper, to give an example one of the key issues emerging is that static and detailed conservation objectives may constitute a problem if a dynamic approach is needed for climate change adaptation. In relation to Natura 2000, the European Court of Justice (Case C-535/07 Commission v Austria) disagreed with the Commission's argument for more specificity in setting the conservation objectives for each species in a SPA and for their inclusion in legally binding form in the designation order. At the domestic level, however, this flexibility is less present. Indeed, in relation to MCZs, under the Marine and Coastal Access Act 2009, the conservation orders must specify the features to be protected. This approach is problematic as it targets protection measures at the individual component features rather than at the ecosystem level (Lieberknecht et al., 2012; Pieraccini, 2013).

5. DISCUSSION
5.1. Evidence versus advocacy

One argument for planning MPA networks is that traditional feature-led conservation efforts (e.g. Natura 2000) are not sufficient for protecting the full range of marine habitats and species and preventing further declines (Jones, 2012a; Wood et al., 2008). However, the planning of MPA networks is challenging not least because of irreconcilable differences in the values and interests of different marine stakeholders (see Jones, 2014 for review); for example, many environmentalists want at a minimum to protect a proportion of representative marine habitats from the effects of bottom-trawling and the fishing industry want to maintain access to fishing grounds. There is also a longstanding scientific debate on the role of MPAs (principally no-take zones) in enhancing fisheries (Caveen et al., 2012; Willis et al., 2003), as well as more generally increasing ecosystem goods and services (Potts et al., 2014). There is, however, significant uncertainty surrounding where, when and how no-take MPAs can be used to achieve such benefits (Caveen et al., 2012), even more so when planning is undertaken at large scales, as the costs of externalities (such as displacement) are often not factored into impact assessments (Hussain et al., 2010). These different values and priorities of conservationists and the fishing industry underpin an important strategic debate running between both camps: that of (1) designating MPAs to protect existing habitat features, or (2) designating MPAs to restore marine ecosystems, and perhaps controversially, increase ecosystem productivity. Some conservationists argue that to achieve the latter, no-trawl MPAs (preferably no-take) must be designated in areas of greatest biological productivity, typically (but not always) where fishing effort would be most concentrated (Roberts and Mason, 2008). However, to date, planning has generally focused on avoiding core fishing areas as it is politically less contentious—this issue is not unique to the North Sea (Devillers et al., 2014).

In such a politically antagonistic arena where competing advocacy coalitions are seeking to maintain (or maximise) their interests (Sabatier, 1988), perhaps unavoidably, science is primarily used as a tool to bolster the credibility of each coalition's cause (Pielke, 2004). Scientific uncertainty had been underplayed by MPA advocates to influence government officials to take action. For example, unsubstantiated claims were made about positive fisheries effects of a no-take MPA network as the Marine and Coastal Access Act was being passed through parliament (e.g. RCEP, 2004: 202, para 8.83; DEFRA, 2008: 60, para 245). This strong fisheries message was an attractive one for government, as officials could take the high ground that they were following the science and promoting a win–win policy, precluding the need for lengthy discussion of different worldviews around the overarching goals of the MPA network. During the planning stage, the fishing industry has continuously emphasised socio–economic uncertainty, and this has clearly influenced where MCZs have been designated (Section 4.1), with some scholars now interpreting the resulting designation of MCZs as reflecting a compromise between conservation and socio-economic interests (Jones, 2012b).

5.2. Realities of planning MPAs

For English MCZs, the vast spatial scale of the North Sea planning area coupled with a restricted timeframe led to several problems regarding information management. Evidence of the location of specific habitats and species entered the planning process haphazardly and, crucially, the short timeframe meant that this evidence was overlooked during the planning of MCZs. This is a sharp contrast with the Scottish MPA process where seabed surveys allowed unique habitat features to be identified, mapped and subsequently put forward for protection. A crucial difference between the English and Scottish MPA processes is that the former was largely driven by targets, abstract planning rules and modelled habitat data, whereas the latter was based on high resolution field-survey data and not target driven.

A major layer of 'wickedness' (see para 2, Section 1) in MPA planning is stakeholder attitudes towards risk (Balint et al., 2011). The incompleteness of the evidence base for planning—i.e. limited fine-scale knowledge of habitat distributions, and fishing gear–habitat interactions—has often meant debates are polarised on the actual restrictions to be implemented in MPAs and on whom the 'burden-of-proof' should fall. Generally speaking, conservationists want fishing activities to be restricted on the basis of the precautionary principle, whereas fishers want evidence that their activities are actually causing damage. Some environmental organisations have invoked inappropriate legal terminology, suggesting that there is a need to prove 'beyond reasonable doubt' that activities in protected areas have no effect. This is the level of proof required to convict criminals in UK law, rather than the more appropriate 'balance of probabilities' required under civil law.[25] Whilst conservationists have accused government of succumbing to short-term socio-economic interests, the uncertainty surrounding the MCZ process both in terms of evidence and objectives of sites has caused the fishing industry to become highly defensive of their historic access rights to fishing grounds.

Despite the recognition of the need for participatory approaches to MPA planning (Christie et al., 2003), the MCZ process was largely reduced to one of bargain and compromise between the ecological and socio-economic interests. This has been likened to an instance of 'thin' proceduralisation as the MCZ process often did not facilitate real dialogue between conservation and industry perspectives (Black, 2000; Pieraccini, n.d.). An interesting contrast in approach to MPA planning was that taken by Scotland. Despite being more top-down, the Scottish process was paradoxically better received by the fishing industry because it was perceived to be more transparent than the MCZ process with less uncertainty posed for industry (NFFO, 2012).

Full no-take zones (or marine reserves) are conspicuously lacking in the North Sea. In English North Sea waters, 13 reference areas (ranging from 0.04 to 52.49 km^2) were initially recommended through the MCZ process for the purpose of scientific monitoring. Their implementation, however, was subsequently contested by the fishing industry as there is no legal mandate for them in the Marine and Coastal Access Act, and DEFRA is not advancing the idea. Generally, the difficulties of establishing no-take zones have arisen due to strong opposition from the fishing industry regarding their perceived costs, and lack of incentives for their establishment (see Jones, 2014 for overview). It should be noted however that no-take MPAs have been successfully deployed in the management of some UK fisheries, such as the Isle of Man scallop fishery (Beukers-Stewart et al., 2005).

Whilst the debate surrounding MPAs is often framed around contests between conservation and fishing interests, the North Sea ecosystem is

---

[25] Letter from Wildlife Trusts, the Marine Conservation Society and Client Earth NEIFCA, 15/1/2013, 'Re: Flamborough ~Head "Red Risk" Fishing Bylaw'.

currently subject to multiple pressures from non-fishing activities such as oil and gas, shipping, aggregate extraction, offshore wind and leisure (Bloomfield et al., n.d.; Knights et al., 2011). Representatives from offshore wind companies had a particularly powerful voice and no MCZs were designated in sites identified by the Crown Estate for potential development of windfarms. The fishing industry, however, has subsequently sought to co-locate MCZs with windfarms to minimise the area of marine space they are restricted from (e.g. Blyth-Skyrme, 2011). The primacy of reaching renewable energy targets was also influential in the UK Government sanctioning a major windfarm development within its Dogger Bank SAC (Section 4.3.1).

A further challenge for planning a North Sea MPA network is that outside a country's 12 nm territorial limit, common EU access means that MPA planning has to be co-ordinated between states. This entails having to deal with the rules and norms of pre-existing governance arrangements (Section 2), and inevitably adds another layer of complexity and uncertainty to planning MPAs, and co-ordinating management of trans-boundary sites such as the Dogger Bank (Section 4.3.1).

5.3. The future of marine conservation in the North Sea

A fundamental problem of the concept of ecological coherence is that it has emerged from a single-disciplinary perspective—the applied ecological sciences (see *Ecological Applications* 2003; 13[1]) and is based on a worldview that believes human activities need to be restricted from large areas of marine ecosystems to achieve maximum benefits (see Roberts, 2007, p. 382). Invariably, benefits are often not explicitly defined and there is often ambiguity over who or what benefits from protection. This is one reason why there is currently an emphasis on research that attempts to provide economic valuations of different types of use/non-use of the marine environment (see Potts et al., 2014 for overview).

The concept of ecological coherence is difficult to implement not least because of the limited evidence base (Jones and Carpenter, 2009); it also underestimates the dynamism of marine ecosystems (i.e. seasonal fish movements, inter-annual variability), overlooks social-ecological complexity (i.e. humans are part of the marine food web) and ignores the irreconcilable difference between utilitarian versus preservationist perspectives (Jones, 2014). Some scientists (e.g. Hiscock, Coastal Futures Conference, 2014) have also criticised the usefulness of the concept for planning because of the confusion that it has caused (see NFFO, 2011). At both strategic and tactical levels, science has too often been used as a political tool for different interests to impose their worldview on policy makers. Whilst this is, to an extent, unavoidable, current governance arrangements have facilitated it by emphasising the importance of science in the planning process without really acknowledging areas of uncertainty and not identifying clear objectives for MPAs. In the case of MCZs, rigid adherence to ecological rules of thumb drove planning, with little thought given to the specific objectives sites were intended to achieve. We suggest that the rush to establish MPA networks to meet abstract planning rules and targets is neither scientific nor pragmatic and has deflected political capital from addressing real risks to different components of the North Sea ecosystem. We also note, however, that there has been a recent re-emphasis on assessing the merits of protection at a much more local scale (e.g. the review of management of EMS, and the phased-in approach of MCZs), and also recognition that management needs to be responsive to ecosystem change—this is critical given the impacts of climate change on the North Sea (see Section 4.4).

Also critical to the planning of MPAs in the North Sea is the management of information. It became apparent during the planning of MCZs that there were many different types of information of potential use to planners, but that there were sensitivities over disclosure (particularly that detailing the activities of fishing vessels due to Data Protection Legislation) and costs (fine resolution habitat data were often collected privately). The scale of the planning process also meant that the rigorous scrutiny of evidence underpinning MCZs could not be carried out. Effective marine conservation can encompass both mitigating current risks to biodiversity and also attempting restoration; however, particularly for the latter, a convincing case must be put forward to stakeholders to justify any potential costs. Whilst to an extent all problems are socially constructed based on people's perceptions of risk (see Burgman, 2005 for overview), we argue that science can have much greater utility in decision-making through frameworks that can deal with uncertainty, subjectivity and different scales of environmental problems. A multi-scaled risk screening process for the North Sea could be carried out that prioritised high risk human–ecosystem interactions for further analysis. Subsequent management objectives and actions (of which MPAs may be one outcome) would be based on the pragmatic application of the precautionary approach within a robust, participatory and integrated decision-making process (e.g. Carey et al., 2007). Not only will this allow time and money to be spent more efficiently, but it will also identify key areas where scientific monitoring would benefit management by reducing uncertainty.

6. CONCLUSION

The North Sea is a complex and dynamic socio-ecological system, due to its ecological variability and the divergent interests of different stakeholders and nations. In addition, an information deficit makes decision-making challenging. Drawing on the case studies presented in this review, we have looked at several MPA processes occurring in the North Sea. We argue that science has been used strategically to make the case for MPA networks, though the tactical use of science to optimise the placement of MPAs for clear purposes is constrained by the quality of real-world data and lack of clear planning objectives.

There is an information deficit that creates considerable uncertainty when MPA networks are planned at large regional-scales to meet abstract planning rules, namely: (1) lack of high resolution spatial and temporal information of the distribution of marine species and habitats and also fishing activity, (2) lack of knowledge of the interaction of human activities with the marine environment, (3) poor understanding of ecosystem connectivity and (4) the inability to predict fisheries displacement effects with confidence. A further challenge for planning a North Sea MPA network is cooperation between states. This inevitably adds another layer of complexity and uncertainty to planning MPAs, particularly the coordination of the management of trans-boundary sites. Finally, global–local debates on food security and environmental justice add additional layers of 'wickedness' to planning MPA networks.

Fundamental differences between the values of conservationists and the fishing industry mean that compromises in protecting the marine environment are inevitable. Crucially, the real-world complexity and dynamism of marine socio-ecological systems, compounded by the interconnectedness of marine and terrestrial food webs through human consumption (Smith et al., 2010) make ecological coherence a practically unattainable technocratic goal. We argue for a more pragmatic approach to marine protection that embraces the complexity of the social and political arena in which decisions are made. Such an approach would draw on evidence at appropriate scales, and facilitate meaningful stakeholder dialogue on real versus perceived threats to marine ecosystems.

ACKNOWLEDGEMENTS

A. C. would like to thank NERC-ESRC for funding that provided the ideas behind this paper. Special thanks to Koen Vanstaen (CEFAS) for providing fishing effort data and Aled Nicholas (Seafish) for mapping. Also thanks to Bill Lart (Seafish) for checking the graphs.

REFERENCES

Anon, 1908. Thirty years of steam trawling. Marlborough Express XLII (25), 2. http://paperspast.natlib.govt.nz/cgi-bin/paperspast?a=d&d=MEX19080130.2.4.

Balint, P., et al., 2011. Wicked Environmental Problems: Managing Uncertainty and Conflict. Island Press, Washington, DC.

Barrett, J., Nicholson, R., Ceron-Cerrasco, R., 1999. Archaeo-ichthyological evidence for long-term socioeconomic trends in northern Scotland: 3500 BC to AD 1500. J. Archaeol. Sci. 26, 353–388.

Barrett, J., et al., 2000. What was the Viking age and when did it happen? A view from Orkney. Norwegian Archaeol. Rev. 39 (1), 1–39.

Barrett, J., Locker, A., Roberts, C., 2004. "Dark Age Economics" revisited: the English fish bone evidence AD 600–1600. Antiquity 78, 618–636.

Beare, D., et al., 2010. An unintended experiment in fisheries science: a marine area protected by war results in Mexican waves in fish numbers-at-age. Naturwissenschaften 97, 797–808.

Beare, D., et al., 2013. Evaluating the effect of fishery closures: lessons learnt from the Plaice Box. J. Sea Res. 84, 49–60.

Beukers-Stewart, B.D., et al., 2005. Benefits of closed area protection for a population of scallops. Mar. Ecol. Prog. Ser. 298, 189–204.

Black, J., 2000. Proceduralizing regulation: part I. Oxf. J. Legal Stud. 20, 597–614.

Bloomfield, H.J., et al., 2012. No-trawl area impacts: perceptions, compliance and fish. Environ. Conserv. 39 (3), 1–11.

Bloomfield, H. et al. (n.d.). How might achieving objectives for seafloor integrity impinge on human use of the marine environment? Under review.

Blyth-Skyrme, R., 2011. Benefits and disadvantages of co-locating windfarms and marine conservation zones. Report to collaborative offshore wind research into the environment Ltd., London, December 2010, 37 pp.

Board of Agriculture and Fisheries, 1908. Annual Report of Proceedings under Acts Relating to Sea Fisheries for Year 1906. His Majesty's Stationary Office, London.

Borley, J., et al., 1923. The plaice fishery and the war. Preliminary report on investigations. Fish. Invest. Ser. II 5, 1–56.

Burgman, M., 2005. Values, history and perception. In: Risks and Decisions for Conservation and Environmental Management. Cambridge University Press, Cambridge, pp. 1–25.

Burns, T., Stahr, C., 2011. Power, knowledge, and conflict in the shaping of commons governance. The case of EU Baltic fisheries. Int. J. Commons 5 (2), 233–258.

Callaway, R., et al., 2007. A century of North Sea epibenthos and trawling: comparison between 1902–1912, 1982–1985 and 2000. Mar. Ecol. Prog. Ser. 346, 27–43.

Cameron, A., Askew, N., 2011. EUSeaMap—preparatory action for development and assessment of a European broadscale seabed habitat map final report. Available at: http://jncc.gov.uk/euseamap.

Cardinale, M., Svedang, H., 2008. Mismanagement of fisheries: policy or science? Fish. Res. 93, 244–247.

Carey, J.M., et al., 2007. Risk-based approaches to deal with uncertainty in a data-poor system: stakeholder involvement in hazard identification for marine national parks and marine sanctuaries in Victoria, Australia. Risk Anal. 27 (1), 271–281.

Caveen, A., 2013. A Critical Examination of the Scientific Credentials of Marine Protected Areas: Sound Science or a Leap of Faith? PhD thesis, Newcastle University, Newcastle.

Caveen, A.J., et al., 2012. Are the scientific foundations of temperate marine reserves too warm and hard? Environ. Conserv. 39 (3), 199–203.

Caveen, A., et al., 2013. MPA policy: what lies behind the science? Mar. Policy 37, 3–10.

Christie, P., et al., 2003. Toward developing a complete understanding: a social science research agenda for marine protected areas. Fisheries 28 (12), 22–26.

Cliquet, A., et al., 2009. Adaptation to climate change. Legal challenges for protected areas. Utrecht Law Rev. 5 (1), 158–175.

Coles, J., 2000. Doggerland: the cultural dynamics of a shifting coastline. Geol. Soc. Lond. Spec. Publ. 175, 393–401.

Crowder, L., Norse, E., 2008. Essential ecological insights for marine ecosystem-based management and marine spatial planning. Mar. Policy 32, 772–778.

Curtis, H., Anderson, J., 2010. 2010 Economic Survey of the UK fishing fleet. The Sea Fish Industry Authority, UK.

Da Rocha, J., Cerviato, S., Villasante, S., 2012. The common fisheries policy: an enforcement problem. Mar. Policy 36 (6), 1309–1314.

Daan, N., et al., 2005. Changes in the North Sea fish community: evidence of indirect effects of fishing? ICES J. Mar. Sci. 62, 177–188.

Davis, F., 1923. An Account of the Fishing Gear of England and Wales, Great Britain. Ministry of Agriculture and Fisheries, UK.

Daw, T., Gray, T., 2005. Fisheries science and sustainability in international policy: a study of failure in the European Union's Common Fisheries Policy. Mar. Policy 29, 189–197.

De Santo, E., 2010. Whose science? Precaution and power-play in European marine environmental decision-making. Mar. Policy 34, 414–420.

DEFRA, 2008. Marine and Coastal Access Bill Impact Assessment. DEFRA, London.

Degnbol, P., et al., 2006. Painting the floor with a hammer: technical fixes in fisheries management. Mar. Policy 30 (5), 534–543. http://dx.doi.org/10.1002/aqc.2445.

Devillers, R., et al., 2014. Reinventing residual reserves in the sea: are we favouring ease of establishment over need for protection? Aquat. Conserv. Mar. Freshwater Ecosyst. http://dx.doi.org/10.1002/aqc.2445.

Douvere, F., 2008. The importance of marine spatial planning in advancing ecosystem-based sea use management. Mar. Policy 32 (5), 762–771.

Dulvy, N.K., Rogers, S.I., Jennings, S., Stelzenmuller, V., Dye, S.R., Skjoldal, H.R., 2008. Climate change and deepening of the North Sea fish assemblage: a biotic indicator of warming seas. J. Appl. Ecol. 45 (4), 1029–1039.

EC, 2004. Council Decision 2004/585/EC of 19 July 2004 establishing Regional Advisory Councils under the Common Fisheries Policy.

EC, 2013. Regulation (EU) No 1380/2013 of the European Parliament and of the Council of 11 December 2013 on the Common Fisheries Policy, amending Council Regulations (EC) No 1954/2003 and (EC) No 1224/2009 and repealing Council Regulations (EC) No 2371/2002 and (EC) No 639/2004 and Council Decision 2004/585/EC.

EC, 2014. EU and Faroe Islands reach agreement on reciprocal exchanges of fishing opportunities. Available at: (accessed: 17.04.2014). http://ec.europa.eu/information_society/newsroom/cf/mare/itemdetail.cfm?item_id=15213.

Engelhard, G., 2009. One hundred and twenty years of change in fishing power of English North Sea trawlers. In: Payne, A., Cotter, J., Potter, T. (Eds.), Advances in Fisheries Science: 50 Years on From Beverton and Holt. Blackwell Publishing, Oxford, pp. 1–25.

Fernandes, P., Cook, R., 2013. Reversal of fish stock decline in the Northeast Atlantic. Curr. Biol. 23 (15), 1432–1437.

Fischer, J., et al., 2014. Land sparing versus land sharing: moving forward. Conserv. Lett. 7 (3), 149–157.

Fleming, N., 2008. Mapping Doggerland. The mesolithic landscapes of the Southern North Sea. Int. J. Naut. Archaeol. 37 (2), 395–398.

Froese, R., Quaas, M., 2013. Rio+20 and the reform of the common fisheries policy in Europe. Mar. Policy 39, 53–55.

Frost, H., Andersen, P., 2006. The common fisheries policy of the European Union and fisheries economics. Mar. Policy 30 (6), 737–746.

FSBI, 2004. In: Effects of fishing on biodiversity in the North Sea. Briefing Paper 3. Fisheries Society of the British Isles, Granta Information Services, 82A High Street, Sawston, Cambridge CB2 4HJ, UK.

Fulton, T., 1895. Fourteenth Annual Report of the Fishery Board for Scotland. House of Commons Parliamentary Papers, UK.

Fulton, E., et al., 2014. An integrated approach is needed for ecosystem based fisheries management: insights from ecosystem-level management strategy evaluation. PLoS One 9 (1), e84242.

Galparsoro, I., Borja, A., Uyarra, M., 2014. Mapping ecosystem services provided by benthic habitats in the European North Atlantic Ocean. Front. Mar. Sci. 1 (23), 1–14.

Gascuel, D., et al., 2012. Towards the implementation of an integrated ecosystem fleet-based management of European fisheries. Mar. Policy 5, 1022–1032.

Gezelius, S., 2008. The arrival of modern fisheries management in the North Atlantic: a historical overview. In: Gezelius, S., Raakjaer, J. (Eds.), Making Fisheries Management Work: Implementation of Policies for Sustainable Fishing. Springer, Dordrecht, pp. 27–40.

Godfray, H., Beddington, J., Crute, I., 2010. Food security: the challenge of feeding 9 billion people. Science 327, 812–818.

Greenstreet, S., et al., 1999. Fishing effects in northeast Atlantic shelf seas: patterns in fishing effort, diversity and community structure. II. Trends in fishing effort in the North Sea by UK registered vessels landing in Scotland. Fish. Res. 40, 107–124.

Greenstreet, S.P.R., Fraser, H.M., Piet, G.J., 2009. Using MPAs to address regional-scale ecological objectives in the North Sea: modelling the effects of fishing effort displacement. ICES J. Mar. Sci. 66 (1), 90–100.

Griffin, L., 2010. The limits to good governance and the state of exception: a case study of North Sea fisheries. Geoforum 41 (2), 282–292.

Halpern, B.S., et al., 2008. A global map of human impact on marine ecosystems. Science 319 (5865), 948–952, Available at: http://www.ncbi.nlm.nih.gov/pubmed/18276889 (accessed 20.1.2014).

Halpern, B.S., Lester, S.E., McLeod, K.L., 2010. Placing marine protected areas onto the ecosystem-based management seascape. Proc. Natl. Acad. Sci. U.S.A. 107 (43), 18312–18317.

Hatchard, J., Gray, T., 2014. From RACs to advisory councils: lessons from north sea discourse for the 2014 reform of the European common fisheries policy. Mar. Policy 47, 87–93.

Heath, M.R., Neat, N.C., Pinnegar, J.K., Reid, D.G., Sims, D.W., Wright, P.J., 2012. Review of climate change impacts on marine fish and shellfish around the UK and Ireland. Aquat. Conservat. Mar. Freshwat. Ecosyst. 22 (3), 337–367.

Hiddink, J., ter Hofstede, R., 2008. Climate induced increases in species richness of marine fishes. Global Change Biol. 14, 453–460.

Hilborn, R., 2007. Managing fisheries is managing people: what has been learned? Fish Fish. 8 (4), 285–296.

Hilborn, R., Ovando, D., 2014. Reflections on the success of traditional fisheries management. ICES J. Mar. Sci. 71 (5), 1040–1046.

Hiscock, S., 2014. Presentation on the meaning of ecological coherence. Coastal Futures Conference, London.

Holden, M., 1994. The building of a conservation policy or the marathon negotiation. In: The Common Fisheries Policy: Origin, Evaluation and Future. Fishing News Books, UK, pp. 1–274.

Hussain, S., et al., 2010. An ex ante ecological assessment of the benefits arising from marine protected areas designation in the UK. Ecol. Econ. 69, 828–838.

ICES, 2013. Multispecies considerations for the North Sea stocks. Section 6.3.1, Book 6, ICES Advice 2013.

Jennings, S., Kaiser, M., 1998. The effects of fishing on marine ecosystems. Adv. Mar. Biol. 34.

Jennings, S., Lee, J., 2012. Defining fishing grounds with vessel monitoring system data. ICES J. Mar. Sci. 69 (1), 51–63.

Jennings, S., et al., 1999. Fishing effects in northeast Atlantic shelf seas: patterns in fishing effort, diversity and community structure. III. International trawling effort in the North Sea: an analysis of spatial and temporal trends. Fish. Res. 40, 125–134.

Jennings, S., Lee, J., Hiddink, J., 2012. Assessing fishery footprints and the trade-offs between landings value, habitat sensitivity, and fishing impacts to inform marine spatial planning and an ecosystem approach. ICES J. Mar. Sci. 69, 1053–1063.

Johnston, D., 1987. The International Law of Fisheries: A Framework for Policy-Orientated Inquiries. New Haven Press, New Haven.

Johnson, M.L., et al., 2014. The behaviour of 60 or so boats from the Scottish prawn fleet between 2010 and 2013. Available at: https://vimeo.com/94814311, (accessed 03.09.14.).

Jones, P.J.S., 2009. Equity, justice and power issues raised by no-take marine protected area proposals. Mar. Policy 33, 759–765.

Jones, P., 2012a. Governing protected areas to fulfil biodiversity conservation obligations: from Habermasian ideals to a more instrumental reality. Environ. Dev. Sustainability 15, 39–50.

Jones, P., 2012b. Marine protected areas in the UK: challenges in combining top-down and bottom-up approaches to governance. Environ. Conserv. 39 (3), 248–258.

Jones, P., 2014. Governing Marine Protected Areas: Resilience Through Diversity (Earthscan). Routledge, Oxon, UK.

Jones, P.J.S., Carpenter, A., 2009. Crossing the divide: the challenges of designing an ecologically coherent and representative network of MPAs for the UK. Mar. Policy 33 (5), 737–743.

Khalilian, S., et al., 2010. Designed for failure: a critique of the Common Fisheries Policy of the European Union. Mar. Policy 34 (6), 1178–1182.

Khan, A., Neis, B., 2010. The rebuilding imperative in fisheries: clumsy solutions for a wicked problem? Prog. Oceanogr. 87, 1–4.

Knights, A. et al., 2011. Sustainable use of European regional seas and the role of the Marine Strategy Framework Directive. Deliverable 1, EC FP7 Project (244273), Options for Delivering Ecosystem-based Marine Management.

Lieberknecht, L., Qui, W., Jones, P., 2012. Finding sanctuary and England's Marine Conservation Zone process: summary and recommendations. University College London, UK.

MacKenzie, B., Myers, R., 2007. The development of the northern European fishery for north Atlantic bluefin tuna Thunnus thynnus during 1900–1950. Fish. Res. 87, 229–239.

Mackinson, S., 2014. Combined analyses reveal environmentally driven changes in the North Sea ecosystem and raise questions regarding what makes an ecosystem model's performance credible? 71 (1), 31–46.

Margetts, A., Holt, S., 1948. Rapports conseil permanent international pour l'exploration de la mer. The effect of the 1939–1945 war on the English North Sea trawl fisheries. 122, 26–46.

Mazur, A., 1981. Media coverage and public opinion on scientific controversies. J. Commun. 31 (2), 106–115.

McBreen, F., et al., 2011. UK SeaMap 2010 predictive mapping of seabed habitats in UK waters. JNCC report no. 446.

McHugh, M., et al., 2011. A century later: long-term change of an inshore temperate marine fish assemblage. J. Sea Res. 65 (2), 187–194.

Mesnil, B., 2012. The hesitant emergence of maximum sustainable yield (MSY) in fisheries policies in Europe. Mar. Policy 36 (2), 473–480.

Molfese, C., Beare, D., Hall-Spencer, J., 2014. Overfishing and the replacement of demersal finfish by shellfish: an example from the English Channel. PLoS One 9(7).

Moorsel, V., 2011. Species and Habitats of the International Dogger Bank. Report ecosub, Doorn, 74 p.

Natural England, JNCC, 2010. Marine Conservation Zone Project: Ecological Network Guidance. Peterborough, UK.

Net Gain, 2011. Final Recommendations: Submission to Natural England and JNCC. Net Gain.

NFFO, 2011. The dubious science of marine conservation zones. http://nffo.org.uk/news/the-dubious-science-of-marine-conservation-zones.html.

NFFO, 2012. Devolution and no take zones. Available at: http://www.nffo.org.uk/news/devolution_2012.html.

Nielsen, J., et al., 2013. Critical report of current fisheries management measures implemented for the North Sea mixed demersal fisheries. DTU Aqua Report No 263–2013.

Olsen, E., et al., 2013. Achieving ecologically coherent MPA networks in Europe: science needs and priorities. In: Larkin, K.E., McDonough, N. (Eds.), Marine Board Position Paper 18. European Marine Board, Ostend, Belgium.

OSPAR, 1992. Convention for the Protection of the marine Environment of the North-East Atlantic.

OSPAR, 2012. 2012 Status report on the OSPAR network of Marine Protected Areas. OSPAR Commission.

Paramor, O.A.L., Allen, K.A., Aanesen, M., Armstrong, C., Hegland, T., Le Quesne, W., Piet, G.J., Raakær, J., Rogers, S., van Hal, R., van Hoof, L.J.W., van Overzee, H.M.J., Frid, C.L.J., 2009. MEFEPO North Sea Atlas. University of Liverpool, UK.

Paust, B., Rice, A.A., 1999. Marketing and shipping live aquatic products. In: Proceedings of the Second International Conference and Exhibition. p. 321.

Pawson, M., 1989. The Coastal Fisheries of England and Wales, part II: A Review of their status in 1988. Rep. MAFF Direct. Fish. Res, Lowestoft, p. 76.

Pawson, M., Benford, T., 1983. The Coastal Fisheries of England and Walesl part I: A Review of their status in 1981. MAFF Directorate of Fisheries Research, Lowestoft.

Pielke Jr., R., 2004. When scientists politicize science: making sense of controversy over The Skeptical Environmentalist. Environ. Sci. Policy 7, 405–417.

Pieraccini, M., 2013. Establishing an ecologically coherent network of marine protected areas in english waters: what does the designation of marine conservation zones under the marine and coastal access act 2009 add to the picture? Environ. Law Rev. 15.

Pieraccini, M., Rethinking participation in environmental decision-making: epistemologies of marine conservation in South-East England. Under review.

Potts, T., et al., 2014. Do marine protected areas deliver flows of ecosystem services to support human welfare? Mar. Policy 44, 139–148, Available at: http://linkinghub.elsevier.com/retrieve/pii/S0308597X13001656 (accessed 26.2.2014).

Raakjaer, J., et al., 2014. Ecosystem-based marine management in European regional seas calls for nested governance structures and coordination—a policy brief. Mar. Policy 50, Part B, 373–381.

Ratz, H.-J., et al., 2010. Complementary roles of European and national institutions under the Common Fisheries Policy and the Marine Strategy Framework Directive. Mar. Policy 34 (5), 1028–1035.

RCEP, 2004. Turning the Tide: Addressing the Impact of Fisheries on the Marine Environment. 25th report of the Royal Commission on Environmental Pollution, UK.

Rijnsdorp, A., et al., 1996. Changes in abundance of demersal fish species in the North Sea between 1906–1909 and 1990–1995. ICES J. Mar. Sci. 53, 1054–1062.

Rijnsdorp, A., et al., 1998. Micro-scale distribution of beam trawl effort in the southern North Sea between 1993 and 1996 in relation to the trawling frequency of the sea bed and the impact on benthic organisms. ICES J. Mar. Sci. 55, 403–419.

Roberts, C.M., 2007. The unnatural history of the sea. Island press, Washington DC.

Roberts, C., Mason, L., 2008. Return to Abundance: A case for Marine Reserves in the North Sea. A report for WWF-UK. University of York, 48 p.

Rogers, S., Ellis, J., 2000. Changes in the demersal fish assemblages of British coastal waters during the 20th century. ICES J. Mar. Sci. 57, 866–881.

Russell, E., 1942. The Overfishing Problem. Cambridge University Press, Cambridge.

Sabatier, P.A., 1988. An advocacy coalition framework of policy change and the role of policy-oriented learning therein. Policy Sci. 21 (2–3), 129–168.

Salomon, A.K., et al., 2011. Bridging the divide between fisheries and marine conservation science. Bull. Mar. Sci. 87 (2), 251–274.

Salomon, M., Markus, T., Dross, M., 2014. Masterstroke or paper tiger? The reform of the EU's Common Fisheries Policy. Mar. Policy 47, 76–84.

SAP, 2011. Science Advisory Panel Assessment of the Marine Conservation Zone Regional Projects Final Recommendations. Department for Environment, Food & Rural Affairs, Published 15 November 2011.

Shellenberger, M., Nordhaus, T., 2004. The death of environmentalism: global warming politics in a post-environmental world. Environmental Grantmakers Association, USA.

Shepherd, J., 2003. Fishing effort control: could it work under the common fisheries policy? Fish. Res. 47, 76–84.

Simpson, S.D., Jennings, S., Johnson, M.P., Blanchard, J.L., Schon, P.-J., Sims, D.W., Genner, M.J., 2011. Continental shelf-wide response of a fish assemblage to rapid warming of the sea. Curr. Biol. 21, 1565–1570.

Smith, M., Roheim, C., Crowder, L., 2010. Sustainability and global seafood. Science 327, 784–786.

Spray, R., Watson, D., 2011. North Norfolk's chalk reef: a report on surveys conducted by Seasearch East. Available at: http://seasearcheast.org.uk.

STEFC, 2012. Scientific, Technical and Economic Committee for Fisheries (STECF)—evaluation of fishing effort regimes in European waters—part 2 (STECF-12-16), 600 p.

Stelzenmuller, V., et al., 2008. Spatial assessment of fishing effort around European marine reserves: implications for successful fisheries management. Mar. Pollut. Bull. 56 (12), 2018–2026.

Suarez de Vivero, J., Rodraguez Mateos, J., Florido del Corral, D., 2008. The paradox of public participation in fisheries governance. The rising number of actors and the devolution process. Mar. Policy 32 (3), 319–325.

Sweeting, C. et al., 2006. A Review of the Similarities and Differences of Planning, Operation, Stakeholder Involvement and Effectiveness of Selected MPAs. INCOFISH, 144p.

Symes, D., Hoefnagel, E., 2010. Fisheries policy, research and the social sciences in Europe: challenges for the 21st century. Mar. Policy 34 (2), 268–275.

Symes, D., Phillipson, J., 2009. Whatever became of social objectives in fisheries policy? Fish. Res. 95, 1–5.

Thurstan, R.H., Brockington, S., Roberts, C.M., 2010. The effects of 118 years of industrial fishing on UK bottom trawl fisheries. Nat. Commun. 1.

Thurstan, R., Hawkins, J., Roberts, C., 2013. Origins of the bottom trawling controversy in the British Isles: 19th century testimonies reveal evidence of early fishery declines. Fish Fisheries 15 (3), 506–522.

Troubworst, A., 2009. International nature conservation law and the adaptation of biodiversity to climate change: a mismatch? J. Int. Law 3, 419–442.

Turrell, W.R., 1992. New hypotheses concerning the circulation of the northern North-Sea and its relation to North-Sea fish stock recruitment. ICES J. Mar. Sci. 49, 107–123.

Urquhart, J., et al., 2011. Setting an agenda for social science research in fisheries policy in Northern Europe. Fish. Res. 108, 240–247.

Valentinsson, D., et al., 1999. Appraisal of the potential for a future fishery on whelks (Buccinium undatum) in Swedish waters: CPUE and biological aspects. Fish. Res. 42 (3), 215–227.

Vanstaen, K., Silva, T., 2010. Integrated inshore and offshore fishing activities data layer in aggregate producing REC areas. MALSF-MEPF Project 09/P116.

Walmsley, S.A., Pawson, M.G., 2005. The Coastal Fisheries of England and Walkes. Part V: A Review of their Status 2005–6(140).

Willis, T.J., et al., 2003. Burdens of evidence and the benefits of marine reserves: putting Descartes before des horse? Environ. Conserv. 30, 97–103.

Wood, L.J., et al., 2008. Assessing progress towards global marine protection targets: shortfalls in information and action. Oryx 42 (3), 340–351.

Worm, B., et al., 2006. Impacts of biodiversity loss on ocean ecosystem services. Science 314 (5800), 787–790.

CHAPTER TEN

Spatial Management of Fisheries in the Mediterranean Sea: Problematic Issues and a Few Success Stories

Carlo Pipitone, Fabio Badalamenti[1], Tomás Vega Fernández, Giovanni D'Anna

CNR–IAMC Sede di Castellammare del Golfo, Castellammare del Golfo, Italy
[1]Corresponding author: e-mail address: fabio.badalamenti@cnr.it

Contents

Abstract

Fishing has been important in the Mediterranean region for many centuries and still has a central role in its economic importance and cultural heritage. A multitude of fishery-oriented marine managed areas have been implemented under a highly complex political and legislative framework to protect fishery resources and sensitive habitats from high impact uses. However, a review of the literature revealed that few data are available to support their effectiveness, except for a few studies on fishery reserves and marine reserves. In these cases, fish biomass has increased and some evidence of ecological and socioeconomic benefits has been documented. The environmental and geopolitical complexity of the Mediterranean region as well as the dominant top-down management approaches, constitute the weakest points in the spatial management of fisheries at regional level. A coordinating role of all national and supranational bodies present in the area is desirable in the near future.

Keywords: Marine protected areas, Fishery reserves, Reserve effect, Fisheries management, Marine policy, Socioeconomic issues

1. INTRODUCTION

In recent decades marine coastal ecosystems have recorded a dramatic growth in human activities with coastal development (Ruiz and Romero, 2003), energy-linked maritime works (Badalamenti et al., 2011), fisheries (Jackson et al., 2001), navigation and tourism (Davenport and Davenport, 2006; UNEP, 2006). Such a growth has in turn resulted in severe impacts on biodiversity and on commercial fishing activities due to loss of fishing grounds and overfishing (Halpern et al., 2008). The international community has tried to counteract these detrimental effects in different ways, the most important of which deal with nature conservation and fisheries policies. The United Nations Convention on the Law of the Sea (UNCLOS) and the Convention on Biological Diversity (CBD) are the main international frameworks from which policies at regional scale have been derived. UNCLOS governs all uses of the oceans but it also coordinates the exclusive economic zones where coastal states have sovereign rights with respect to the use of natural resources and environmental protection. While the main objective of the CBD is the conservation of biological diversity it also focuses on the relationships between conservation and fishing. The CBD, implemented by the Jakarta Mandate on Marine and Coastal Biodiversity of 1995, singles out marine protected areas (MPAs) as an important tool to reconcile nature conservation and fishing, as well as to promote the sustainable use of marine and coastal living resources.

The Mediterranean Sea has not escaped to the trend described above. The whole basin is subject to pervasive human impacts that range from a medium level in the vast majority of the basin to a very high level in densely populated coastal areas (Halpern et al., 2008). The management of human activities is particularly complex due to the political and administrative fragmentation of the area, and fisheries management is no different (Papaconstantinou and Farrugio, 2000). Additional difficulties arise from the conflicts between trawling, tourism and artisanal fisheries (Gómez et al., 2006), the high biological diversity (Coll et al., 2010) and the oligotrophic nature of most of the waters (Margalef, 1985). The political and cultural heterogeneity is wide, encompassing 21 coastal countries belonging to three continents, eight of which are European Union (EU) members while four are EU candidates or potential candidates. In such regional context the international Conventions above overlap with the regulations of the EU and of the single states resulting in a complex, often disconnected framework of legal and administrative instruments created to manage nature conservation and the sustainable use of fishery resources (Arceo et al., 2013; Katsanevakis et al., 2011).

The two main sources of inspiration for conservation and fisheries policies within the Mediterranean region are the EU and the United Nations, the latter mainly through the Food and Agriculture Organization (FAO) and the Regional Activity Center for Specially Protected Areas (RAC-SPA), which was established in Tunis in 1985 by decision of the Contracting Parties to the Barcelona Convention. The EU emanates legally binding directives as well as voluntarily adopted instruments. The FAO is especially active in generating guidelines and recommendations, particularly through its General Fisheries Commission for the Mediterranean (GFCM), a FAO regional body specifically tailored to tackle Mediterranean and Black Sea fishery issues with 23 member countries plus the EU. The picture is further complicated by the fact that the GFCM recommendations become legally binding for EU member states in the Mediterranean. The main EU legislative acts regarding conservation are the Birds Directive (2009/147/EC), the Habitats Directive (92/43/EEC), the Environmental Impact Assessment Directive (85/337/EEC), the Strategic Environmental Assessment Directive (2001/42/EC), and the Marine Strategy Framework Directive (2008/56/EC). Fisheries are regulated at the EU level by the Common Fisheries Policy (CFP) (2371/2002/EC), which is currently under a reform process after been criticised by fisheries stakeholders for failing to achieve the main goal of reducing overfishing (Daw and Grey, 2005). The reform of the CFP was approved on December 2013 and the new EU fisheries legislation will start in 2014. The Integrated Maritime Policy (EU Reg. 1255/2011) tries to harmonise the different approaches underpinning these policies through the concept of Marine Spatial Planning, of which MPAs are particular cases. On the other hand, the FAO Technical Guidelines on Fisheries Management (FAO, 2011) state that MPAs can play an important role in the achievement of sustainable fishing and in preserving critical habitats or sensitive life stages of marine species. Moreover, the agreement for the establishment of the GFCM and the International Convention for the Conservation of Atlantic Tunas (ICCAT) are other applications of the above FAO Guidelines at a regional scale. Within GFCM, Geographical Sub-

Areas have been established to facilitate fisheries management (through, i.e., trawl fisheries management plans, provisions for the establishment of large offshore no-trawl areas devoted to the permanent or temporal protection of essential fish habitats, etc.). In addition, the EU legislation regulates the minimum depth and distance offshore for trawling (EC Reg. 1967/2006).

Another management tool for fisheries widespread in the Mediterranean is represented by artificial reefs (Fabi et al., 2011; FAO, 1995). Even if no explicit legislation exists, specific guidelines referring to the deployment of artificial reefs in the marine environment have been adopted (UNEP-MAP, 2005).

Following the CBD input, MPAs have become a very common tool in many countries to achieve conservation targets. Additionally, in many cases conservation instruments have worked synergistically with other legal instruments specifically oriented to the management of fisheries. In this context, MPAs have sometimes been deliberately established with the purpose of acting as fisheries management tools on the basis of frequent reports on their potential effects such as e.g., increased fish biomass and spillover, i.e., a net movement of adult fish to the outside of a protected area possibly due to a density-dependent mechanism (Garcia-Charton et al., 2008; PISCO, 2011; Roberts, 1997). As such, MPAs are a powerful tool able to contribute to ecosystem-based marine spatial management and to marine spatial planning in general, and in turn, they can benefit from spatial planning concepts (Ehler and Douvere, 2009; Katsanevakis et al., 2011).

In the Mediterranean region, MPAs have been established following different rationales. For example, in EU countries, MPAs have often been proposed by governments or governmental agencies (Badalamenti et al., 2000), in a few cases following a bottom-up process (Arceo et al., 2013). In non-EU countries, international organisations—notably the RAC-SPA and MedPAN[1] have played a major role. RAC-SPA in particular coordinates initiatives and activities for the implementation of the Specially Protected Areas and Biological Diversity (SPA/BD) protocol, that is the main tool for implementing the CBD in the Mediterranean for the *in situ* sustainable management of coastal and marine biodiversity. RAC-SPA also formulates recommendations for guidelines and common criteria for the selection of marine and coastal protected areas that could be included in the list of Specially Protected Areas of Mediterranean Importance (SPAMI) as well as

---

[1] Network of marine protected area managers in the Mediterranean: http://www.medpan.org/.

guidelines for the establishment and management of protected areas and any other technical tool relevant to the implementation of the SPA/BD protocol.

Currently 681 MPAs *sensu lato* exist and account for 132,300 km^2, i.e., 5.26% of the total surface of the Mediterranean. These figures drop to 44,743 km^2 and 1.78% without the major contribution of the Pelagos Sanctuary, which is a large offshore MPA in the northern Tyrrhenian Sea aimed at the protection of cetaceans. In addition, the distribution of MPAs is markedly clumped, being concentrated mainly in the northwestern sector (Gabrié et al., 2012).

The awareness that the impact of human activities (mostly fisheries) was causing adverse effects on the marine environment led Mediterranean countries to implement active protection measures for nature conservation and fisheries enhancement (Badalamenti et al., 2000). Along with traditional MPAs, which are established mainly for biodiversity conservation, a number of other marine managed areas (MMAs) created under the above mentioned EU and national legislation and policies have been implemented in the Mediterranean, although the documented examples of their success are scanty.

The aim of this chapter is therefore to summarise the main typologies of fishery-oriented MMAs existing in the Mediterranean and to discuss their efficacy in the light of the results reported in the academic literature.

2. TYPES OF MEDITERRANEAN MMAs

To our knowledge the first case of a spatial fishery exclusion adopted in the Mediterranean Sea as a measure to counteract excessive fishing pressure dates back to 1627 and concerns the Kingdom of Naples in southern Italy (Lentini, 2010). The first MPAs were created in the 1960s. The list of 681 Mediterranean MPAs includes 170 national and international MPAs, 507 Natura 2000 sites,[2] the Pelagos Sanctuary and 4 GFCM fisheries restricted areas, but not other types of MMAs created specifically for fisheries management (Gabrié et al., 2012). Hereafter we examine all types of fishery-oriented MMAs present in the Mediterranean. Their declared purpose is to rebuild depleted stocks and often to benefit artisanal fisheries but their implicitly expected effects may well include the protection of essential fish

---

[2] Network of protected sites established under the European 92/43/EEC Directive (Habitat Directive).

habitats and more generally habitats and areas important as nursery, spawning or feeding grounds.

2.1. Fishery reserves

These are spatially delimited areas created to manage fisheries, in which fishing is restricted or forbidden permanently or for limited periods to one or more fishing gears (Agardy, 1997; Auster and Shackell, 1997). Within this broad category we group four types of MMAs:

2.1.1 Établissements de pêche *(fishery establishments)* and cantonnements de pêche *(fishery reserves)*

These are small inshore managed areas created along the French Mediterranean coast during the late 1970s and early 1980s (Arceo et al., 2013; Meinesz et al., 1983). There are three *établissements de pêche* and one *cantonnement de pêche* in continental France plus nine *cantonnements de pêche* in Corsica. Their aim is to protect the populations of commercial species and to provide benefits to adjacent artisanal fisheries. In both categories fishing is totally excluded, furthermore they can also include artificial reef areas.

2.1.2 Off-shore managed areas

The "Mammellone", a high seas area located between the Italian island of Lampedusa and the Tunisian coast, although unilaterally considered by Tunisia an exclusive fishing zone since 1951 has always been fished by Italian trawlers due to its high productivity, raising frequent and harsh conflicts between the two countries (MRAG Ltd., 2013). After an Italy–Tunisia bilateral agreement signed in 1971, the Mammellone has been formally considered a part of the Tunisian exclusive economic zone. In 1979, Italy declared the area a biological protection zone and has formally prohibited Italian trawlers to fish therein.

2.1.3 No-trawl areas

Year-round no-trawl areas have been established in the gulfs of South Evoikos, Pagassitikos and Orei along the Aegean coast of Greece and in eastern Lefkada, Laganas Gulf, Messolonghi Lagoon and in three bays inside the Korinthiakos Gulf along the Greek Ionian coast (Tsikliras and Stergiou, 2007 and authors' unpublished data). Italy has established three no-trawl areas along the northern and eastern coasts of Sicily in the gulfs of Castellammare, Patti and Catania (Pipitone et al., 2000). In these three gulfs, all uses including artisanal fishing are permitted except trawling. It is worth noting that a trawl ban had been imposed in the gulfs of Castellammare and Termini Imerese (northern Sicily) at the end of the 1890s as a mean for rebuilding the already depleted fish stocks and to reduce conflicts between fisheries sectors (Anon., 1899).

The EC Reg. 1967/2006, which provides a framework for the sustainable exploitation of fisheries resources in the Mediterranean, prohibits the use of bottom-towed gears on (1) seagrass beds, coralligenous habitats and at depths over 1000 m, (2) at less than 1.5 miles from the coastline irrespective of the depth (or less than 0.7 miles with a special derogation upon a justified request from a member state). Prior to this regulation most Mediterranean countries had a ban on trawling within 3 miles from the coast or less if the depth was over 50 m, which is still valid where not in contrast with the above regulation. Also in 2005 the GFCM produced a recommendation prohibiting the use of dredges and trawl nets at depths over 1000 m which extend over 58% of the Mediterranean seabed.[3]

2.1.4 No-take zones

Small no-take zones have been created in certain cases to protect fishery resources and rebuild locally depleted populations of commercial species. In Sardinia, the first no-take zone was established off the western coast in 1998 as a protection measure for the spiny lobster, *Palinurus elephas* (Follesa et al., 2007) followed in 2009 by the creation of a further five zones with the same objective all around the island (Follesa, 2010). In southwestern Turkey five no-take zones were established in 2010 in Gökova Bay with the purpose of fish population restocking and protection (Bann and Basak, 2011; Ünal, 2010).

2.2. Fisheries restricted areas

Areas where fishing activities are regulated or prohibited through a recommendation issued by GFCM. A fisheries restricted areas (FRA) may be requested to GFCM by any stakeholder involved with management, research or production, and can be established (after evaluation of the proposal) to protect any kind of marine resource and environment from the impact of fishing (GFCM, 2014). As with all GFCM recommendations, member states are expected to adopt the measures foreseen by the FRA.

---

[3] Recommendation GFCM/2005/1 on the management of certain fisheries exploiting demersal and deepwater species.

Of the four Mediterranean FRAs only the one created in 2009 in the eastern Gulf of Lions has explicit fisheries management purposes, more specifically the protection of spawning aggregation areas of hake (*Merluccius merluccius*) and other demersal species from excessive fishing.[4] The others, namely the Lophelia reef off Capo Santa Maria di Leuca, the Nile delta area cold hydrocarbon seeps and the Eratosthenes Seamount, mention the benefits to sustainable fisheries amongst their effects.[5] The FRAs are considered a valuable tool for the Mediterranean since they are (i) easy to establish, (ii) open to different types of stakeholders, (iii) complementary with other protection measures (GFCM, 2014).

2.3. Marine protected areas

Of the 170 Mediterranean MPAs, 161 have been created under a national designation while 9 have an international designation. Their size spans from less than 1 to over 4000 km^2, with the Pelagos Sanctuary 87,500 km^2 in extent (Gabrié et al., 2012). Established worldwide for multiple uses of which nature conservation is the principal, most MPAs in the Mediterranean are composed of one or more of the following zones characterised by different levels of protection: a fully protected core zone, a buffer zone and a peripheral zone (Abdulla et al., 2008). Such a spatial zoning scheme allows for the possibility of practising both artisanal fishing, i.e., small-scale fishing operations conducted with small boats using fixed or drifting gear and recreational fishing in the buffer and peripheral zones (with the exclusion of spearfishing). All in all Mediterranean MPAs are used also as a fisheries management tool and as a way to assign user rights to local fishers. Spain exemplifies this concept very well: most Spanish MPAs are established and managed by the fisheries departments of the local or national government.[6]

2.4. Biological protection zones

Twelve offshore zones ranging in size from 50 to over 2200 km^2 were established between 1993 and 2009 by the Italian government in areas

---

[4] Recommendation GFCM/33/2009/1 on the establishment of a fisheries restricted area in the Gulf of Lions to protect spawning aggregations and deep sea sensitive habitats.

[5] Recommendation GFCM/30/2006/3 on the establishment of fisheries restrictive areas in order to protect the deep sea sensitive habitats.

[6] Spanish Ministry of Agriculture, Food and Environment: http://www.magrama.gob.es/en/pesca/temas/proteccion-recursos-pesqueros/reservas-marinas-de-espana/rmarinas-intro.asp

deemed critical for the reproduction or growth of commercial species. Seven such zones are in the Adriatic while five are in the Tyrrhenian. Fishing is strictly regulated and permitted only to set gears, with the exception of a few zones where trawling is permitted for part of the year.

2.5. Artificial reef areas

About 250 artificial reef areas were created in 10 Mediterranean countries after the first scientifically planned reefs deployed in 1974 off the Adriatic coast of central Italy (Bombace, 1989; Fabi et al., 2011). Artificial reefs have been planned as a fisheries management tool able to increase fishing yields and reduce conflicts among different fisheries sectors and, to a lesser extent, as a protection device for coastal sensitive habitats, namely *Posidonia oceanica* beds. It is the intention that these targets are reached essentially through antitrawling action. While pyramids of concrete blocks are the most common submerged artificial structures, many other different manmade structures have been used to fulfil different specific purposes (Fabi et al., 2011). A few Mediterranean artificial reefs have been created for research purposes and for recreational fishing and diving.

2.6. Exclusive fishing zones

In the last few decades several Mediterranean countries, namely Malta, Algeria, Spain and Libya, have created an enlarged exclusive fishing zone extending the offshore limits of their territorial waters under the UNCLOS framework. Malta has managed a 25-mile exclusive fishing zone since 1971 with the purpose of protecting its own artisanal fisheries against the impact of foreign industrial fishing (Dimech et al., 2009). With EC Regulations 1626/1994 and 813/2004 Malta has been allowed a 25-mile fisheries management zone where fishing effort, vessel size and type of gear are strictly regulated.

3. THE EFFECTS OF MMAs ON FISHERIES

Research on MMAs in the Mediterranean region has addressed to a large extent the effect of protection on ecological processes and on the structure of fish and benthic assemblages inside marine reserves (Bevilacqua et al., 2006; Claudet et al., 2010, 2011; Coll et al., 2012; Consoli et al., 2013; Di Franco et al., 2009; Fouzai et al., 2012; Guidetti et al., 2014; Seytre and Francour, 2009; Valls et al., 2012; Villamor and Becerro, 2012). The academic literature on the effects of MMAs on fisheries is still relatively scarce and is concentrated more on the species (abundance, size structure) than on human components (fisheries yields, fishing patterns, fishermen perceptions and attitudes, governance issues).

Table 10.1 provides a summary of the literature on MMAs that relate to commercial fish populations and to their fisheries. This is only a subset of the wider literature on Mediterranean MMAs. The references cited in the table are based on the results of fishing surveys, tag-and-recapture surveys and visual census, which provided data on the abundance, population structure and distribution of fish, and of socioeconomic surveys, which provided data on fishing yields, revenues and fishers' attitudes and perceptions.

The MMAs in the table were arbitrarily categorised as small (<100 km^2) and large (>100 km^2). The first group includes 15 MMAs located in 4 countries (3 in the western basin and 1 in the eastern basin) and belonging to three different typologies: marine protected areas, no-take zones and artificial reef areas. They were established with the main objective of either fisheries enhancement or conservation, sometimes both. The habitats included within their borders are mainly rocky bottoms, *P. oceanica* meadows, sandy bottoms and to a lesser extent detritic and coralligenous bottoms. Their mean age is 28.14 ± 11.79 years (SD). These MMAs are managed by the local or national government or by an association of local authorities that sometimes involve fishermen. The second group includes seven MMAs located in four countries (three in the western basin and one in the eastern basin) and belonging to three different typologies: marine protected areas, no-trawl areas and exclusive fishing zones. Their mean age is 23.71 ± 9.20 years (SD). The two MPAs are among the largest in the Mediterranean and are typical multipurpose protected areas devoted to biodiversity conservation and fisheries management. The Strait of Bonifacio nature reserve is managed under a France–Italy bilateral agreement. The other MMAs were created as fisheries management tools and extend over a wide portion of soft bottoms; their management is granted by local or national authorities. The Greek no-trawl areas served as a case study for the first research carried out in a large fishery exclusion area in the Mediterranean (Vassilopoulou and Papaconstantinou, 1999), although no further report followed this first one.

From Table 10.1 it is apparent that Mediterranean MMAs differ a lot as regards size, habitat, declared objectives, type and amount of data collected. We have selected four different case studies that are representative examples of such diversity and give an insight on the effects of MMAs on fishery

Table 10.1 Summary of the literature on Mediterranean fishery-oriented MMAs

Location (country)	Typology	Main objectives	Type of data	Size	Habitat	Age	References
Côte Bleue marine park with Carry-le-Rouet and Cap Couronne NTZs (France)	MPA	Conservation, fisheries enhancement, research	Socioeconomic data. Spatial pattern of fishing operations. Commercial fishing yields. Fish abundance (CPUE, UVC)	S	Po, SB, RB	31	Goñi et al. (2008), Harmelin Vivien et al. (2008), Forcada et al. (2009), and Leleu et al. (2012)
Côte Bleue marine park (France)	ARA	Fisheries enhancement, research	Fish abundance (UVC). Fish assemblage structure	S	SB	31	Charbonnel and Bachet (2011)
Port-Cros marine reserve (France)	MPA	Conservation	Socioeconomic data. Spatial pattern of fishing operations. Commercial fishing yields	S	Po, RB	51	Cadiou et al. (2009)
Cerbère–Banyuls marine reserve (France)	MPA	Conservation	Socioeconomic data. Spatial pattern of fishing operations. Commercial fishing yields. Fish abundance and size (CPUE, UVC)	S	RB	40	Bell (1983), Stelzenmüller et al. (2007, 2009), Forcada et al. (2008, 2009), Goñi et al. (2008, 2009), Harmelin Vivien et al. (2008), and Garcia Rubies et al. (2013)
Cabrera Archipelago national park (Spain)	MPA	Conservation		S	Po, SB, RB, CD, C	23	

Continued

Table 10.1 Summary of the literature on Mediterranean fishery-oriented MMAs—cont'd

Location (country)	Typology	Main objectives	Type of data	Size	Habitat	Age	References
Cabo de Palos—Islas Hormigas marine reserve (Spain)	MPA	Fisheries enhancement		S	Po, SB, RB, CD, C	19	
Tabarca Island marine reserve (Spain)	MPA	Fisheries enhancement		S	Po, RB	31	
Medes Islands marine reserve (Spain)	MPA	Conservation		S	Po, RB	31	
Columbretes Islands marine reserve (Spain)	MPA	Fisheries enhancement	Commercial fishing yields, abundance (CPUE) and size of fish and spiny lobster. Movement of spiny lobster	S	RB	24	Goñi et al. (2001, 2006, 2010) and Stobart et al. (2009)
Alicante (Spain)	ARA	Fisheries enhancement, antitrawling	Abundance and distribution of fish (UVC)	S	SB, Po	37	Ramos-Esplá et al. (2000) and Bayle Sempere et al. (2001)
Plemmirio marine reserve (Italy)	MPA	Conservation	Abundance and diversity of fish (UVC)	S	RB	24	Consoli et al. (2013)
Su Pallosu no-take zone (Italy)	NTZ	Fisheries enhancement	Commercial fishing yields. Abundance (CPUE), size and movement of spiny lobster	S	RB	16	Follesa et al. (2007, 2008, 2009, 2011)

Torre Guaceto marine reserve (Italy)	MPA	Conservation	Socioeconomic data. Commercial fishing yields	S	Po, RB	23	Guidetti and Claudet (2009) and Guidetti et al. (2010)
Central Adriatic Sea (Italy)	ARA	Fisheries enhancement, offshore aquaculture	Commercial fishing yields of fish and sessile bivalves. Abundance (CPUE, UVC) and diversity of fish	S	SB	40	Bombace et al. (2000)
Gökova Bay no-take zones (Turkey)	NTZ	Fisheries enhancement	Socioeconomic data	S	RB	4	Ünal (2010) and Bann and Basak (2011)
Fisheries Management Zone (Malta)	EFZ	Fisheries enhancement	Socioeconomic data. Abundance (CPUE) and size of demersal assemblage	L	SB, MB, MD	43	Dimech et al. (2008, 2009, 2012)
Gulf of Castellammare no-trawl area (Italy)	NTA	Fisheries enhancement	Socioeconomic data. Commercial landings data. Abundance (CPUE), size, distribution of fish. Fish assemblage structure and diversity	L	SB, MB, MD	24	Anon. (1899), Pipitone et al. (2000, 2001), D'Anna et al. (2001), Badalamenti et al. (2002, 2008), Whitmarsh et al. (2002, 2003), Himes (2003), Mardle et al. (2004), and Fiorentino et al. (2008)

Continued

Table 10.1 Summary of the literature on Mediterranean fishery-oriented MMAs—cont'd

Location (country)	Typology	Main objectives	Type of data	Size	Habitat	Age	References
Gulf of Patti no-trawl area (Italy)	NTA	Fisheries enhancement	Fish abundance (CPUE)	L	SB, MB, MD	24	Potoschi et al. (2006)
Egadi Islands marine reserve (Italy)	MPA	Fisheries enhancement, conservation	Socioeconomic data	L	RB, Po, SB	23	Himes (2003)
Tavolara–Punta Coda Cavallo marine reserve (Italy)	MPA	Conservation, fish enhancement	Socioeconomic data	L	RB, Po, SB, C	17	Micheli and Niccolini (2013)
Strait of Bonifacio nature reserve (France)	MPA	Fisheries enhancement, conservation	Commercial fishing yields and fish abundance (only a rough in/out comparison)	L	Po, SB, RB	15	Mesnildrey et al. (2010)
Aegean Sea no-trawl areas (Greece)	NTA	Fisheries enhancement	Fish abundance (CPUE), size and diversity	L	MB, SB	Uncertain (>20)	Vassilopoulou and Papaconstantinou (1999)

MPA, marine protected area; ARA, artificial reefs area; NTZ, no-take zone; EFZ, exclusive fishing zone; NTA, no-trawl area; CPUE, catch per unit effort from experimental fishing surveys; UVC, underwater visual census; Po, *Posidonia oceanica*; SB, sandy bottom; RB, rocky bottom; CD, coastal detritus; C, coralligenous; MD, muddy detritus; MB, muddy bottom. Age in years.

resources: (1) the Gulf of Castellammare, a fishery reserve that occupies a vast portion of soft bottoms in northwestern Sicily; (2) Torre Guaceto, a small inshore MPA with rocky habitats along the Adriatic coast of Apulia; (3, 4) Su Pallosu and Columbretes Islands, two offshore fishery reserves off western Sardinia and eastern Spain, respectively.

(1) Gulf of Castellammare fishery reserve (Italy)

The Gulf of Castellammare, located in northwestern Sicily, was declared a no-trawl area in 1990 by the Sicilian parliament with the purpose of rebuilding the fishery resources and of mitigating the conflict between set-gear and trawl fishing fleets (Pipitone et al., 2000). Except bottom-towed gears, all other professional and recreational fishing techniques are permitted and regulated under national legislation. The no-trawl area is $200 \ km^2$ and extends over the continental shelf and beyond starting from the coastline. As a consequence of protection, the demersal resources of the Gulf shelf increased eightfold 4 years after the start of the ban, with even higher increments of target species such as red mullet *Mullus barbatus*, pandora *Pagellus erythrinus*, hake *Merluccius merluccius* and monkfish *Lophius budegassa* among others (Pipitone et al., 2000). Figure 10.1 shows the increase in demersal biomass after the start of the ban inside the no-trawl area (IN) and in an adjacent trawled area (OUT) as evidenced by a BACI (Before After Control Impact; Green, 1979) comparison. Additionally an ACI (After Control Impact; Glasby, 1997) approach was applied to trawl survey data collected

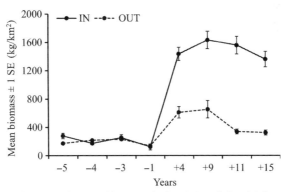

Figure 10.1 Trend in mean demersal biomass before (−) and after (+) the start of the trawl ban (1990) inside the Gulf of Castellammare no-trawl area (IN) and in an adjacent trawled area (OUT). Trawl survey data, all seasons, $N = 522$. Vertical bars: 1 standard error (SE). The results of the three-way ANOVA on the root-tranformed demersal biomass (Before vs. After, BA; IN vs. OUT, IO; Year, Ye (BA × IO)) showed a significant effect of the interaction term BA × IO ($F_{1, 25.8} = 24.36; p = 0.0041$). The subsequent pairwise tests showed a significant difference between IN and OUT after the trawl ban ($t = 9.92, p = 0.0291$) and between Before and After in both IN ($t = 11.08, p = 0.0278$) and OUT ($t = 2.95, p = 0.0314$).

in 2004 and 2005 in IN, OUT and in two control areas subject to intense trawling, one located 75 km (Gulf of Termini Imerese) and the other 130 km (Gulf of Sant'Agata) eastwards from the Gulf of Castellammare. The analysis showed that IN had more biomass than the three trawled areas, and that OUT had more biomass than the controls (Figure 10.2). The results of both analyses suggest that OUT benefited from the adjacent no-trawl area, possibly through spillover or larval export. Data from the 2004–2005 survey show an inflexion in the pattern of biomass rebuilding in line with the abolition of subsidies to excluded trawlers, the decrease of enforcement and the increase of anecdotal reports of poaching along the inner side of the no-trawl border (Stefanoni et al., 2008).

A trammel net survey conducted from 1990 to 1998 in the no-trawl area (D'Anna et al., 2001) showed that the biomass of shallow-water species that are commonly targeted by the small-scale fishery increased gradually since the ban (Figure 10.3).

The ecosystem effects of the trawl ban assessed through investigation of the trophodynamics of two commercial fish species in the Gulf of Castellammare and in the two controls did not yield clear and homogeneous results (Fanelli et al., 2010; Sinopoli et al., 2012). Badalamenti et al. (2008) found only limited effects of trawling on the trophodynamics of three

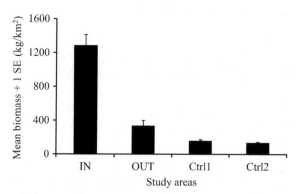

Figure 10.2 Mean demersal biomass in the Gulf of Castellammare no-trawl area (IN), in an adjacent trawled area (OUT), and in two trawled controls: the Gulf of Termini Imerese (Ctrl1) and the Gulf of Sant'Agata (Ctrl2). Trawl survey data, autumn 2004 and spring 2005 cumulated, $N = 180$. Vertical bars: 1 standard error (SE). The one-way ANOVA with a planned comparison performed on the root-transformed demersal biomass showed a significant difference among areas ($F_{3, 176} = 142.79$; $p = 0.0001$). The planned comparisons showed higher biomass in IN than in OUT ($F_{1, 58} = 28.44$; $p = 0.0001$) and in OUT than in the controls ($F_{1, 131} = 31.23$; $p = 0.0001$), which in turn did not display any difference between them ($F_{1, 118} = 1.07$; $p = 0.3001$).

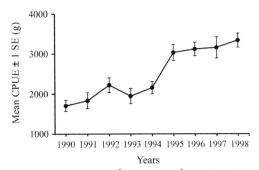

Figure 10.3 Mean CPUEs (gr × 500 m^{-1} net × day^{-1}) in the Gulf of Castellammare no-trawl area. Trammel net survey data, $N = 680$. Vertical bars: 1 standard error (SE). Data collected at the start of the trawl ban and in the successive 8 years (from D'Anna et al., 2001). Spearman's correlation coefficient $r = 0.95$, $p < 0.0001$.

commercial fish species in the Gulf of Castellammare compared to the Gulf of Termini Imerese, i.e., no difference in the source of carbon and different trophic level in two species. The analysis of biomass spectra in the same areas gave an unexpected result, i.e., higher midpoint height coupled with steeper slope in the Gulf of Castellammare. This could have been caused possibly by higher overall biomass and higher abundance of smaller fish due to size-selective artisanal fishing in the untrawled gulf, and by recruitment over-fishing in the trawled gulfs (Sweeting et al., 2009).

(2) Torre Guaceto MPA (Italy)

Torre Guaceto is a 22 km^2 coastal MPA established in 1991 along the eastern coast of Apulia (southern Adriatic Sea). Like all Italian MPAs it is divided in three zones with different level of protection: zone A (integral protection), zone B (partial protection) and zone C (buffer zone). The first recorded effect of protection was the higher abundance and size of two predators highly sought after by professional and recreational fishermen—the seabreams *Diplodus sargus* and *D. vulgaris*—inside zone A (Guidetti, 2006). In 2005, after 5 years of total fishing ban inside the whole MPA, a com-anagement initiative was set based on the limitation of key predators removal and of fishing effort and on the use of selective gear. The buffer zone was reopened to controlled fishing activity and commercial catches per unit effort (CPUEs) were collected inside the buffer zone and outside the MPA from January 2005 to April 2008 (Guidetti and Claudet, 2009; Guidetti et al., 2010). Data analysis showed that the CPUEs of total catch and of the most abundant target species were always higher inside the buffer zone than outside, and that inside the MPA, after an initial decrease, densities stabilised at about twice the value than outside. Further research based on a multidisciplinary approach (Di Franco et al., 2012) demonstrated that the MPA hosted more and larger *D. sargus* spawners than the outside fished area and that their propagules were able to reach areas as far as 100 km down-current, to replenish distant fishing grounds.

(3, 4) Su Pallosu (Italy) and Columbretes Islands (Spain) fishery reserves
Su Pallosu and the Columbretes Islands provide strong evidence of a positive effect of a fishery reserve on the European spiny lobster, *Palinurus elephas* population and fishery. Su Pallosu is a 4 km^2 area located 6 km off the central-western coast of Sardinia (Italy). It is characterised by precoralligenous and coralligenous detritic bottoms with sand and mud patches at a depth of 50–100 m. In 1998 this traditional fishing ground for local artisanal fishermen was selected by the Sardinian parliament as a no-take restocking zone for the lobster. The Columbretes Islands marine reserve is located 55 km off the coast of eastern Spain between Valencia and the Balearic Islands, and was declared a no-take marine reserve by the Fisheries Department of the Spanish government in 1990. The protected area is about 44 km^2 and is characterised by a coralligenous rocky habitat with sand and mud patches down to a maximum depth of 80 m. Like Su Pallosu, the Columbretes used to be a traditional fishing ground for lobster.

Follesa et al. (2008, 2011) carried out studies on the spiny lobster in Su Pallosu based on experimental and commercial CPUE data and on recapture data of lobsters that were released inside the no-take zone after tagging. Data were collected in 1997, i.e., 1 year before the fishing ban, and in each of the successive 12 years. A progressive increase of lobster abundance was recorded inside and outside the no-take zone. After 12 years of protection the inside biomass was 500% higher than in 1998 while the abundance inside was 4.7-fold higher than outside. Spatial trends in commercial CPUEs and in the recapture of tagged specimens gave evidence of adult spillover: 39% of tagged lobsters were recaptured within 2.5 km from the reserve border and 34% between 2.5 and 5 km from the border, with occasional recaptures occurring up to 50 km away. The area of highest productivity extended up to 6 km from the border with a dramatic decline in catches beyond that distance. Also an increase in mean body size was recorded inside and outside the reserve that contributed to compensate commercial catches against the loss of fishing ground. The data collected in Su Pallosu suggest that, where there has been severe overfishing previously, for a species with moderate mobility like *P. elephas*, even a very small MMA may provide benefits to the resource soon after the start of protection.

Goñi et al. (2006, 2010) collected the same type of data as Follesa et al. (2008, 2011) in the Columbretes Islands 9–17 years after the start of protection. CPUEs of lobsters were higher inside (4–154 lobsters $\times$ 600 m^{-1} net $\times$ day^{-1}) than outside (0–10 lobsters $\times$ 600 m^{-1} net $\times$ day^{-1}) the reserve, with a nonlinear decrease of catches with increasing distance from the reserve centre as well as from the reserve boundary. A depression in CPUEs was found at the boundary and was followed by a slight increase and a plateau up to 1500 m from the border, which indicated a concentration of fishing effort at the border and stable CPUEs up to a certain distance. This distance is shorter than in Su Pallosu due to "fishing the line" effect and to higher fishing intensity in the Columbretes. A similar trend was found in the recapture rate of tagged lobsters released inside the reserve. No recaptures occurred beyond 1500 m from the reserve boundary. Using tag-recapture data and CPUEs collected from 1998 to 2007 from the local artisanal fishery along with estimates of lobster abundance, a net 10% in weight mean annual benefit to the catch was produced by spillover that compensated for the loss of fishing ground.

4. SOCIOECONOMIC ISSUES

Socioeconomic issues in the conservation and management of nature have rarely been addressed by managers and policy makers in the Mediterranean region (Arceo et al., 2013). Most decision processes in the fields of fisheries management and conservation have followed a top-down approach and stakeholders are rarely involved in the preparatory and planning phases of MMAs. As a result the creation of marine protected and managed areas often faces local incomprehension and lack of acceptance, if not open hostility and protest, which affect management success (Badalamenti et al., 2000). The same lack of interest is observed in the academic literature: the very few papers on this topic are the outcome of independent research based on interviews to stakeholders.

Himes (2003) interviewed the artisanal fishermen of two Sicilian MMAs: the Egadi Islands MPA and the Gulf of Castellammare fishery reserve. The results show how different the situation can be a few tens of kilometres apart. While the Egadi MPA—one of the largest in the Mediterranean—is perceived as unsuccessful and ineffective by fishermen and managers, and yet it is optimistically supported by both of them, the Gulf of Castellammare fishery reserve is considered useful and successful by artisanal fishermen who firmly believe that it has helped the local fishery.

The Gulf of Castellammare social and economic issues have also been investigated by Whitmarsh et al. (2002, 2003), whose research showed that (1) the revenues of artisanal fishermen based inside the no-trawl area have grown since the ban, (2) these fishers foster the maintenance of the trawl ban and would quit the fishery if the ban was lifted, (3) the artisanal fishermen based outside the no-trawl area are hampered by the higher costs incurred to reach the protected area and by the displacement of trawlers onto their traditional fishing grounds.

Contrasting attitudes were displayed by fishermen in Malta (Dimech et al., 2009), where a 25-mile exclusive fishing zone has been in place since 1971. While pelagic fishers did not recognise any benefit coming from the exclusive zone due to the erratic nature of their target species, demersal fishers had a more neutral perception of the management benefits. A contradictory statement among Maltese fishers was that, while admitting an overall economic benefit from the exclusive zone, the abundance of fish did not increase inside the managed area.

A similar contradictory attitude towards MPAs was recorded among Balearic artisanal fishermen who, while recognising the benefits to biodiversity conservation and fisheries, did not consider protected areas as a good mean to minimise user conflicts (Maynou et al., 2013).

A promising experience comes from the Torre Guaceto MPA, where a comanagement approach has been agreed between MPA managers, artisanal fishermen and scientists (Guidetti and Claudet, 2009; Guidetti et al., 2010). Fishermen have accepted limitations to their gear, technique and fishing power with the aim of attaining sustainable fishing and better economic performance under strict patrolling and yield monitoring.

A good level of involvement of fishermen in the decisional and management process is also present along the French Mediterranean coast, where associations of artisanal fishermen called *prud'homies* have had a role in the planning and functioning of MMAs (Arceo et al., 2013). Their involvement in the management of the Côte Bleue Marine Park is one of the reasons why the zonation and fishing limitations in that park have been accepted by fishermen and why their perception of the benefits deriving from protection is so positive (Leleu et al., 2012).

5. DISCUSSION

Our brief analysis of fishery-oriented MMAs highlights some peculiar aspects: (i) in the Mediterranean several different types of MMAs exist,

(ii) the literature addressing their effects on fisheries is scanty especially with respect to the human component of fisheries, and (iii) the recorded effects are positive and sometimes dramatic in terms of fish biomass increase like in the Gulf of Castellammare no-trawl area.

Where MMAs have been efficiently enforced our review showed substantial effects on the density of previously overexploited populations and less clear-cut results on the trophodynamics of the system. While the ban on trawling imposed in the Gulf of Castellammare resulted in increased biomass of demersal fish (Pipitone et al., 2000), it had no (Fanelli et al., 2009; Sinopoli et al., 2012) or little (Badalamenti et al., 2008; Fanelli et al., 2010) effect on the trophic level and on the source of carbon in the diet of single species. A spillover effect has been detected in Su Pallosu and in the Columbretes Islands and to a lesser extent in the Gulf of Castellammare. The lack of a clear spillover effect for rock-dwelling target fish species abundant within some MMAs could be due to strong site fidelity and long residency time within no-take zones (B. Hereu, personal communication; Di Lorenzo et al., 2014; G. Maza, personal communication). In coastal rocky areas, groupers and seabreams consistently provided short-term positive responses to the exclusion of fishing activities even in small MPAs together with a few other species like scorpionfish (*Scorpaena porcus*), striped mullet (*Mullus surmuletus*) and some wrasses (La Mesa and Vacchi, 1999). On soft bottom areas, red mullet (*M. barbatus*), hake (*M. merluccius*), pandora (*P. erithrynus*) and anglerfish (*L. budegassa*) increased their number and biomass in area where trawl fishing had been permanently removed (Badalamenti et al. 2008; Pipitone et al., 2000). In all cases stock rebuilding and spillover were evident after surprisingly few years if compared to patterns found in tropical areas (Russ, 2002).

Six hundred and eighty-one MPAs is a very high number for a relatively small region such as the Mediterranean, which testifies to the attention of coastal countries towards the conservation of natural resources. It should be noted that the largest number are found along the coasts of wealthier countries such as Spain, France and Italy (Gabrié et al., 2012). Yet most of these MPAs are tiny as suggested by the small protected surface: 5.26% that drops to 1.78% if we subtract the Pelagos Sanctuary. If we consider only the fishery-oriented MMAs this value goes roughly around less than 1%: in all cases very far from the 10% suggested by CBD in order to maintain functional biodiversity and even farther from the 20% to 30% suggested by Bohnsack et al. (2002) to attain sustainable fisheries. Therefore the surface currently covered by MMAs is clearly inadequate to meet stock rebuilding targets and attain fisheries sustainability at regional level. It is also quite obvious from Table 10.1 that there is only limited published evidence of the effects of fishery-oriented MMAs (again, mostly from the western basin and only very few from the eastern basin) if compared to the huge literature on Mediterranean MPAs in general as can be drawn through a search in the Web Of Science™ (http://wokinfo.com/).

It is hardly arguable that Mediterranean MMAs form a functional network, as has been largely advocated elsewhere (e.g. Murray et al., 1999). The reason is the uneven distribution and excessive distances between MMAs, which pose barriers to the dispersion abilities of adults and larvae of fish (Andrello et al., 2013; Di Lorenzo et al., 2014). Additionally, scientific understanding of the demographic dynamics of Mediterranean stocks is very limited. The genetic structure, which can be used to infer reproductive connectivity patterns, is known only for a few species (e.g. Cimmaruta et al., 2005). However even for the few species for which such information is available, management actions do not match the demographic patterns of reproduction (Reiss et al., 2009). Moreover, insight about potential reproductive exchanges is severely impaired by the reduced information about the early life history traits of species (Abaunza et al., 2008) and the circulation of water masses except at the mesoscale (e.g. Cimmaruta et al., 2005; González-Wangüemert et al., 2012).

Few among the reviewed MMAs provide scientific evidence of management effects on the social and economic dimensions of fisheries. The effects appear to be dependent on either the local context or the focus of the study: while Whitmarsh et al. (2002, 2003), Leleu et al. (2012) and Maynou et al. (2013) found a positive perception of the effects of protection among the artisanal fishermen of the Gulf of Castellammare, the Côte Bleue marine park and the Baleares Islands respectively, Gómez et al. (2006) found the effects of the Cap de Creus MPA to be detrimental for artisanal fisheries, due among other reasons to the excessive development of tourism (Badalamenti et al., 2000). Where fishing restrictions are not perceived as able to resolve at least some of the fishery problems like, e.g., competition with foreign fleets as is the case in the Maltese fisheries management zone (Dimech et al., 2009), fishermen complain about the MMA regardless of the positive biological effects observed (Dimech et al., 2008). The inconsistency of the results is further amplified by anecdotal evidence suggesting that a large proportion of artisanal catches remains unreported in small or remote MMAs, as well as those inhabited by small communities, and where cultural traditions—reshaped by tourist demand—prevail (Badalamenti et al., 2000).

To make the assessment of the MMAs effects harder there is little information that exists on their benefits to adjacent fisheries, with the exception of the examples provided of the Columbretes and Su Pallosu lobster fisheries—possibly a result of the sedentary nature of the resource—and of the indirect evidence of spillover in the Gulf of Castellammare.

Also, poaching and illegal fishing practises are suspected to pass largely unnoticed. Although not exclusive to the region, enforcement is definitely a weak point in the management of Mediterranean MMAs. Shortage of financial resources impairs the efficacy of patrolling, and poaching is a frequent plague that nullifies actual or potential benefits of protection (Guidetti et al., 2008). In the Gulf of Castellammare enforcement dropped drastically about 10 years after the start of the ban. This change, coupled to the cessation of the subsidies that were granted to local trawlers excluded from the Gulf (Stefanoni et al., 2008), were the main reasons for the increase in poaching activity, which is the likely cause of the decrease of fish biomass observed inside and outside the no-trawl border (Figure 10.1). Management failures are always likely when a top-down approach is followed in both the planning and functioning phases and when stakeholders are not involved, as it is almost the rule in the Mediterranean (Arceo et al., 2013). Such failures are extremely difficult to counteract in the absence of compensating mechanisms and of comanagement approaches in order to ensure equity (Badalamenti et al., 2012). The case of Torre Guaceto demonstrates clearly how successful an adaptive comanagement strategy based on the collaboration between local authorities, fishermen and scientists can be (Guidetti and Claudet, 2009; Guidetti et al., 2010).

The cultural aspects of fisheries in the region are often disregarded within management approaches (Badalamenti et al., 2000). To date, all efforts aimed at supporting fishermen cultural heritage and welfare—that is rooted in well preserved ancient practises (Bekker-Nielsen and Casasola, 2010)—have been concentrated on the promotion of artisanal fishing practises, which are favoured by MMAs due to their lower environmental impact and higher selectivity. Artisanal fisheries are economically relevant and socially important to their localities in the Mediterranean region. Despite the low productivity of the Mediterranean Sea (Margalef, 1985), the number of fishermen is high in most coastal communities (Papaconstantinou and Farrugio, 2000; Tzanatos et al., 2006). In terms of a stable occupation, artisanal fishing is the main long-term occupation in the fishery sector and often the only viable job. Actually, although increasing tourism has provided new and diversified opportunities (e.g., scuba diving and eco-tourism), only a tiny fraction of fishers and their relatives have taken advantage of them switching to such new jobs (Badalamenti et al., 2000). The majority of arti-sanal fishermen operate from numerous, small and low-powered vessels, tar-get a wide variety of species and use multiple fishing gears according to the fishing ground and season. This conduct puts them among the best practi-tioners of balanced sustainable fishing. Their fishing capacity is low and the economic sustainability of the activity is based on relatively high market prices obtained for their fresh, high-quality product, which is marketed locally with the benefits of additional demand from tourists. Often, the behaviour of local fishermen is regulated by traditional rules (norms) that grant rights sometimes inherited through family lines (Arceo et al., 2013; Gómez et al., 2006; Papaconstantinou and Farrugio, 2000). Nevertheless, artisanal fisheries suffer a recession that MMAs cannot halt or revert. While industrial fishing is being severely reduced, some forms of tourism and the relocation and regulation of human uses within MMAs, which often ignore the traditional forms of regulation among local fishers pose unforeseen threats to the economic viability and preservation of their activity (Arceo et al., 2013; Gómez et al., 2006 and references therein). Mediterranean arti-sanal fishermen are getting progressively fewer and older and hardly able to exert any lobbying in the fisheries policy context (Gómez et al., 2006). New, innovative and more specific actions are needed to reinvigorate artisanal fisheries in Mediterranean MMAs.

Equity issues complicate the above picture. We are not aware of any mechanism that provide for an even distribution of costs and benefits derived from the creation of MMAs among stakeholders, although such mechanisms could make people ignore the bigger picture and the non-monetary cultural value of artisanal fishing. Equity and competition issues occur between different fishery sectors (relocation of fishing effort being a clear example: Whitmarsh et al., 2002) as well as between fisheries and other users like tourist operators, energy industries, and commercial ship-ping. Tourism in particular represents a powerful economic driver in the Mediterranean, but local communities often fail to take advantage of the full array of development opportunities created by the tourist flow (Badalamenti et al., 2000; Font and Lloret, 2011). These issues are further complicated by the lack of precisely defined property rights on the mari-time domain, although the zonation and fishing regulations inside MPAs, which generally allow exclusive fishing rights to local fishermen, actually work to some extent as a user rights system.

6. CONCLUSIONS

Fishing has been important to the Mediterranean region since ancient times and it still plays a central role in terms of economic importance and cultural heritage (Farrugio et al., 2003). The review of the current knowledge on fishery-oriented MMAs served as a starting point to explore their functioning and main issues. Although benefits to fishery resources inside and sometimes outside the MMAs have been recorded, most such areas and in particular their no-take zones are generally too small to produce evident and durable benefits. Large MMAs such as no-trawl areas that drastically limit fishing mortality while keeping lower impact fishing practises are a very promising tool, but they should be maintained through effective appropriate enforcement and possibly with compensation measures for the negatively affected fishery sectors (Badalamenti et al., 2012).

Lack of compliance is still a major issue. A harmonised and integrated process that looks at fisheries as a component of a social–ecological system (Berkes, 2003) is urgently needed. Moreover the increase of stakeholder participation, the adoption of bottom-up policies and adaptive comanagement schemes should be pursued (Guidetti and Claudet, 2009; Qiu and Jones, 2013). However the geopolitical scenario of the Mediterranean is complex and adopting a common management view for marine and coastal areas may not be an easy task.

The protection of ecologically important large areas that may include also the high seas could produce indirect positive effects to fisheries (De Juan and Lleonart, 2010), although such projects are not readily feasible due to patrolling and enforcement difficulties. International agreements among scientific and management bodies can help in this respect (GFCM, 2014). A common policy at regional level that encompasses conservation and fishery issues is desirable in the near future with a coordinating role played by the EU along with the Arab League and other supranational organisations such as UN bodies and NGOs.

ACKNOWLEDGEMENTS

We wish to thank all those who kindly contributed with valuable bibliographic information: Cristina Follesa, Stelios Katsanevakis, José Luis Sanchez-Lizaso, Kostas Stergiou, Vahdet Ünal and Celia Vassilopoulou.

REFERENCES

Abaunza, P., Gordo, L.S., Santamaría, M.T., Iversen, S.A., Murta, A.G., Gallo, E., 2008. Life history parameters as basis for the initial recognition of stock management units in horse mackerel (*trachurus trachurus*). Fish. Res. 89, 167–180.

Abdulla, A., Gomei, M., Maison, E., Piante, C., 2008. Status of Marine Protected Areas in the Mediterranean Sea. IUCN, Malaga and WWF, Paris.

Agardy, T., 1997. Marine Protected Areas and Ocean Conservation. Academic Press, San Diego, CA.

Andrello, M., Mouillot, D., Beuvier, J., Albouy, C., Thuiller, W., Manel, S., 2013. Low connectivity between Mediterranean marine protected areas: a biophysical modeling approach for the dusky grouper *Epinephelus marginatus*. PLoS ONE 8, e68564.

Anon., 1899. Documenti ufficiali sugli splendidi effetti delle zone di esperimento per la pesca nei Golfi di Termini Imerese e di Castellammare (Sicilia). Giovanni Villa, Palermo.

Arceo, H.O., Cazalet, B., Alino, P.M., Mangialajo, L., Francour, P., 2013. Moving beyond a top-down fisheries management approach in the northwestern Mediterranean: some lessons from the Philippines. Mar. Policy 39, 29–42.

Auster, P.J., Shackell, N.L., 1997. Fishery reserves. In: Boreman, J.G., Nakashima, B.S., Wilson, J.A., Kendall, R.L. (Eds.), Northwest Atlantic groundfish: perspectives on a fishery collapse. American Fisheries Society, Bethesda, Maryland, pp. 159–166.

Badalamenti, F., Ramos, A.A., Voultsiadou, E., Sanchez Lizaso, J.L., D'Anna, G., Pipitone, C., Mas, J., Ruiz Fernandez, J.A., Whitmarsh, D., Riggio, S., 2000. Cultural and socio-economic impacts of Mediterranean marine protected areas. Environ. Conserv. 27, 110–125.

Badalamenti, F., D'Anna, G., Pinnegar, J.K., Polunin, N.V.C., 2002. Size-related trophodynamic changes in three target fish species recovering from intensive trawling. Mar. Biol. 141, 561–570.

Badalamenti, F., Sweeting, C.J., Polunin, N.V.C., Pinnegar, J., D'Anna, G., Pipitone, C., 2008. Limited trophodynamics effects of trawling on three Mediterranean fishes. Mar. Biol. 154, 765–773.

Badalamenti, F., Alagna, A., D'Anna, G., Terlizzi, A., Di Carlo, G., 2011. The impact of dredge-fill on *Posidonia oceanica* seagrass meadows: regression and patterns of recovery. Mar. Pollut. Bull. 62, 483–489.

Badalamenti, F., Pipitone, C., Fiorentino, F., D'Anna, G., 2012. The trawling ban in Hong Kong's inshore waters—a round of applause and a plea to learn from others' mistakes. Mar. Pollut. Bull. 64, 1513–1514.

Bann, C., Basak, E., 2011. Economic analysis of Gökova Bay Special Environmental Protection Area. Ministry of Environment and Urbanization, General Directorate of Natural Assets Protection, Ankara, 60 pp.

Bayle Sempere, J.T., Ramos Esplá, A.A., Palazon, J.A., 2001. Análisis del efecto producción - atracción sobre la ictiofauna litoral de un arrecife artificial alveolar en la reserva marina de Tabarca (Alicante). Bol. Inst. Esp. Oceanog. 17, 73–85.

Bekker-Nielsen, T., Casasola, D.B. (Eds.), 2010. Ancient nets and fishing gear. In: Proceedings of The International Workshop on Nets and Fishing Gear in The Classical Antiquity: A First Approach. Servicio de Publicaciones de la Universidad de Cádiz & Aarhus University Press, Cádiz, Aarhus.

Berkes, F., 2003. Alternatives to conventional management: lessons from small-scale fisheries. Environments 31 (1), 5–19.

Bell, J.D., 1983. Effects of depth and marine reserve fishing restrictions on the structure of a rocky reef fish assemblage in the N/W Mediterranean Sea. J. Appl. Ecol. 20, 357–369.

Bevilacqua, S., Terlizzi, A., Fraschetti, S., Russo, G.F., Boero, F., 2006. Mitigating human disturbance: can protection influence trajectories of recovery in benthic assemblages? J. Anim. Ecol. 75, 908–920.

Bohnsack, J.A., Causey, B., Crosby, M.P., Griffis, R.B., Hixon, M.A., Hourigan, T.F., Koltes, K.H., Maragos, J.E., Simons, A., Tilmant, J.T., 2002. A rationale for minimum 20-30% no-take protection. In: Proceedings of the Ninth International Coral Reef Symposium 2, Bali, 23-27 October 2000, pp. 615–619.

Bombace, G., 1989. Artificial reefs in the Mediterranean Sea. Bull. Mar. Sci. 44, 1023–1032.

Bombace, G., Fabi, G., Fiorentini, L., 2000. Artificial reefs in the Adriatic Sea. In: Jensen, A.C., Collins, K.J., Lockwood, A.P.M. (Eds.), Artificial reefs in European Seas. Kluwer Academic Publishers, Dordrecht, pp. 31–63.

Cadiou, G., Boudouresque, C.F., Bonhomme, P., Le Direach, L., 2009. The management of artisanal fishing within the Marine Protected Area of the Port-Cros National Park (northwest Mediterranean Sea): a success story? ICES J. Mar. Sci. 66, 41–49.

Cimmaruta, R., Bondanelli, P., Nascetti, G., 2005. Genetic structure and environmental heterogeneity in the European hake (Merluccius merluccius). Mol. Ecol. 14, 2577–2591.

Charbonnel, E., Bachet, F., 2011. Artificial reefs in the Cote Bleue Marine Park: assessment after 25 years of experiments and scientific monitoring. In: Ceccaldi, H.J., Dekeyser, I., Girault, M., Stora, G. (Eds.), Global Change: Mankind-Marine Environment Interactions. Proceedings of the 13th French-Japanese Oceanography Symposium. Springer, pp. 73–79.

Claudet, J., Osenberg, C.W., Domenici, P., Badalamenti, F., Milazzo, M., Falcon, J.M., Bertocci, I., Benedetti Cecchi, L., Garcia Charton, J.A., Goni, R., Borg, J.A., Forcada, A., De Lucia, G.A., Perez Ruzafa, A., Afonso, P., Brito, A., Guala, I., Le Direach, L., Sanchez Jerez, P., Somerfield, P.J., Planes, S., 2010. Marine reserves: fish life history and ecological traits matter. Ecol. Appl. 20, 830–839.

Claudet, J., Garcia-Charton, J.A., Lenfant, P., 2011. Combined effects of levels of protection and environmental variables at different spatial resolutions on fish assemblages in a marine protected area. Conserv. Biol. 25, 105–114.

Coll, M., Piroddi, C., Steenbeek, J., Kaschner, K., Ben Rais Lasram, F., Aguzzi, J., Ballesteros, E., Bianchi, C.N., Corbera, J., Dailianis, T., Danovaro, R., Estrada, M., Froglia, C., Galil, B.S., Gasol, J.P., Gertwagen, R., Gil, J., Guilhaumon, F., Kesner-Reyes, K., Kitsos, M.S., Koukouras, A., Lampadariou, N., Laxamana, E., Lopez-Fe De La Cuadra, C.M., Lotze, H.K., Martin, D., Mouillot, D., Oro, D., Raicevich, S., Rius-Barile, J., Saiz-Salinas, J.I., San Vicente, C., Somot, S., Templado, J., Turon, X., Vafidis, D., Villanueva, R., Voultsiadou, E., 2010. The biodiversity of the Mediterranean Sea: estimates, patterns, and threats. PLoS ONE 5, e11842.

Coll, J., Garcia Rubies, A., Morey, G., Grau, A.M., 2012. The carrying capacity and the effects of protection level in three marine protected areas in the Balearic islands (NW Mediterranean). Sci. Mar. 76, 809–826.

Consoli, P., Sarà, G., Mazza, G., Battaglia, P., Romeo, T., Incontro, V., Andaloro, F., 2013. The effects of protection measures on fish assemblage in the Plemmirio marine reserve (central Mediterranean Sea, Italy): a first assessment 5 years after its establishment. J. Sea Res. 79, 20–26.

D'Anna, G., Badalamenti, F., Pipitone, C., 2001. Rendimenti di pesca sperimentale con tramaglio nel Golfo di Castellammare dopo otto anni di divieto della pesca a strascico. Biol. Mar. Mediterr. 8, 704–707.

Davenport, J., Davenport, J.L., 2006. The impact of tourism and personal leisure transport on coastal environments: a review. Estuar. Coast. Shelf Sci. 67, 280–292.

Daw, T., Grey, T., 2005. Fisheries science and sustainability in international policy: a study of failure in the European Union's common fisheries policy. Mar. Policy 29, 189–197.

De Juan, S., Lleonart, J., 2010. A conceptual framework for the protection of vulnerable habitats impacted by fishing activities in the Mediterranean high seas. Ocean Coast. Manag. 53, 717–723.

Di Franco, A., Bussotti, S., Navone, A., Panzalis, P., Guidetti, P., 2009. Evaluating effects of total and partial restrictions to fishing on Mediterranean rocky-reef fish assemblages. Mar. Ecol. Prog. Ser. 387, 275–285.

Di Franco, A., Coppini, G., Pujolar, J.M., De Leo, G.A., Gatto, M., Lyubartsev, V., Melia, P., Zane, L., Guidetti, P., 2012. Assessing dispersal patterns of fish propagules from an effective Mediterranean marine protected area. PLoS ONE 7, e52108.

Di Lorenzo, M., D'Anna, G., Badalamenti, F., Giacalone, V.M., Starr, R.M., Guidetti, P., 2014. Fitting the size of no-take zones to species movement patterns: a case study on a Mediterranean seabream. Mar. Ecol. Prog. Ser. 502, 245–255.

Dimech, M., Camilleri, M., Hiddink, J.G., Kaiser, M.J., Ragonese, S., Schembri, P.J., 2008. Differences in demersal community structure and biomass size spectra within and outside the Maltese Fishery Management Zone (FMZ). Sci. Mar. 72, 669–682.

Dimech, M., Darmanin, M., Smith, I.P., Kaiser, M.J., Schembri, P.J., 2009. Fishers' perception of a 35-year old exclusive Fisheries Management Zone. Biol. Conserv. 142, 2691–2702.

Ehler, C., Douvere, F., 2009. Marine Spatial Planning: A Step-By-Step Approach Toward Ecosystem-Based Management. UNESCO, Paris.

Fabi, G., Spagnolo, A., Bellan-Santini, D., Charbonnel, E., Cicek, B.A., Goutayer Garcia, J.J., Jensen, A.C., Kallianotis, A., Neves Santos, M., 2011. Overview on artificial reefs in Europe. Braz. J. Oceanogr. 59, 155–166.

Fanelli, E., Badalamenti, F., D'Anna, G., Pipitone, C., 2009. Diet and trophic level of scaldfish *Arnoglossus laterna* in the southern Tyrrhenian Sea (western Mediterranean): contrasting trawled versus untrawled areas. J. Mar. Biol. Ass. U.K. 89, 817–828.

Fanelli, E., Badalamenti, F., D'Anna, G., Pipitone, C., Romano, C., 2010. Trophodynamic effects of trawling on the feeding ecology of Pandora, *Pagellus erythrinus*, off the northern Sicily coast (Mediterranean Sea). Mar. Freshw. Res. 61, 408–417.

FAO, 1995. Code of Conduct for Responsible Fisheries. FAO, Rome.

FAO, 2011. Fisheries management. 4, Marine protected areas and fisheries. FAO Technical Guidelines for Responsible Fisheries. 4(Suppl. 4), FAO, Rome.

Farrugio, H., Oliver, P., Biagi, F., 2003. An overview of the history, knowledge, recent and future research trends in the Mediterranean fisheries. Sci. Mar. 57, 105–119.

Fiorentino, F., Badalamenti, F., D'Anna, G., Garofalo, G., Gianguzza, P., Gristina, M., Pipitone, C., Rizzo, P., Fortibuoni, T., 2008. Changes in spawning-stock structure and recruitment pattern of red mullet, *Mullus barbatus*, after a trawl ban in the Gulf of Castellammare (central Mediterranean Sea). ICES J. Mar. Sci. 65, 1175–1183.

Follesa, M.C., 2010. La duplice valenza delle aree di ripopolamento attivo di *Palinurus elephas*: utili siti di sperimentazione e potenti strumenti gestionali. PhD thesis, Università degli Studi di Cagliari, Cagliari.

Follesa, C., Cuccu, D., Cannas, R., Sabatini, A., Cau, A., 2007. Emigration and retention of *Palinurus elephas* (Fabricius, 1787) in a central western Mediterranean marine protected area. Sci. Mar. 71, 279–285.

Follesa, M.C., Cuccu, D., Cannas, R., Cabiddu, S., Murenu, M., Sabatini, A., Cau, A., 2008. Effects of marine reserve protection on spiny lobster (*Palinurus elephas* Fabr., 1787) in a central western Mediterranean area. Hydrobiologia 606, 63–68.

Follesa, M.C., Cannas, R., Cau, A., Cuccu, D., Gastoni, A., Ortu, A., Pedoni, C., Porcu, C., Cau, A., 2011. Spillover effects of a Mediterranean marine protected area on the European spiny lobster *Palinurus elephas* (Fabricius, 1787) resource. Aquat. Conserv. Mar. Freshwat. Ecosyst. 21, 564–572.

Font, T., Lloret, J., 2011. Socioeconomic implications of recreational shore angling for the management of coastal resources in a Mediterranean marine protected area. Fish. Res. 108, 214–217.

Forcada, A., Bayle Sempere, J.T., Valle, C., Sanchez Jerez, P., 2008. Habitat continuity effects on gradients of fish biomass across marine protected area boundaries. Mar. Envir. Res. 66, 536–547.

Forcada, A., Valle, C., Bonhomme, P., Criquet, G., Cadiou, G., Lenfant, P., Sanchez Lizaso, J.L., 2009. Effects of habitat on spillover from marine protected areas to artisanal fisheries. Mar. Ecol. Prog. Ser. 379, 197–211.

Fouzai, N., Coll, M., Palomera, I., Santojanni, A., Arneri, E., Christensen, V., 2012. Fishing management scenarios to rebuild exploited resources and ecosystems of the Northern-Central Adriatic (Mediterranean Sea). J. Mar. Syst. 102, 39–51.

Gabrié, C., Lagabrielle, E., Bissery, C., Crochelet, E., Meola, B., Webster, C., Claudet, J., Chassanite, A., Marinesque, S., Robert, P., Goutx, M., Quod, C., 2012. The Status of Marine Protected Areas in the Mediterranean Sea. MedPAN Collection, Marseille.

Garcia-Charton, J.A., Perez-Ruzafa, A., Marcos, C., Claudet, J., Badalamenti, F., Benedetti Cecchi, L., Falcon, J.M., Milazzo, M., Schembri, P.J., Stobart, B., Vandeperre, F., Brito, A., Chemello, R., Dimech, M., Domenici, P., Guala, I., Le Direach, L., Maggi, E., Planes, S., 2008. Effectiveness of European Atlanto-Mediterranean MPAs: do they accomplish the expected effects on populations, communities and ecosystems? J. Nat. Conserv. 16, 193–221.

Garcia Rubies, A., Hereu, B., Zabala, M., 2013. Long-term recovery patterns and limited spillover of large predatory fish in a Mediterranean MPA. PLoS One 8, e73922.

GFCM, 2014. GFCM SAC, Subcommittee on Marine Environment and Ecosystems (SCMEE): Final report of working group on marine protected areas, Bar, Montenegro, 3 February 2014. FAO, Rome, 31 pp.

Glasby, T.M., 1997. Analysing data of post-impact studies using asymmetrical analysis of variance: a case study of epibiota on marinas. Aust. J. Ecol. 22, 448–459.

Gómez, S., Lloret, J., Demestre, M., Riera, V., 2006. The decline of the artisanal fisheries in Mediterranean coastal areas: the case of Cap de Creus (Cape Creus). Coast. Manag. 34, 217–232.

Goñi, R., Reñones, O., Quetglas, A., 2001. Dynamics of a protected Western Mediterranean population of the European spiny lobster Palinurus elephas (Fabricius, 1787) assessed by trap surveys. Mar. Freshwat. Res. 52, 1577–1587.

Goñi, R., Quetglas, A., Reñones, O., 2006. Spillover of spiny lobsters Palinurus elephas from a marine reserve to an adjoining fishery. Mar. Ecol. Prog. Ser. 308, 207–219.

Goñi, R., Adlerstein, S., Alvarez-Berastegui, D., Forcada, A., Reñones, O., Criquet, G., Polti, S., Cadiou, G., Valle, C., Lenfant, P., Bonhomme, P., Perez Ruzafa, A., Sanchez Lizaso, J.L., Garcia Charton, J.A., Bernard, G., Stelzenmüller, V., Planes, S., 2008. Spillover from six western Mediterranean marine protected areas: evidence from artisanal fisheries. Mar. Ecol. Prog. Ser. 366, 159–174.

Goñi, R., Hilborn, R., Diaz, D., Mallol, S., Adlerstein, S., 2010. Net contribution of spillover from a marine reserve to fishery catches. Mar. Ecol. Prog. Ser. 400, 233–243.

González-Wangüemert, M., Vega Fernández, T., Pérez-Ruzafa, A., Giacalone, M., D'Anna, G., Badalamenti, F., 2012. Genetic considerations on the introduction of farmed fish in marine protected areas: the case of study of white seabream restocking in the gulf of Castellammare (southern Tyrrhenian Sea). J. Sea Res. 68, 41–48.

Green, R.H., 1979. Sampling Design and Statistical Methods for Environmental Biologists. John Wiley & Sons, New York, NY.

Guidetti, P., 2006. Marine reserves reestablish lost predatory interactions and cause community changes in rocky reefs. Ecol. Appl. 16, 963–976.

Guidetti, P., Claudet, J., 2009. Comanagement practices enhance fisheries in marine protected areas. Conserv. Biol. 24, 312–318.

Guidetti, P., Milazzo, M., Bussotti, S., Molinari, A., Murenu, M., Pais, A., Spanò, N., Balzano, R., Agardy, T., Boero, F., Carrada, G., Cattaneo-Vietti, R., Cau, A., Chemello, R., Greco, S., Manganaro, A., Notrbartolo Di Sciara, G., Russo, G.F., Tunesi, L., 2008. Italian marine reserve effectiveness: does enforcement matter? Biol. Conserv. 141, 699–709.

Guidetti, P., Bussotti, S., Pizzolante, F., Ciccolella, A., 2010. Assessing the potential of an artisanal fishing co-management in the marine protected area of Torre Guaceto (southern Adriatic Sea, SE Italy). Fish. Res. 101, 180–187.

Guidetti, P., Baiata, P., Ballesteros, E., Di Franco, A., Hereu, B., Macpherson, E., Micheli, F., Pais, A., Panzalis, P., Rosenberg, A.A., Zabala, M., Sala, E., 2014. Large-scale assessment of Mediterranean marine protected areas effects on fish assemblages. PLoS ONE 9, e91841.

Halpern, B.S., Walbridge, S., Selkoe, K.A., Kappel, C.V., Micheli, F., D'agrosa, C., Bruno, J.F., Casey, K.S., Ebert, C., Fox, H.E., Fujita, R., Heinemann, D., Lenihan, H.S., Madin, E.M. P., Perry, M.T., Selig, E.R., Spalding, M., Steneck, R., Watson, R., 2008. A global map of human impact on marine ecosystems. Science 319, 948–952.

Harmelin Vivien, M., Le Direach, L., Bayle Sempere, J., Charbonnel, E., Garcia Charton, J.A., Ody, D., Perez Ruzafa, A., Reñones, O., Sanchez Jerez, P., Valle, C., 2008. Gradients of abundance and biomass across reserve boundaries in six Mediterranean marine protected areas: evidence of fish spillover? Biol. Cons. 141, 1829–1839.

Himes, A.H., 2003. Small-scale Sicilian fisheries: opinions of artisanal fishers and sociocultural effects in two MPA case studies. Coast. Manag. 31, 389–408.

Jackson, J.B.C., Kirby, M.X., Berger, W.H., Bjorndal, K.A., Duarte, C.M., Botsford, L.W., Bourque, B.J., Bradbury, R.H., Cooke, R., Erlandson, J., Estes, J.A., Hughes, T.P., Kidewell, S., Lange, C.B., Lenihan, H.S., Pandolfi, J.M., Peterson, C.H., Steneck, R.S., Tegner, M.J., Warner, R., 2001. Historical over-fishing and the recent collapse of coastal ecosystems. Science 293, 629–638.

Katsanevakis, S., Stelzenmuller, V., South, A., Kirk Sorensen, T., Jones, P.J.S., Kerr, S., Badalamenti, F., Anagnostou, C., Breen, P., Chust, G., D'Anna, G., Duijn, M., Filatova, T., Fiorentino, F., Hulsman, H., Johnson, K., Karageorgis, A.P., Kroncke, I., Mirto, S., Pipitone, C., Portelli, S., Qiu, W., Reiss, H., Sakellariou, D., Salomidi, M., Van Hoof, L., Vassilopoulou, V., Vega Fernandez, T., Voge, S., Weber, A., Zenetos, A., Ter Hofstede, R., 2011. Ecosystem-based marine spatial management: review of concepts, policies, tools, and critical issues. Ocean Coast. Manag. 54, 807–820.

La Mesa, G., Vacchi, M., 1999. An analysis of the fish assemblage of the ustica island marine reserve (Mediterranean Sea). Mar. Ecol. 20, 147–165.

Leleu, K., Alban, F., Pelletier, D., Charbonnel, E., Letourneur, Y., Boudouresque, C.F., 2012. Fishers' perceptions as indicators of the performance of marine protected areas (MPAs). Mar. Policy 36, 414–422.

Lentini, R., 2010. Tra frodi e legalità: pesca a strascico e pesca con la dinamite nei compartimenti marittimi di Palermo e di Trapani tra Ottocento e Novecento. In: D'Arienzo, V., Di Salvia, B. (Eds.), Pesci, barche, pescatori nell'area mediterranea dal medioevo all'età contemporanea. Franco Angeli, Milano, pp. 255–286.

Mardle, S., James, C., Pipitone, C., Kienzle, M., 2004. Bioeconomic interactions in an established fishing exclusion zone: the Gulf of Castellammare, NW Sicily. Nat. Res. Model. 17, 287–316.

Margalef, R. (Ed.), 1985. Western Mediterranean. Pergamon Press, Oxford.

Maynou, F., Morales-Nin, B., Cabanellas-Reboredo, M., Palmer, M., Garcia, E., Grau, A.M., 2013. Small-scale fishery in the Balearic islands (W Mediterranean): a socio-economic approach. Fish. Res. 139, 11–17.

Meinesz, A., Lefevre, J.R., Beurier, J.P., Boudouresque, C.F., Miniconi, R., O'neill, J., 1983. Les zones marines protegees des cotes francaises de méditerranée. Bulletin d'Ecologie 14, 35–50.

Mesnildrey, L., Gascuel, D., Lesueur, M., Le Pape, O., 2010. Analyse des effets des réserves de pêche. Agrocampus Ouest, Rennes.

Micheli, F., Niccolini, F., 2013. Achieving success under pressure in the conservation of intensely used coastal areas. Ecol. Soc. 18, 19.

MRAG Ltd., 2013. Costs and Benefits Arising from the Establishment of Maritime Zones in the Mediterranean Sea. Final Report. European Commission, DG MARE, Brussels.

Murray, S.N., Ambrose, R.E., Bohnsack, J.A., Botsford, L.W., Carr, M.H., Davis, G.E., Dayton, P.K., Gotshall, D., Gunderson, D.R., Hixon, M.A., Lubchenco, J., Mangel, M., MacCall, A., McArdle, D.A., Ogden, J.C., Roughgarden, J., Starr, R.M., Tegner, M.J., Yoklavich, M.M., 1999. No-take reserve networks: sustaining fishery populations and marine ecosystems. Fish. Manag. 24, 11–25.

Papaconstantinou, C., Farrugio, H., 2000. Fisheries in the Mediterranean. Mediterr Mar. Sci. 1, 5–18.

Pipitone, C., Badalamenti, F., D'Anna, G., Patti, B., 2000. Fish biomass increase after a four-year trawl ban in the gulf of Castellammare (NW Sicily, Mediterranean Sea). Fish. Res. 48, 23–30.

PISCO: Partnership for Interdisciplinary Studies of Coastal Oceans, 2011. The science of marine reserves. www.piscoweb.org.

Potoschi, A., Battaglia, P., Rinelli, P., Perdichizzi, F., Manganaro, A., Greco, S., 2006. Variazione dei rendimenti con rete a strascico in un'area a parziale protezione nel golfo di Patti (Sicilia settentrionale) in 20 anni di monitoraggio. Biol. Mar. Medit. 13, 149–157.

Qiu, W., Jones, P.J.S., 2013. The emerging policy landscape for marine spatial planning in Europe. Mar. Policy 39, 182–190.

Ramos-Esplá, A.A., Guillen, J.E., Bayle, J.T., Sanchez Jerez, P., 2000. Artificial anti-trawling reefs off Alicante, south-eastern Iberian Peninsula: evolution of reef block and set design. In: Jensen, A.C., Collins, K.J., Lockwood, A.P.M. (Eds.), Artificial reefs in European seas. Kluwer Academic Publishers, Dordrecht, pp. 195–218.

Reiss, H., Hoarau, G., Dickey-Collas, M., Wolff, W.J., 2009. Genetic population structure of marine fish: mismatch between biological and fisheries management units. Fish Fish. 10, 361–395.

Roberts, C.M., 1997. Ecological advice for the global fisheries crisis. Trends Ecol. Evol. 12, 35–38.

Ruiz, J.M., Romero, J., 2003. Effects of disturbances caused by coastal constructions on spatial structure, growth dynamics and photosynthesis of the seagrass Posidonia oceanica. Mar. Pollut. Bull. 46, 1523–1533.

Russ, G.R., 2002. Yet another review of marine reserves as reef fishery management tools. Coral Reef Fishes: Dynamics and Diversity in a Complex Ecosystem. Academic Press, New York, NY.

Seytre, C., Francour, P., 2009. The Cap roux MPA (Saint-Raphael, French Mediterranean): changes in fish assemblages within four years of protection. ICES J. Mar. Sci. 66, 180–187.

Sinopoli, M., Fanelli, E., D'Anna, G., Badalamenti, F., Pipitone, C., 2012. Assessing the effects of a trawling ban on diet and trophic level of hake, Merluccius merluccius, in the southern Tyrrhenian Sea. Sci. Mar. 76, 677–690.

Stefanoni, S., D'Anna, G., Pipitone, C., Badalamenti, F., 2008. Analisi economica delle politiche di gestione della pesca nel Golfo di Castellammare. In: Pipitone, V., Cognata, A. (Eds.), La valutazione delle risorse ambientali. Approcci multidisciplinari al Golfo di Castellammare. Franco Angeli, Milano, pp. 178–201.

Stelzenmüller, V., Maynou, F., Martin, P., 2007. Spatial assessment of benefits of a coastal Mediterranean Marine Protected Area. Biol. Cons. 136, 571–583.

Stelzenmüller, V., Maynou, F., Martin, P., 2009. Patterns of species and functional diversity around a coastal marine reserve: a fisheries perspective. Aquat. Conservat. Mar. Freshwat. Ecosyst. 19, 554–565.

Stobart, B., Warwick, R., Gonzalez, C., Mallol, S., Diaz, D., Reñones, O., 2009. Long-term and spillover effects of a marine protected area on an exploited fish community. Mar. Ecol. Prog. Ser. 384, 47–60.

Sweeting, C.J., Badalamenti, F., D'Anna, G., Pipitone, C., Polunin, N.V.C., 2009. Steeper biomass spectra of demersal fish communities after trawler exclusion in Sicily. ICES J. Mar. Sci. 66, 195–202.

Tsikliras, A.C., Stergiou, K.I., 2007. Fisheries management and marine protected areas. In: Papaconstantinou, C., Zenetos, A., Vassilopoulou, V., Tserpes, G. (Eds.), State of Hellenic Fisheries. Hellenic Centre for Marine Research, Athens, pp. 306–314.

Tzanatos, E., Dimitriou, E., Papaharisis, L., Roussi, A., Somarakis, S., Koutsikopoulos, C., 2006. Principal socio-economic characteristics of the Greek small-scale coastal fishermen. Ocean Coast. Manage. 49, 511–527.

Ünal, V., 2010. Fishery management in Gökova Special Environment Protection Area. In: Kırac, C.O., Veryeri, N.O. (Eds.), Putting peen to practice in marine and coastal areas; Gokova integrated coastal and marine management planning project. Final report. BBI Matra, Netherlands, pp. 1–92.

UNEP, 2006. Marine and Coastal Ecosystems and Human Well-Being: A Synthesis Report Based on the Findings of the Millennium Ecosystem Assessment. UNEP, Nairobi.

UNEP-MAP, 2005. Guidelines for the Placement at Sea of Matter for Purpose Other than Mere Disposal (Construction of Artificial Reefs), UNEP(DEC)/MED WG.270/10.

Valls, A., Gascuel, D., Guenette, S., Francour, P., 2012. Modeling trophic interactions to assess the effects of a marine protected area: case study in the NW Mediterranean Sea. Mar. Ecol. Prog. Ser. 456, 201–214.

Vassilopoulou, V., Papaconstantinou, C., 1999. Marine protected areas as reference points for precautionary fisheries: a case study of trawl reserves in Greek waters. In: CIESM Worksop on Precautionary Approach to Local Fisheries in the Mediterranean Sea. Kerkenna Island (Tunisia), 23-26 September 1999. CIESM, Monaco, pp. 67–70.

Villamor, A., Becerro, M.A., 2012. Species, trophic, and functional diversity in marine protected and non-protected areas. J. Sea Res. 73, 109–116.

Whitmarsh, D., James, C., Pickering, H., Pipitone, C., Badalamenti, F., D'Anna, G., 2002. Economic effects of fisheries exclusion zones: a Sicilian case study. Mar. Resour. Econ. 17, 239–250.

Whitmarsh, D., Pipitone, C., Badalamenti, F., D'Anna, G., 2003. The economic sustainability of artisanal fisheries: the case of the trawl ban in the Gulf of Castellammare, NW Sicily. Mar. Policy 27, 489–497.

SUBJECT INDEX

Note: Page numbers followed by "*f*" indicate figures and "*t*" indicate tables.

TAXONOMIC INDEX

Note: Page numbers followed by "*f*" indicate figures and "*t*" indicate tables.

CPI Antony Rowe
Eastbourne, UK
October 29, 2014